Harnessing the Abyss:
AI and the Future of Deep Sea Mining

Oliver-Andreas Leszczynski is a global leader in AI-driven industrial transformation, with over 15 years of experience integrating artificial intelligence into essential industries. His academic background in Political Science, Business Management, and Applied AI enables him to bridge economic theory with cutting-edge technology, driving innovation across the maritime and deep-sea mining sectors.

As Director for AI at the Institute of Northern-European Economic Research (INER) and a leader in private maritime enterprises, Leszczynski spearheads large-scale AI initiatives that modernize manufacturing and enhance operational efficiencies. A pioneer in AI-supported deep-sea mining, he advocates for the responsible adoption of emerging technologies to strengthen Western economic resilience.

A staunch proponent of transatlantic partnerships, he fosters U.S.-European cooperation in AI to maintain global competitiveness and strategic leadership. His vision combines economic liberalism with targeted AI investments, ensuring that digitalization and automation sustain long-term industrial and geopolitical stability. Through his leadership, he is shaping the future of AI in global maritime industries.

Oliver-Andreas Leszczynski

Harnessing the Abyss: AI and the Future of Deep Sea Mining

Innovations, Challenges, and Ethical Pathways in Ocean Resource Governance

Bibliographic information from the German National Library:
The German National Library listed this publication in the German National Bibliography; detailed bibliographic data are available on the Internet at http://dnb.dnb.de.

Mail: leszczynski@iner.nbs.de

INER Institute of Northern-European Economic Research

Graphic: Rnko7 / Shutterstock.com

Publisher: BoD · Books on Demand GmbH, Überseering 33, 22297 Hamburg, Germany, bod@bod.de

Print: Libri Plureos GmbH, Friedensallee 273, 22763 Hamburg, Germany

ISBN: 978-3-7693-3708-2

Contents

To my father, Andreas Leszczynski,

My greatest role model, my best friend, and the epitome of what a father should be. You have been my constant source of inspiration, my unwavering motivator, and my steadfast anchor through every challenge. Your guidance taught me resilience, your wisdom instilled ambition, and your belief in me gave me the courage to reach for the stars.

This book is as much a reflection of your lessons as it is of my journey. I am where I am today because of your presence, your support, and your enduring love. Thank you for showing me the power of perseverance and the beauty of striving for greatness.

This work, in all its depth and ambition, is dedicated to you.

Prolog

A New Frontier Beneath the Waves

Beneath the ocean's surface lies a world that is at once ancient and unexplored, a vast expanse that holds secrets as profound as those of the stars. The deep sea, covering more than two-thirds of our planet, is Earth's final frontier, a realm of extraordinary biodiversity, complex ecosystems, and untapped mineral wealth. For centuries, humanity has viewed the ocean with a mix of wonder and trepidation, relying on its bounty for sustenance, trade, and inspiration. Yet, its greatest depths have remained largely out of reach; until now.

The convergence of global challenges and technological breakthroughs has brought the deep ocean into sharper focus. The transition to renewable energy, the rise of advanced manufacturing, and the increasing demand for digital technologies have created an insatiable appetite for critical minerals. Cobalt, nickel, manganese, and rare earth elements, essential components of batteries, wind turbines, and semiconductors, are driving a new race for resources. At the same time, artificial intelligence (AI) and robotics have revolutionized humanity's ability to explore and interact with the deep sea, making what was once unimaginable not only possible but imminent.

This book explores the profound implications of this moment, where the promise of deep sea mining collides with the imperative for sustainability, equity, and stewardship. It is a story of technological ingenuity and ethical responsibility, of innovation and restraint, of progress and preservation. At its heart lies a question that transcends the technical and the economic: How can humanity harness the potential of the ocean without compromising its integrity?

Artificial intelligence plays a central role in this narrative. With its capacity to process vast datasets, predict complex patterns, and optimize decision-making, AI is transforming the way resources are explored, extracted, and managed. It enables precision in exploration, efficiency in operations, and rigor in environmental monitoring, offering a pathway to minimize ecological impacts while maximizing economic benefits. Yet, the deployment of AI in the deep ocean also raises profound questions about governance, equity, and the ethical use of technology. Its power must be wielded thoughtfully, with a recognition of its limitations and a commitment to the greater good.

The stakes are high. The deep sea is not merely a repository of minerals; it is a vital component of Earth's climate system, a reservoir of carbon, a cradle of biodiversity, and a source of untold scientific knowledge. Its ecosystems, delicate and largely unexplored, are as fragile as they are fascinating. The decisions made in the pursuit of deep sea mining will shape the ecological and social landscapes of the future, with implications that extend far beyond the immediate goals of resource extraction.

This book is not only a chronicle of the opportunities and challenges of deep sea mining but also a call to action. It invites policymakers, industry leaders, researchers, and citizens to engage with the complexities of this new frontier, to navigate its uncertainties with care, and to align its development with principles of sustainability, equity, and transparency. It challenges readers to think deeply about the values that should guide humanity's relationship with the ocean and to envision a future where innovation serves as a bridge to stewardship, not a justification for exploitation.

As the exploration of the deep sea accelerates, humanity stands at a crossroads. The choices made today will echo for generations, shaping the legacy of the ocean and its role in sustaining life on Earth. This book is a journey into the depths of that decision-making process, an exploration of how we might balance the promise of technology with the wisdom of responsibility. It is a reflection on what it means to lead, to innovate, and to care for a world that is as mysterious as it is vital, a world beneath the waves, waiting to be understood.

1. Introduction and Context

1.1. Historical Overview of Deep Sea Mining

The quest to harvest the riches of the seabed is as old as human curiosity about the mysteries of the ocean. Yet, the formal history of deep sea mining is marked by a blend of scientific exploration, geopolitical ambition, and technological innovation that has unfolded over the past century. It is a history defined by moments of bold ambition, tempered by daunting challenges, and punctuated by groundbreaking technological and legal milestones.

Long before the advent of modern deep sea mining, the ocean was seen as an inexhaustible reservoir of mystery and resources. In the 19th century, oceanographic expeditions, such as the famous HMS Challenger voyage (1872–1876), revealed the first glimpses of the deep seabed's potential. Although the mission's primary focus was scientific exploration, it uncovered vast plains of polymetallic nodules, dark, rock-like deposits rich in manganese, nickel, cobalt, and copper, scattered across the abyssal plains.

These early discoveries hinted at the potential economic value of the deep seabed but remained beyond the technological grasp of the era. Industrial extraction methods were confined to terrestrial mining, and the sheer depth and pressure of the ocean made deep sea resource recovery a theoretical rather than practical pursuit.

The post-World War II period marked a turning point in humankind's relationship with the ocean. Driven by advances in sonar, underwater robotics, and submarine technology, oceanographers and engineers began to map and explore the seafloor in unprecedented detail. The discovery of hydrothermal vents in the late 1970s, rich with metallic sulfides, provided a new impetus for deep sea mining. These vents, formed by underwater volcanic activity, deposit highly concentrated metals such as gold, silver, and zinc, making them an attractive target for future exploitation.

During this period, industrial players and governments began to see the ocean floor as a potential solution to the growing demand for strategic metals. The Cold War further fueled this interest, as access to rare and strategic minerals became a cornerstone of national security. Nations like the United States and the Soviet Union invested heavily in oceanographic research, spurred not only by economic motives but also by the strategic importance of undersea territory.

The 1960s and 1970s were transformative decades for deep sea mining. For the first time, technological advancements made seabed resource extraction seem achievable. The development of remotely operated vehicles (ROVs) and deep-sea drilling platforms, pioneered by companies like Global Marine and organizations such as the National Oceanic and Atmospheric Administration (NOAA), marked significant milestones.

At the same time, private entities like the multinational consortium Deepsea Ventures began experimental efforts to harvest polymetallic nodules from the Pacific Ocean. Although these attempts were limited in scale and marred by technical and logistical difficulties, they underscored the growing appetite for exploiting deep ocean resources.

In parallel, nations began to grapple with the question of how to govern the resources of the deep seabed. The 1970s saw the birth of the United Nations Convention on the Law of the Sea (UNCLOS), a groundbreaking treaty that sought to establish a legal framework for the world's oceans, including provisions for deep sea mining. The International Seabed Authority (ISA), established under UNCLOS, was tasked with overseeing mining activities and ensuring that benefits derived from seabed resources were shared equitably among nations.

By the 1980s, enthusiasm for deep sea mining waned. The collapse of metal prices and the technical challenges of working at extreme depths made commercial operations economically unviable. For decades, deep sea mining remained a niche interest, confined to academic research and occasional exploratory missions. The technology required to access deep ocean resources was prohibitively expensive, and terrestrial mining operations continued to dominate the global supply of strategic metals.

However, the turn of the 21st century brought a resurgence of interest. As global population growth and industrialization accelerated, the demand for rare earth elements and other strategic metals surged. These materials, critical for the production of electronics, renewable energy technologies, and advanced weaponry, became a linchpin of modern economies. At the same time, terrestrial reserves of these metals faced depletion, geopolitical instability, and environmental degradation.

The new millennium marked a renaissance for deep sea mining, driven by technological advancements, increasing demand for critical resources, and a shifting geopolitical landscape. The advent of high-resolution satellite imaging, advanced AI-driven geospatial analytics, and autonomous underwater systems revolutionized the exploration and assessment of the seabed.

Private companies like Nautilus Minerals and DeepGreen Metals emerged as pioneers in the field, securing exploration licenses in mineral-rich regions of the Pacific and Indian Oceans. Simultaneously, state-backed entities in China, Russia, and India began ramping up their investments in deep sea mining technologies, recognizing the strategic importance of these resources in securing long-term economic and military advantages.

China, in particular, has become a dominant force in the field, leveraging its expertise in AI, robotics, and manufacturing to develop cutting-edge subsea mining systems. This has sparked concerns in the West, as China's control over critical supply chains, coupled with its ambitions in the South China Sea, could lead to a new era of resource-driven geopolitical tension.

As of the 2020s, deep sea mining remains a field poised on the brink of large-scale commercialization. While some view it as the key to unlocking a sustainable supply of critical minerals, others warn of the potential for irreversible environmental harm and the exacerbation of global inequalities.

What is clear, however, is that the future of deep sea mining will be inextricably linked to advances in technology, particularly artificial intelligence. AI promises to address many of the industry's longstanding challenges, from improving resource discovery to optimizing extraction methods and mitigating environmental impacts.

1.2. Scope and Challenges

The immense potential of deep sea mining is undeniable, offering access to vast untapped reserves of critical minerals that are essential for the functioning of advanced economies. Yet this promise is intricately intertwined with a web of complex challenges that must be addressed if this emerging industry is to succeed. The following exploration seeks to define the scope of deep sea mining in its entirety, from the extraordinary resources it aims to harness to the technological, economic, regulatory, and environmental hurdles that lie in its path. This is not merely an academic exercise, but an urgent and necessary analysis of an industry poised to redefine global resource dynamics.

The seabed, covering over 70% of the Earth's surface, is a vast repository of mineral wealth. Beneath the crushing depths of the oceans lie three primary categories of resources: polymetallic nodules, cobalt-rich ferromanganese crusts, and seafloor massive sulfides (SMS). Each of these presents a unique opportunity and a distinct set of extraction challenges. Polymetallic nodules, scattered across the abyssal plains, are a trove of manganese, nickel, cobalt, and copper. These metals are indispensable for industries ranging from steel production to battery technology, particularly in the context of the accelerating global transition to renewable energy systems. Found at depths of 4,000 to 6,000 meters, such nodules are especially abundant in the Clarion-Clipperton Zone (CCZ) in the Pacific Ocean, a region increasingly viewed as the focal point of future mining operations.

Cobalt-rich ferromanganese crusts, in contrast, are located on underwater ridges and seamounts. These crusts, which adhere to rocky substrates, are rich in cobalt, platinum, and rare earth elements. Their potential economic value is immense, given the demand for rare earths in high-tech applications, yet their attachment to the seabed renders extraction a formidable technical challenge. The third major resource category, seafloor massive sulfides, forms around hydrothermal vents, where volcanic activity has deposited rich concentrations of gold, silver, zinc, and copper. These deposits, often boasting metal grades far superior to terrestrial ores, offer an enticing target for mining operations but carry profound ecological risks due to the fragile and unique ecosystems they host.

Geographically, the resources of the deep sea are concentrated in specific zones of economic and strategic interest. The Pacific Ocean is home not only to the Clarion-Clipperton Zone but also to other resource-rich areas such as the Mid-Pacific Mountains and the East Pacific Rise. The Indian Ocean has similarly attracted attention for its SMS deposits and ferromanganese crusts, while the South Atlantic offers untapped potential with implications for the resource strategies of nations in Africa and South America. These resources are distributed across regions governed by international law, as well as within the exclusive economic zones (EEZs) of coastal states, presenting both opportunities and conflicts as nations seek to assert control over undersea wealth.

Transforming these latent resources into economically viable products, however, involves navigating an intricate and highly interdependent value chain. Exploration forms the critical first step, requiring advanced technologies capable of mapping and analyzing the seabed with unprecedented precision. The integration of autonomous underwater vehicles (AUVs), machine learning algorithms, and high-resolution geospatial analytics has begun to revolutionize this phase, enabling the precise identification of mineral-rich zones. Extraction, the next phase, demands innovations in robotics and subsea engineering to overcome the extreme pressures and temperatures of the deep ocean. From robotic platforms designed to harvest nodules to advanced cutting tools for crusts and sulfides, each technological solution must be customized to the unique demands of its target resource.

Processing and refining present further challenges, as marine minerals differ in composition from their terrestrial counterparts, necessitating new metallurgical techniques. This step is followed by transportation, requiring sophisticated logistics networks to move extracted materials from remote oceanic sites to processing hubs. Each stage of this value chain carries its own costs, risks, and technological demands, underscoring the complexity of developing a cohesive and efficient industry.

The physical challenges of deep sea mining are among the most daunting. The extreme environments of the deep ocean, characterized by crushing pressures, near-freezing temperatures, and perpetual darkness, test the limits of human engineering. Conventional machinery is ill-equipped to operate under such conditions, necessitating the development of advanced materials and robust design principles. Human presence in these depths is impractical, making remote and autonomous systems essential. These systems must integrate real-time decision-making capabilities, relying on artificial intelligence to navigate, assess, and perform operations in environments that defy direct human oversight.

Data scarcity exacerbates these physical challenges. Despite decades of oceanographic research, our knowledge of the deep seabed remains limited. Geological and environmental data are sparse, and the high costs of acquiring additional data hinder comprehensive resource assessments. Advanced probabilistic modeling and adaptive learning systems powered by AI hold promise for addressing these gaps, but even these require sufficient initial data to build reliable models. The uncertainty surrounding resource estimates and environmental impacts introduces significant financial risk, complicating investment decisions and project planning.

Economic and regulatory barriers further compound the challenges. The capital intensity of deep sea mining is immense, encompassing the costs of exploration, technological development, and infrastructure deployment. The extended payback periods and volatility of metal prices add layers of financial complexity, deterring private investment and limiting participation to well-funded state-backed initiatives or multinational corporations. Compounding this, terrestrial mining operations, with their established supply chains and economies of scale, continue to dominate the global market for strategic metals, challenging the economic competitiveness of undersea ventures.

Regulatory uncertainties loom large. The United Nations Convention on the Law of the Sea (UNCLOS) provides a legal framework for deep sea mining, yet ambiguities persist, particularly regarding enforcement and the allocation of benefits from seabed resources. The International Seabed Authority (ISA), established under UNCLOS, has issued exploration licenses to various entities but has yet to finalize a comprehensive legal regime for commercial mining. This regulatory vacuum creates uncertainty for investors and complicates the development of industry standards.

Beyond technical and economic concerns, deep sea mining faces intense scrutiny over its environmental and social implications. The ecosystems of the deep ocean are among the least understood and most vulnerable on the planet. Hydrothermal vents, for instance, are home to unique species adapted to extreme conditions, many of which could face extinction if their habitats are disrupted. Similarly, the sediment plumes generated by mining activities could have far-reaching impacts, affecting marine life well beyond the immediate vicinity of extraction sites. Public perception and ethical considerations add further dimensions of complexity. The principle of the "common heritage of mankind," enshrined in UNCLOS, calls for equitable sharing of benefits derived from seabed mining. Yet balancing this ideal with the profit motives of corporations and the strategic interests of nations remains an unresolved challenge.

Deep sea mining represents a bold and transformative frontier, poised to address the growing demand for critical resources while grappling with profound physical, economic, and ethical challenges. Central to overcoming these challenges is the integration of artificial intelligence, a technology that offers unprecedented capabilities for navigating the complexities of this emerging industry. By addressing the scope and challenges outlined here, the potential of deep sea mining can be realized in a manner that aligns technological ambition with responsible stewardship of the planet's last great frontier.

1.3. Role of Artificial Intelligence

Artificial Intelligence (AI) represents a transformative tool in deep sea mining, capable of addressing the technical, logistical, and environmental challenges inherent in exploring and exploiting the ocean's vast mineral wealth. At the heart of this transformation lies a suite of advanced AI methodologies, each tailored to solve specific problems and optimize processes in the complex operational theater of the deep ocean. Understanding these methodologies, machine learning, deep learning, evolutionary algorithms, and reinforcement learning, provides the foundation for appreciating their role in revolutionizing this nascent industry.

Machine Learning: The Cornerstone of Predictive Intelligence

Machine learning (ML), a subfield of AI, excels at extracting patterns from data and making predictions based on historical trends. In deep sea mining, ML is particularly valuable for exploration and resource estimation. Geological and geophysical datasets, often vast and heterogeneous, are processed using algorithms that can detect correlations and infer the likelihood of mineral deposits in specific regions of the seabed.

Supervised learning, one of the most common ML approaches, is instrumental in analyzing labeled datasets obtained from prior explorations. For example, sonar readings, sediment samples, and bathymetric maps can be used to train models to recognize areas likely to contain polymetallic nodules or seafloor massive sulfides. Unsupervised learning, on the other hand, enables clustering of unstructured data to identify patterns without pre-labeled outcomes. This is especially useful in initial reconnaissance, where exploratory data may not yet be categorized.

Applications of ML extend beyond prospecting. Predictive maintenance of subsea equipment, anomaly detection in operational parameters, and real-time optimization of mining activities are all enhanced by the algorithmic capabilities of machine learning. By incorporating historical performance data and sensor readings, ML models can predict equipment failures before they occur, reducing costly downtimes and improving operational efficiency.

Deep Learning: Unlocking Complex Patterns

Deep learning (DL), a subset of machine learning inspired by the architecture of the human brain, involves the use of artificial neural networks to model complex, non-linear relationships in data. Its strength lies in its ability to process high-dimensional and unstructured data types, such as images, audio, and video, making it indispensable in underwater exploration.

Convolutional neural networks (CNNs), a prominent form of deep learning, are particularly adept at processing visual data. In deep sea mining, CNNs can analyze images captured by autonomous underwater vehicles (AUVs) to identify mineral deposits with unprecedented accuracy. These networks can distinguish polymetallic nodules from surrounding sediment, assess the quality of crust formations, and even identify signs of environmental disturbance, such as sediment plumes or damage to benthic habitats.

Deep learning also excels in combining multiple data modalities to create comprehensive models of the seabed. For example, by integrating sonar readings, optical imagery, and geochemical data, DL systems can generate highly accurate three-dimensional maps of resource-rich zones. Recurrent neural networks (RNNs), another form of DL, are used for sequential data analysis, such as interpreting time-series data from oceanographic sensors to predict environmental changes or assess the stability of mining operations.

The self-learning capabilities of deep learning are particularly valuable in unstructured and dynamic environments like the deep ocean. Unlike traditional algorithms that require manual adjustment, deep learning models continuously adapt and improve as they process new data. This adaptability is critical in deep sea mining, where conditions are often unpredictable and rapidly changing.

Evolutionary Algorithms: Mimicking Natural Selection for Optimization

Evolutionary algorithms (EAs) take inspiration from the principles of biological evolution, such as mutation, crossover, and selection, to solve optimization problems. These algorithms are particularly suited to addressing the complex logistical and engineering challenges of deep sea mining, where multiple competing objectives must be balanced.

For instance, EAs can optimize the layout and navigation paths of robotic mining equipment to maximize resource recovery while minimizing environmental impact. By simulating thousands of potential solutions, these algorithms iteratively refine operational strategies to identify the most effective and efficient approaches. This is particularly valuable in environments where traditional optimization techniques struggle due to the high-dimensional and non-linear nature of the problem space.

In addition to route planning, evolutionary algorithms are used in the design of mining tools and systems. For example, they can optimize the configuration of dredging or cutting equipment to balance durability, energy efficiency, and extraction performance. By mimicking the trial-and-error process of natural selection, EAs accelerate the development of innovative solutions that would be infeasible to achieve through conventional engineering methods alone.

Reinforcement Learning: Decision-Making in Dynamic Environments

Reinforcement learning (RL) is a unique branch of AI in which agents learn to make decisions by interacting with their environment and receiving feedback in the form of rewards or penalties. This approach is particularly well-suited to the challenges of autonomous operation in the deep sea, where dynamic and uncertain conditions demand adaptive decision-making.

In the context of deep sea mining, RL agents can control robotic systems, such as AUVs and remotely operated vehicles (ROVs), to navigate complex terrain, avoid obstacles, and perform extraction tasks with minimal human intervention. For example, an RL-based navigation system can learn optimal paths through rugged seafloor landscapes by trial and error, improving its performance with each mission.

One of the key strengths of RL is its ability to handle environments where pre-defined rules are insufficient. For example, during a mining operation, unforeseen events such as equipment malfunctions or unexpected environmental changes can disrupt predefined plans. An RL agent, however, can adapt in real time, recalibrating its strategy to continue operations effectively. This capability is enhanced when combined with multi-agent reinforcement learning, where fleets of robotic systems collaborate to complete tasks efficiently, such as coordinating the collection and transport of mined materials.

Beyond physical operations, reinforcement learning is also applied to resource management and logistics. For instance, RL algorithms can optimize the allocation of energy resources among different subsystems, ensuring efficient power distribution for mining, navigation, and data transmission in energy-constrained environments.

Integrating AI Methods for Synergistic Solutions

While each AI methodology offers distinct advantages, their true potential lies in their integration. For example, machine learning models can provide initial resource predictions, which are then refined by deep learning systems analyzing high-resolution imagery. Evolutionary algorithms can optimize equipment configurations and operational strategies, while reinforcement learning enables autonomous systems to adapt dynamically to real-time conditions. This synergistic approach ensures that the complexities of deep sea mining are addressed holistically, leveraging the strengths of each AI technique to overcome challenges and unlock the full potential of this transformative industry.

As we delve deeper into the role of AI in the following sections, these methodologies will be explored in the context of specific applications, illustrating how they drive innovation and efficiency across every stage of the deep sea mining value chain. From exploration and resource assessment to environmental monitoring and autonomous operation, the integration of AI stands as the linchpin of a new era in subsea resource exploitation.

Artificial intelligence (AI) has emerged as a transformative force capable of addressing the multifaceted challenges within the value chain of deep sea mining. From exploration to extraction, logistics to environmental monitoring, AI-driven innovations are revolutionizing every stage of this emerging industry. By seamlessly integrating data analytics, machine learning, robotics, and decision-making algorithms, AI offers unprecedented opportunities to optimize operations, minimize costs, and ensure sustainability.

AI in Exploration: Mapping the Uncharted

Exploration is the foundation of deep sea mining, demanding precise identification of mineral-rich zones across vast and poorly understood seabed environments. Traditional exploration methods rely on a combination of sonar mapping, sediment sampling, and direct observation, all of which are time-consuming, expensive, and prone to uncertainty. AI is transforming this stage by enabling faster, more accurate, and cost-effective resource identification.

AI-powered geospatial analytics play a pivotal role in processing large datasets obtained from autonomous underwater vehicles (AUVs), remotely operated vehicles (ROVs), and satellite imagery. Machine learning algorithms can analyze bathymetric data to identify patterns indicative of polymetallic nodules, ferromanganese crusts, or seafloor massive sulfides. These models not only pinpoint likely resource locations but also predict their size, composition, and economic value with high precision.

Deep learning techniques further enhance exploration by interpreting high-resolution imagery and multi-sensor data. Convolutional neural networks (CNNs), for instance, can automatically classify geological features in subsea photographs, distinguishing mineral deposits from surrounding sediments or biological matter. Reinforcement learning algorithms enable AUVs to autonomously navigate complex underwater terrains, optimizing their trajectories for maximum data collection.

Moreover, AI-driven predictive modeling helps address one of exploration's most significant challenges: data scarcity. Using sparse datasets from prior expeditions, AI systems employ probabilistic techniques to extrapolate potential resource distributions, reducing the need for extensive field surveys. This not only saves time and resources but also minimizes environmental disturbance in unexplored regions.

AI in Extraction: Enhancing Precision and Efficiency

Once resources are identified, the extraction phase presents its own set of challenges, including extreme pressure, low temperatures, and rugged seafloor conditions. AI is central to overcoming these obstacles, enabling precise, efficient, and adaptive extraction processes that operate autonomously under inhospitable conditions.

Autonomous robotic systems, guided by AI, form the backbone of subsea mining operations. Reinforcement learning algorithms allow these systems to adapt to dynamic conditions in real time, whether navigating uneven terrain or responding to equipment malfunctions. Swarm intelligence techniques, inspired by the collective behavior of biological systems, enable fleets of robots to coordinate their activities, such as clearing paths, collecting nodules, or cutting ferromanganese crusts, while avoiding collisions and optimizing resource recovery.

AI also plays a critical role in equipment optimization. Machine learning models analyze sensor data to monitor the performance and wear of cutting tools, dredging equipment, and hydraulic systems. Predictive maintenance algorithms ensure that potential failures are identified and addressed before they disrupt operations, reducing downtime and extending the lifespan of costly machinery.

In addition, evolutionary algorithms are used to optimize the design and configuration of extraction systems. By simulating thousands of potential designs, these algorithms identify configurations that maximize efficiency while minimizing energy consumption and environmental impact. This approach has proven especially valuable in designing dredging systems for polymetallic nodules and cutting technologies for cobalt-rich crusts.

AI in Logistics: Streamlining Transportation and Resource Flow

The logistical demands of deep sea mining are immense, requiring the efficient transportation of extracted materials from remote oceanic sites to processing facilities. AI is reshaping this phase of the value chain by optimizing transportation routes, fleet management, and supply chain integration.

Route optimization algorithms, powered by machine learning and graph theory, determine the most efficient paths for transport vessels and mining support ships. These algorithms account for variables such as ocean currents, weather conditions, and fuel consumption, ensuring timely and cost-effective delivery of extracted materials. In addition, AI systems can dynamically adjust routes in response to changing conditions, such as adverse weather or equipment delays, minimizing disruptions.

AI also enhances fleet management by enabling real-time monitoring and coordination of transport vessels. Autonomous ships equipped with AI systems can operate with minimal human intervention, reducing operational costs and increasing safety. These systems integrate data from GPS, oceanographic sensors, and onboard cameras to make informed decisions about navigation, collision avoidance, and docking.

Supply chain optimization is another critical area where AI excels. By integrating data from mining sites, transport networks, and processing facilities, AI systems enable end-to-end visibility of resource flows. Predictive analytics models forecast demand for specific minerals, allowing operators to align production schedules with market trends. Blockchain technology, augmented by AI, ensures traceability and provenance, enhancing transparency and compliance with international regulations.

AI in Monitoring: Safeguarding the Environment

Environmental monitoring is a cornerstone of responsible deep sea mining, ensuring that operations do not irreparably harm delicate marine ecosystems. AI-driven systems are revolutionizing this aspect of the value chain by enabling continuous, real-time monitoring of environmental conditions and facilitating adaptive management strategies.

Sensor networks deployed on the seabed collect vast amounts of data on water quality, sediment composition, and biological activity. AI algorithms process this data to detect anomalies, such as increased turbidity or changes in chemical composition, that may indicate environmental disturbances caused by mining activities. By identifying these impacts early, operators can adjust their methods to mitigate harm, ensuring compliance with environmental regulations and minimizing ecological footprints.

Computer vision techniques, powered by deep learning, enable the automatic identification and classification of marine species captured in video footage. This is particularly valuable for monitoring biodiversity in areas surrounding mining operations. AI systems can track population dynamics over time, providing critical insights into the long-term impacts of resource extraction on local ecosystems.

AI also supports the development of predictive models for environmental impact assessments. By simulating different mining scenarios, these models estimate potential effects on marine habitats and guide the design of mitigation measures. For instance, sediment plume dispersion models, enhanced by machine learning, predict the spread of particulate matter generated during extraction, allowing operators to implement strategies to contain or reduce its impact.

Moreover, AI-driven platforms enable transparent reporting and stakeholder engagement. Real-time data visualizations and dashboards provide regulators, environmental organizations, and local communities with access to monitoring results, fostering trust and accountability. These platforms can also facilitate adaptive management, allowing operators to modify their practices in response to feedback from stakeholders or changes in environmental conditions.

AI as the Unifying Force

The integration of AI into the value chain of deep sea mining represents a paradigm shift, enabling operators to tackle the immense technical, logistical, and environmental challenges of this industry. By enhancing exploration accuracy, improving extraction efficiency, optimizing logistics, and safeguarding the environment, AI provides a unifying force that ensures the viability and sustainability of deep sea mining operations.

Through this lens, deep sea mining emerges not just as an industrial activity but as a pioneering field where human ingenuity, technological innovation, and ecological stewardship converge.

2. Geological and Oceanographic Fundamentals for Deep Sea Mining

2.1. Geological Distribution of Deep Sea Mineral Resources

The formation and distribution of mineral resources in the deep ocean are intimately tied to the dynamic processes of the Earth's tectonic system. Plate tectonics, the fundamental driver of geological activity on our planet, shapes the seafloor and creates the conditions necessary for the accumulation of valuable mineral deposits. Among the most significant of these are polymetallic nodules, hydrothermal vent deposits, and cobalt-rich ferromanganese crusts. Each of these resources forms through unique processes that are both geologically intricate and economically significant. Understanding these processes provides a critical foundation for the exploration and exploitation of deep sea minerals.

Hydrothermal Vents and Seafloor Massive Sulfides

Hydrothermal vents are among the most striking geological features on the seafloor, forming primarily along mid-ocean ridges and volcanic arcs. These vents arise from the interaction between seawater and the hot, newly formed crust produced at divergent plate boundaries or in subduction zones. The process begins as cold seawater infiltrates fractures in the oceanic crust, descending to depths where it encounters magma-heated rock. This interaction heats the water to temperatures exceeding 350°C and facilitates a complex chemical exchange between the seawater and the surrounding rock.

The heated, mineral-laden water eventually rises back to the seafloor, where it is expelled through vent openings. Upon contact with the cold seawater, the dissolved minerals precipitate, forming deposits known as seafloor massive sulfides (SMS). These sulfides are rich in valuable metals such as copper, zinc, lead, gold, and silver. The geochemical composition of SMS deposits varies based on the local geology and the specific minerals present in the underlying crust. For example, vents along mid-ocean ridges often yield copper-rich sulfides, while those associated with volcanic arcs may produce higher concentrations of gold and silver.

Hydrothermal vent fields often host complex ecosystems, with organisms that derive energy from chemosynthesis rather than photosynthesis. This biological richness adds an additional layer of complexity and controversy to the exploitation of SMS deposits, as mining could disrupt these fragile ecosystems. Nevertheless, the high metal grades found in SMS deposits, often far exceeding those of terrestrial ores, make them an enticing target for deep sea mining.

Polymetallic Nodules: The Treasures of the Abyssal Plains

Polymetallic nodules, also known as manganese nodules, are potato-sized concretions that litter the vast abyssal plains of the deep ocean. These nodules form over millions of years through a slow process of accretion, where metal ions from seawater and pore fluids in sediments precipitate around a nucleus, often a small fragment of rock or fossil. The resulting nodules are composed of manganese, iron, nickel, cobalt, and copper, with minor amounts of rare earth elements and other trace metals.

The most extensive and economically significant deposits of polymetallic nodules are found in the Clarion-Clipperton Zone (CCZ) of the Pacific Ocean, a region stretching between Hawaii and Mexico. This area is characterized by vast expanses of relatively flat seafloor, which provides an ideal setting for nodule formation. The high concentrations of critical metals in CCZ nodules, combined with their abundance, make them a cornerstone of future deep sea mining prospects.

The formation of polymetallic nodules is influenced by a combination of biological, chemical, and physical factors. Microbial activity plays a significant role in catalyzing the precipitation of manganese and other metals, while the slow sedimentation rates in abyssal environments ensure that nodules remain exposed on the seafloor rather than being buried. Currents and turbulence also play a role, as they prevent excessive sediment accumulation and promote the growth of nodules by allowing continuous access to metal-rich water.

Cobalt-Rich Ferromanganese Crusts: Seamount Guardians

Cobalt-rich ferromanganese crusts form on the flanks and summits of seamounts, ridges, and other underwater elevations. These crusts are thin, hard layers of metal oxides that adhere to rocky substrates, growing at an extraordinarily slow rate, typically less than a millimeter per million years. The primary metals in these crusts include cobalt, manganese, and iron, with significant concentrations of rare earth elements, platinum, and other valuable trace metals.

The formation of these crusts is driven by the precipitation of metals from seawater, a process that occurs under specific oceanographic conditions. Areas with strong currents, low sedimentation rates, and high biological productivity are particularly conducive to crust development. The presence of volcanic rock as a substrate further enhances the formation of ferromanganese crusts, as it provides a stable surface for mineral deposition.

Seamounts in the central Pacific Ocean, particularly those within the EEZs of small island nations such as Kiribati and the Cook Islands, are among the most promising targets for cobalt-rich crust mining. These formations are located at depths of 1,000 to 3,000 meters, making them more accessible than polymetallic nodules or SMS deposits. However, their attachment to the underlying rock presents a significant technical challenge, requiring the development of specialized cutting and collection technologies.

The Role of Tectonic Settings in Resource Distribution

The distribution of these mineral resources is intricately linked to the tectonic processes that shape the Earth's crust. Mid-ocean ridges, where tectonic plates diverge, are the primary sites for hydrothermal vent formation and SMS deposits. These ridges represent the most volcanically active regions on Earth, producing new oceanic crust that interacts with seawater to create the conditions for vent ecosystems and mineralization.

In contrast, the abyssal plains, home to polymetallic nodules, are vast, stable regions of the seafloor located far from tectonic plate boundaries. These areas are characterized by low biological activity and slow sedimentation rates, which allow nodules to grow undisturbed over millions of years. The plains often lie adjacent to tectonic plate margins, where upwelling currents bring metal-rich water to the surface.

Seamounts and ridges, where cobalt-rich crusts form, are associated with both volcanic activity and tectonic uplift. These features create topographic highs that interact with ocean currents, promoting the precipitation of metals from seawater. Many of these seamounts are remnants of extinct volcanoes, providing a stable substrate for crust formation over geological timescales.

Geological Masterpieces of the Deep

The formation processes and tectonic settings of deep sea mineral resources reveal a remarkable interplay of geological forces, chemical reactions, and biological activity. Hydrothermal vents, polymetallic nodules, and cobalt-rich ferromanganese crusts each represent a distinct geological masterpiece, shaped by the Earth's dynamic crustal processes and the unique conditions of the deep ocean.

Understanding these processes is not merely an academic pursuit but a practical necessity for the future of deep sea mining. The mineral wealth of the seafloor holds immense economic potential, but its distribution and formation are deeply tied to the geological contexts in which these resources are found. This knowledge serves as the foundation for exploration strategies, guiding efforts to locate, assess, and ultimately extract these treasures of the deep with precision and care.

Mapping the seafloor is one of the most critical and challenging aspects of deep sea mining, as the success of exploration and subsequent resource extraction depends on the ability to accurately understand and visualize the terrain, mineral distributions, and geological structures of the seabed. Despite advances in technology, significant challenges remain in achieving the necessary resolution, coverage, and accuracy to map vast, remote, and complex underwater landscapes. These challenges are amplified by the unique physical and environmental conditions of the deep ocean, making seafloor mapping a demanding endeavor requiring sophisticated tools and methodologies.

Resolution: The Quest for Detail in an Unknown World

The resolution of seafloor maps refers to the level of detail with which features can be discerned and measured. High-resolution maps are crucial for identifying mineral-rich areas, characterizing geological formations, and planning extraction operations. However, achieving such resolution in the deep sea is a formidable challenge due to the vast scale of the ocean floor and the limitations of current mapping technologies.

Satellite-based methods, such as satellite altimetry, provide global coverage of the seafloor by measuring variations in the sea surface caused by gravitational anomalies. While these methods are invaluable for large-scale mapping, they offer only low-resolution data, typically on the order of kilometers. This is insufficient for the detailed exploration required for deep sea mining, where individual mineral deposits or seafloor features may span only a few meters.

In contrast, ship-based multibeam sonar systems are capable of producing high-resolution maps with resolutions of tens to hundreds of meters, depending on depth. These systems emit acoustic waves and measure the time it takes for the waves to reflect back from the seafloor, creating detailed bathymetric profiles. However, even with multibeam sonar, achieving fine resolution over large areas is time-intensive and resource-intensive. For example, mapping the Clarion-Clipperton Zone (CCZ) alone, a prime area for polymetallic nodules, would require years of dedicated survey time due to its immense size.

At the highest resolutions, autonomous underwater vehicles (AUVs) and remotely operated vehicles (ROVs) are used to conduct close-range surveys of specific areas. Equipped with high-frequency sonar, cameras, and laser scanners, these platforms can achieve centimeter-scale resolution, capturing fine details of seafloor topography and sediment characteristics. However, their limited operational range and high deployment costs restrict their use to focused surveys rather than large-scale mapping.

Coverage: The Immensity of the Ocean Floor

The sheer size of the ocean floor presents a monumental challenge to comprehensive mapping efforts. Covering more than 70% of the Earth's surface, the seafloor is the largest unexplored frontier on the planet. Despite centuries of oceanographic exploration, less than 25% of the global seabed has been mapped in detail, and even this coverage is unevenly distributed, with most high-resolution data concentrated in coastal and economically important areas.

Achieving sufficient coverage for deep sea mining operations requires a balance between efficiency and detail. Large-scale surveys are typically conducted using ship-based multibeam sonar, which can cover thousands of square kilometers in a single mission. However, this approach is limited by the trade-off between coverage and resolution: the deeper the water, the lower the resolution of the resulting maps due to the wider spread of acoustic beams.

Expanding coverage further requires integrating data from multiple sources, including satellites, ships, and underwater vehicles. AI and machine learning are increasingly used to combine these datasets, filling gaps and extrapolating missing information to create more comprehensive maps. For example,

AI algorithms can analyze patterns in existing bathymetric data to predict the topography of unmapped regions, reducing the need for extensive field surveys.

Despite these advances, large areas of the deep ocean remain unmapped, particularly in remote regions beyond national jurisdictions. The logistical challenges of deploying survey vessels to these areas, coupled with the high costs of operation, limit the ability to achieve comprehensive coverage. This lack of data poses significant risks for deep sea mining, as unmapped areas may contain unforeseen geological hazards or environmental sensitivities.

Bathymetric Complexity: The Challenges of Uneven Terrain

The seafloor is far from a uniform landscape; it is a complex and dynamic environment shaped by tectonic activity, sedimentation, and biological processes. This complexity presents significant challenges for mapping and interpreting bathymetric data, particularly in regions with rugged terrain, such as mid-ocean ridges, seamounts, and volcanic arcs.

Mid-ocean ridges, where tectonic plates diverge, are among the most geologically active regions on Earth. These ridges are characterized by steep slopes, fissures, and lava flows, creating a highly irregular topography that is difficult to map with conventional sonar systems. The presence of hydrothermal vents and associated sulfide deposits further complicates the terrain, requiring high-resolution imaging to identify and assess potential mining targets.

Seamounts, underwater mountains formed by volcanic activity, pose similar challenges. These features often rise thousands of meters above the surrounding seafloor and are covered with cobalt-rich ferromanganese crusts. Mapping seamounts requires advanced sonar systems capable of capturing steep gradients and irregular surfaces. The ruggedness of these features also necessitates careful navigation by survey vessels and AUVs to avoid collisions and ensure accurate data collection.

In addition to geological complexity, sediment dynamics add another layer of difficulty. Abyssal plains, which are the primary habitat for polymetallic nodules, may appear flat on a large scale but often contain subtle variations in sediment thickness and composition. These variations can affect the distribution of nodules and must be accounted for in mapping efforts. High-resolution sediment mapping, often achieved through sub-bottom profilers and sediment coring, is essential for understanding these micro-topographical features.

The interpretation of bathymetric data is further complicated by the limitations of acoustic technology in complex environments. Acoustic shadowing, caused by steep slopes or overhangs, can obscure parts of the seafloor, creating gaps in the data. Advanced processing techniques, including machine learning algorithms, are increasingly used to reconstruct missing information and enhance the accuracy of bathymetric models.

The Role of Technology and Innovation

Overcoming the challenges of resolution, coverage, and bathymetric complexity requires continual advancements in mapping technologies. The integration of multiple data collection platforms, from satellites to AUVs, is key to achieving comprehensive and accurate seafloor maps. Artificial intelligence is playing a transformative role in processing and analyzing the vast datasets generated by these platforms, enabling more efficient and precise mapping.

Emerging technologies, such as synthetic aperture sonar and multi-modal imaging systems, promise to further enhance mapping capabilities. Synthetic aperture sonar, for example, can achieve higher resolutions over larger areas than traditional sonar systems by combining multiple acoustic signals into a single, detailed image. Multi-modal systems, which integrate sonar, optical, and magnetic sensors, provide a more complete picture of the seafloor by capturing different types of data simultaneously.

As these technologies evolve, the cost and logistical barriers to seafloor mapping are likely to decrease, enabling more extensive surveys and higher-quality data. However, the challenges of mapping the deep ocean will remain significant, requiring ongoing investment, innovation, and collaboration among governments, industries, and research institutions.

Navigating the Unknown

Seafloor mapping is the gateway to deep sea mining, providing the critical information needed to locate, assess, and extract valuable mineral resources. Yet the challenges of achieving sufficient resolution, coverage, and accuracy in the vast and complex environment of the deep ocean cannot be understated. The interplay of technological, logistical, and environmental factors makes mapping a demanding and resource-intensive process.

Despite these challenges, the advances in sonar systems, autonomous vehicles, and AI-driven data analytics offer a promising path forward. By addressing the limitations of current mapping technologies and embracing innovative solutions, the deep sea mining industry can unlock the potential of the seafloor while minimizing risks and uncertainties.

2.2. Oceanographic and Environmental Conditions

The environmental conditions of the deep ocean are as alien as any extraterrestrial landscape, presenting unique challenges and opportunities for deep sea mining. The interplay of temperature, pressure, ocean currents, and the sub-seafloor environment creates a complex operational theater, where the extremes of nature test the limits of technology and ingenuity. Understanding these conditions is essential not only for designing effective mining systems but also for assessing the potential impacts of resource extraction on the fragile ecosystems that thrive under such extreme circumstances.

Temperature: The Abyssal Chill and Thermal Extremes

The deep ocean is a domain of perpetual cold. At depths below 1,000 meters, where sunlight no longer penetrates, temperatures hover just above freezing, typically ranging between 1°C and 4°C. This thermal uniformity is broken only in proximity to hydrothermal vents, where superheated fluids, enriched with dissolved minerals, can exceed 350°C upon expulsion. The stark contrast between vent temperatures and the surrounding water creates steep thermal gradients, which pose both challenges and opportunities for mining operations.

For equipment and robotics, the near-freezing ambient temperature of the deep sea necessitates materials and components capable of withstanding prolonged exposure to cold without compromising structural integrity or functionality. Thermal expansion and contraction must be carefully managed, particularly in systems that transition between the cold deep ocean and the warmer surface environment. Electronics must be insulated against condensation and frost, while hydraulic systems require low-temperature fluids to maintain performance.

Conversely, hydrothermal vents represent a source of localized thermal energy that could potentially be harnessed to power mining operations. Advances in thermoelectric technology might enable the conversion of heat differentials into usable energy, providing a sustainable power source for autonomous mining platforms. However, the volatility of vent systems, which can exhibit rapid temperature fluctuations and corrosive chemical environments, demands robust and adaptive engineering solutions.

Pressure: The Crushing Depths

Pressure in the deep ocean increases by approximately one atmosphere (1 bar) for every 10 meters of depth, reaching over 600 bars (6,000 atmospheres) at depths of 6,000 meters. This immense pressure is one of the greatest engineering challenges of deep sea mining, as it imposes extraordinary demands on equipment and systems operating at such depths.

The structural integrity of mining equipment is paramount. Materials must be capable of withstanding compressive forces that would crush conventional machinery. Titanium alloys, high-strength steels, and advanced composites are commonly used for their ability to endure extreme pressure without deforming or failing. Additionally, the design of pressure-resistant housings for electronics, sensors, and cameras is critical to ensuring the functionality of these components in the deep ocean.

Pressure also affects fluid dynamics and mechanical systems. Hydraulic actuators, pumps, and valves must be specifically engineered to function reliably under high-pressure conditions. Similarly, the behavior of gases, such as those used in buoyancy systems, is altered by pressure, requiring precise control mechanisms to ensure stability and predictability during operations.

For autonomous underwater vehicles (AUVs) and remotely operated vehicles (ROVs), pressure-resistant design is a fundamental requirement. Advanced computational modeling and testing are used to optimize the shape and structure of these vehicles, minimizing stress concentrations and enhancing their ability to navigate and perform tasks at great depths.

Ocean Currents: Dynamic Forces of the Deep

Ocean currents in the deep sea are less intense than those in surface waters but still exert a significant influence on mining operations. These currents, driven by thermohaline circulation and influenced by topographical features, can vary in speed and direction, creating dynamic forces that affect navigation, stability, and sediment dispersal.

At depths beyond 1,000 meters, currents are typically slower, averaging between 0.02 and 0.1 meters per second. However, localized phenomena, such as internal waves and eddies, can generate stronger flows, particularly around seamounts, ridges, and other bathymetric features. These currents pose challenges for the precise positioning of mining equipment and vehicles, as even small forces can disrupt stability and increase energy consumption for station-keeping.

Currents also play a critical role in the dispersal of sediments and plumes generated by mining activities. Understanding the behavior of these flows is essential for predicting the environmental impact of resource extraction. Computational fluid dynamics (CFD) models, enhanced by AI, are used to simulate current patterns and assess the potential spread of sediments, enabling operators to design mitigation measures that minimize ecological disruption.

On the positive side, ocean currents offer opportunities for energy generation. Kinetic energy harvesting systems, such as underwater turbines, could be deployed to convert current flow into electricity, providing a renewable power source for deep sea mining operations. Such systems would need to be carefully integrated into mining platforms to ensure compatibility with operational requirements and environmental constraints.

The Sub-Seafloor Environment: Hidden Complexities

Beneath the seabed lies the sub-seafloor environment, a largely unexplored frontier that holds both geological mysteries and operational challenges. The sub-seafloor is composed of sediment layers, basaltic crust, and, in some regions, mineral-rich deposits associated with hydrothermal systems. The physical and geochemical properties of these layers have significant implications for mining operations.

Sediment layers vary in thickness and composition, influencing the accessibility of resources such as polymetallic nodules. In areas like the Clarion-Clipperton Zone, nodules rest on or just below the sediment surface, requiring minimal disturbance for collection. However, variations in sediment density and cohesion can complicate extraction, necessitating adaptive mining technologies capable of handling diverse substrate conditions.

In regions targeted for seafloor massive sulfides, the sub-seafloor environment is characterized by intricate networks of mineralized veins and fractures. These formations result from the circulation of hydrothermal fluids, which deposit metals as they cool. Extracting these resources requires detailed geological mapping to identify high-grade zones and avoid unnecessary disruption to the surrounding rock.

Additionally, the sub-seafloor environment hosts microbial communities that play a role in biogeochemical cycles. These microorganisms, adapted to extreme conditions, may influence the stability and geochemistry of mining sites. Understanding their interactions with minerals and fluids is critical for predicting the long-term effects of resource extraction and for designing operations that minimize ecological disturbance.

The Challenges of an Alien Frontier

The environmental conditions of the deep ocean, its frigid temperatures, immense pressures, dynamic currents, and complex sub-seafloor geology, create one of the most challenging operational environments on Earth. Yet these same conditions hold immense promise for technological innovation and resource discovery. To succeed in this frontier, deep sea mining operations must integrate cutting-edge engineering, advanced materials, and sophisticated modeling tools to navigate these extremes.

The availability and distribution of mineral resources in the deep ocean are shaped by a dynamic interplay of biotic (biological) and abiotic (non-biological) factors. This relationship is crucial to understanding the geological processes that create valuable deposits such as polymetallic nodules, seafloor massive sulfides (SMS), and cobalt-rich ferromanganese crusts. While abiotic processes dominate resource formation on a macro scale, biotic influences play an often-overlooked but critical role in modifying, stabilizing, and even initiating certain mineral deposits. A nuanced understanding of these interactions is essential for developing effective exploration and extraction strategies.

Abiotic Factors: The Geophysical Engine of Resource Formation

The primary forces driving the formation of deep sea mineral deposits are abiotic in nature, rooted in geological, chemical, and physical processes that operate over vast timescales. Plate tectonics, hydrothermal activity, and sedimentation govern the formation and distribution of mineral-rich environments across the ocean floor.

Hydrothermal vent systems, for example, are the quintessential product of abiotic processes. Located along mid-ocean ridges and subduction zones, these systems are powered by magma-induced heat that drives seawater circulation through the crust. As seawater interacts with the hot rock, it leaches metals and other elements, which then precipitate to form SMS deposits upon cooling.

The composition of these deposits, rich in copper, zinc, gold, and silver, is determined by the geochemistry of the underlying rock and the fluid dynamics of the vent system.

Similarly, the slow accretion of polymetallic nodules on abyssal plains is driven by abiotic factors such as the precipitation of manganese, iron, nickel, and cobalt from seawater and pore fluids. The nodules grow over millions of years as metals accumulate layer by layer around a nucleus, typically a fragment of rock or shell. The sedimentation rate, oceanic currents, and the chemistry of the water column are key abiotic parameters that influence the size, distribution, and composition of these nodules.

Cobalt-rich ferromanganese crusts, which form on the flanks and summits of seamounts, are also the product of long-term abiotic processes. These crusts develop through the precipitation of metals from seawater under specific conditions of temperature, pressure, and current dynamics. The morphology and thickness of these crusts are closely tied to the geological features of the seamounts, as well as the prevailing oceanographic conditions.

Biotic Factors: Life's Role in Shaping Mineral Resources

Although abiotic processes dominate the formation of deep sea mineral deposits, biotic factors exert a significant influence, particularly in modifying and enhancing resource availability. Microbial activity, in particular, plays a pivotal role in the geochemical cycles that contribute to mineral deposition.

In hydrothermal vent systems, microbial communities thrive in extreme conditions, utilizing chemical energy from vent fluids to sustain themselves in the absence of sunlight. These chemosynthetic organisms play an active role in mineral precipitation by altering the chemical composition of vent fluids. For example, microbial mats near vents can influence the oxidation state of metals such as iron and manganese, promoting their precipitation as sulfides or oxides. The presence of microbial biofilms on SMS deposits can also impact the stability and geochemistry of these resources, potentially affecting their extractive value.

In the case of polymetallic nodules, microbial activity within the surrounding sediments has been shown to facilitate the accumulation of metals. Microorganisms mediate redox reactions that control the mobility and availability of elements such as manganese and cobalt. Over geological timescales, this microbial influence contributes to the growth and chemical composition of the nodules, underscoring the symbiotic relationship between biotic and abiotic processes.

The formation of cobalt-rich crusts is similarly influenced by biological factors. The surfaces of seamounts, where these crusts develop, often host microbial communities that catalyze metal precipitation. These microbes can interact with seawater to promote the deposition of manganese and other metals, enhancing the growth of ferromanganese crusts. Additionally, the biological productivity of the overlying water column, driven by planktonic organisms, can influence the availability of organic matter, which in turn affects the geochemistry of the seafloor.

Interdependence of Biotic and Abiotic Influences

The interplay between biotic and abiotic factors is particularly evident in environments where both forces operate simultaneously, creating feedback loops that influence resource formation. For instance, the activity of hydrothermal vents is fundamentally abiotic, driven by tectonic and magmatic processes. However, the biological communities that colonize these vents can modify the local geochemistry, impacting the composition and structure of the resulting mineral deposits.

In abyssal plains, the slow sedimentation rates that expose polymetallic nodules to metal-rich water columns are influenced by abiotic factors such as ocean currents. Yet, the microbial ecosystems within the sediments play a crucial role in cycling nutrients and metals, ensuring the continuous availability of elements necessary for nodule growth.

Seamounts, where cobalt-rich crusts are found, exemplify this interdependence. The abiotic conditions of strong currents and low sedimentation create an ideal environment for crust formation. At the same time, the biological activity of microbes and higher organisms contributes to the geochemical conditions that sustain metal precipitation. These intertwined processes highlight the necessity of understanding both biotic and abiotic influences when assessing resource potential.

Challenges and Implications for Deep Sea Mining

The dual influence of biotic and abiotic factors on resource availability presents unique challenges for deep sea mining. While abiotic processes determine the bulk formation and distribution of resources, biotic influences can introduce variability in their composition and accessibility. For instance, microbial activity might enhance the concentration of certain metals in a deposit, making it more economically valuable, but it could also create geochemical heterogeneities that complicate extraction.

Moreover, the role of biotic factors underscores the importance of preserving the ecological integrity of deep sea environments. The microbial communities that influence mineral deposits are integral to the broader ecosystem and may play roles in nutrient cycling and carbon sequestration that are not yet fully understood. Disrupting these communities through mining activities could have unintended consequences, both locally and globally.

From an operational perspective, understanding the biotic and abiotic interplay is critical for designing effective exploration and extraction strategies. Advanced technologies, such as AI-driven geochemical modeling and in situ microbial analysis, can provide insights into the factors shaping resource availability, enabling more precise targeting of valuable deposits while minimizing environmental impact.

A Complex Interplay

The availability of deep sea mineral resources is governed by a complex interplay of abiotic and biotic factors, each contributing to the formation, distribution, and composition of these deposits. While geological and physical processes provide the primary framework for resource formation, biological influences add an additional layer of complexity, modifying and enhancing the characteristics of mineral deposits.

For the deep sea mining industry, understanding these interactions is not merely an academic exercise but a practical necessity. The integration of biotic and abiotic insights into exploration and extraction methodologies will be essential for unlocking the potential of the deep ocean's mineral wealth while ensuring the sustainability of its ecosystems.

2.3. Data Acquisition and Sensor Technologies

The exploration and exploitation of deep sea mineral resources hinge on the ability to collect accurate and comprehensive data from one of the most challenging environments on Earth. Advanced marine instrumentation forms the backbone of this effort, enabling the detection, analysis, and assessment of valuable deposits at depths that exceed the reach of conventional methods. Central to these capabilities are autonomous underwater vehicles (AUVs), remotely operated vehicles (ROVs), sensor suites, and sampling tools, each of which plays a unique role in overcoming the technical and environmental challenges of deep sea exploration.

Autonomous Underwater Vehicles (AUVs): The Vanguard of Autonomous Exploration

Autonomous underwater vehicles (AUVs) have revolutionized deep sea exploration by providing a platform for high-resolution mapping and data acquisition in environments where human presence is impractical. These untethered, self-propelled systems are equipped with sophisticated onboard sensors and navigation systems, allowing them to operate independently for extended periods. AUVs are particularly well-suited to large-scale surveys of mineral-rich regions, such as the Clarion-Clipperton Zone (CCZ), where their ability to cover vast areas efficiently is invaluable.

Modern AUVs are equipped with multibeam sonar systems for bathymetric mapping, side-scan sonar for imaging seafloor textures, and sub-bottom profilers for penetrating sediment layers to reveal subsurface structures. These instruments enable detailed characterization of the seabed, identifying potential mineral deposits and geological features of interest. AUVs also incorporate inertial navigation systems, Doppler velocity logs, and acoustic positioning systems, ensuring precise navigation and data georeferencing even in areas without GPS coverage.

The modular design of AUVs allows them to be customized for specific missions. For example, advanced models are equipped with high-resolution cameras and laser scanners for visual mapping, magnetometers for detecting ferromagnetic minerals, and water samplers for analyzing chemical signatures in the water column. These capabilities make AUVs indispensable for pre-extraction surveys, providing the high-resolution data needed to assess the economic potential of target sites.

Remotely Operated Vehicles (ROVs): Precision and Control in Extreme Depths

Remotely operated vehicles (ROVs) complement AUVs by providing the precision and control required for tasks that demand direct human oversight.

These tethered systems are controlled in real-time from a surface vessel, allowing operators to perform detailed inspections, sampling, and intervention tasks. ROVs are particularly valuable in deep sea mining for tasks such as collecting polymetallic nodules, sampling hydrothermal vent deposits, and deploying monitoring equipment.

ROVs are equipped with a wide array of tools and sensors tailored to specific applications. High-definition cameras and lighting systems provide real-time video feeds for visual inspection, while robotic arms with interchangeable end-effectors enable precise manipulation of objects and sampling of sediments or mineral deposits. Many ROVs also carry specialized instruments, such as spectrometers for in situ chemical analysis, and corers for extracting sediment samples.

The tethered nature of ROVs offers the advantage of unlimited power and data transmission, enabling the use of high-energy instruments and real-time data processing. However, the reliance on a tether also limits their operational range and mobility compared to AUVs. Despite these limitations, ROVs remain essential for tasks that require human decision-making and adaptability, such as selecting specific areas for sampling or responding to unexpected challenges during a mission.

Sensor Suites: The Eyes and Ears of Deep Sea Exploration

Sensor suites form the core of marine instrumentation, providing the data necessary to locate and characterize mineral deposits. These suites typically combine multiple sensors, each designed to measure specific physical, chemical, or biological parameters of the deep sea environment. The integration of these sensors into AUVs, ROVs, and other platforms enables comprehensive data acquisition across multiple dimensions.

Acoustic sensors, including multibeam and side-scan sonar, are fundamental for mapping the seafloor and identifying geological features indicative of mineral deposits. Sub-bottom profilers extend this capability by imaging subsurface layers, revealing buried structures and sediment compositions that are not visible on the surface.

Optical sensors, such as high-resolution cameras and laser scanners, provide detailed visual and topographical data. These instruments are particularly useful for identifying the distribution and density of polymetallic nodules, assessing the morphology of hydrothermal vent structures, and detecting biotic features that may influence resource availability or environmental impact.

Chemical sensors, including spectrometers and water samplers, measure the concentrations of dissolved elements and compounds in the water column. These sensors are critical for detecting the geochemical signatures associated with hydrothermal activity, such as elevated levels of iron, manganese, and sulfides, which indicate the presence of seafloor massive sulfides.

Magnetometers and gravimeters are used to detect variations in the Earth's magnetic and gravitational fields, which can signal the presence of ferromagnetic minerals or dense ore bodies. These sensors are particularly valuable for exploring areas with complex geology, such as volcanic arcs and seamounts.

Biological sensors, including environmental DNA (eDNA) samplers and plankton nets, provide data on the biotic components of the deep sea environment. While not directly related to mineral extraction, these sensors help assess the potential ecological impacts of mining and inform strategies for mitigating harm to deep sea ecosystems.

Sampling Tools: Ground Truthing the Data

While sensors provide critical remote data, direct sampling is essential for validating and refining the interpretations drawn from sensor measurements. Sampling tools collect physical samples of sediments, rocks, and water, enabling detailed laboratory analysis that complements in situ data.

Core samplers, including gravity corers and box corers, are used to extract cylindrical sections of seafloor sediment. These samples reveal the stratigraphy, composition, and physical properties of the substrate, providing insights into the processes that formed and shaped mineral deposits. Core samples are also analyzed for the presence of microfossils, organic matter, and trace elements, which can indicate the environmental history of the site.

Grab samplers and dredges are used to collect larger samples of surface sediments and rocks, including polymetallic nodules and fragments of seafloor massive sulfides. These tools are particularly valuable for assessing the mineral content and economic potential of deposits.

Water samplers, such as Niskin bottles and CTD rosettes, collect water from specific depths for chemical and physical analysis. These samples provide information on temperature, salinity, and dissolved element concentrations, which are critical for understanding the geochemical processes that influence resource formation.

Advanced sampling tools, such as in situ microdrills and laser ablation systems, are capable of collecting small, precise samples directly from mineral surfaces. These tools are used in conjunction with ROVs and AUVs to target specific features of interest, minimizing environmental disturbance while maximizing the quality of the data collected.

The Technological Backbone of Deep Sea Mining

Advanced marine instrumentation, encompassing AUVs, ROVs, sensor suites, and sampling tools, represents the technological backbone of deep sea mining. These systems enable the collection of high-resolution, multi-dimensional data that is essential for identifying, characterizing, and assessing mineral resources in the deep ocean. Each platform and tool brings unique capabilities to the exploration process, ensuring that operators can navigate the immense challenges of this environment with precision and efficiency.

The integration of these technologies into a cohesive data acquisition strategy is critical for the success of deep sea mining. As the field continues to evolve, innovations in sensor technology, vehicle design, and data processing will further enhance the capabilities of marine instrumentation, paving the way for a new era of resource exploration and sustainable extraction. Through these advancements, the mysteries of the deep ocean can be unlocked, transforming its mineral wealth into a cornerstone of the global economy.

The integration of geophysical, geochemical, and biological data is at the core of advanced deep sea mining strategies. This interdisciplinary approach ensures that the full complexity of the seafloor and its resources is understood, enabling precise resource identification, effective extraction planning, and mitigation of environmental impacts. By combining diverse data streams into cohesive models, operators can address the unique challenges posed by the deep ocean, unlocking its mineral wealth while maintaining ecological stewardship.

Geophysical Data: Mapping the Physical Framework

Geophysical data provides the foundational understanding of the seafloor's physical structure, including its topography, subsurface layers, and tectonic features. These datasets are primarily obtained through acoustic and electromagnetic techniques, such as multibeam sonar, side-scan sonar, sub-bottom profiling, and magnetometry.

Multibeam sonar is indispensable for creating detailed bathymetric maps, revealing the seafloor's topography and identifying features associated with mineral deposits, such as hydrothermal vent fields or abyssal plains rich in polymetallic nodules. Sub-bottom profilers complement this by imaging the sediment layers beneath the surface, highlighting buried deposits or geological structures indicative of mineralization processes. Magnetometry detects variations in the Earth's magnetic field, pinpointing areas with high concentrations of ferromagnetic minerals, such as cobalt and nickel, often associated with massive sulfide deposits or ferromanganese crusts.

Geophysical data serves as the skeleton of integrated resource models, defining the physical context within which geochemical and biological processes operate. Accurate mapping of faults, ridges, and volcanic structures also aids in assessing geohazards, such as slope instability or potential seismic activity, which could impact mining operations.

Geochemical Data: Unlocking the Composition of Resources

Geochemical data reveals the elemental and molecular composition of the seafloor and its surrounding environment, offering insights into the processes that form and concentrate valuable minerals. This data is collected through both in situ sensors and direct sampling, including water samplers, sediment cores, and rock dredges.

Chemical sensors, such as mass spectrometers and spectrophotometers, measure concentrations of dissolved elements like manganese, cobalt, copper, and rare earth elements in the water column and pore fluids. These measurements help identify active hydrothermal systems or sedimentary environments conducive to nodule growth. Subsurface sampling, combined with laboratory analysis, further refines this understanding, providing detailed mineralogical profiles of specific deposits.

The geochemical signature of a site often acts as a fingerprint, linking resource potential to specific geological processes. For instance, elevated levels of sulfides and metals in hydrothermal fluids indicate active vent systems capable of producing seafloor massive sulfides. Similarly, the presence of certain isotopes can reveal the age and growth rate of polymetallic nodules, guiding exploration efforts toward regions with higher economic potential.

Biological Data: The Living Landscape of the Deep

Biological data adds an essential dimension to resource modeling by documenting the diversity, distribution, and ecological roles of deep sea organisms.

This information is critical for understanding the interplay between biological processes and resource availability, as well as for assessing the environmental impacts of mining activities.

Environmental DNA (eDNA) analysis has emerged as a powerful tool for characterizing biological communities without the need for direct sampling. By extracting DNA from water or sediment samples, researchers can identify the species present in an area, including those that are difficult or impossible to observe through traditional methods. eDNA analysis is particularly useful in regions surrounding hydrothermal vents and abyssal plains, where microbial activity plays a role in mineral formation and cycling.

Visual surveys conducted with high-resolution cameras and video systems provide additional biological data, capturing the distribution of megafauna, such as deep sea corals and fish, as well as smaller organisms that inhabit sediments and mineral deposits. These observations help delineate habitats and identify potential areas of ecological sensitivity.

Microbial data is especially relevant in the context of mineral formation. Chemosynthetic microbes near hydrothermal vents, for example, influence the precipitation of metals like iron and manganese, directly contributing to the formation of seafloor massive sulfides. Similarly, microbial communities in abyssal sediments play a role in the redox reactions that control the availability of elements like cobalt and nickel, essential for nodule growth.

Data Integration: Building a Holistic Understanding

Integrating geophysical, geochemical, and biological data into a unified framework requires advanced computational tools and analytical techniques. Artificial intelligence (AI) and machine learning play a pivotal role in this process, enabling the analysis of complex, multi-dimensional datasets to identify patterns and correlations that might not be apparent through traditional methods.

Geospatial modeling software combines geophysical maps, geochemical profiles, and biological observations to create three-dimensional representations of potential mining sites. These models allow operators to visualize the relationships between physical features, mineral concentrations, and ecological zones, supporting more informed decision-making.

For example, in regions targeted for polymetallic nodules, integrated models can overlay bathymetric maps with sediment composition data and biological distributions. This helps identify areas with the highest resource density while avoiding ecologically sensitive zones. Similarly, in hydrothermal vent fields, combining geochemical signatures with biological observations can distinguish between active and inactive systems, prioritizing deposits that are less likely to impact living ecosystems.

Predictive analytics further enhance this integration by simulating future scenarios. Models can predict the spread of sediment plumes generated by mining activities, estimate the regeneration rates of biological communities, and assess the potential economic yield of resource extraction. These simulations provide a dynamic and adaptive framework for managing operations in real time.

Challenges and Innovations in Data Integration

Despite its potential, integrating geophysical, geochemical, and biological data presents significant challenges. The sheer volume and heterogeneity of data collected from the deep sea require robust storage, processing, and analysis infrastructure. Ensuring compatibility between datasets, collected at different scales and resolutions, is another obstacle that demands sophisticated interpolation and standardization techniques.

To address these challenges, cloud-based platforms and edge computing solutions are increasingly being employed. These technologies enable real-time data processing and sharing, ensuring that operators have access to up-to-date information from multiple sources. Additionally, advancements in sensor technology, such as multi-modal instruments that collect geophysical, geochemical, and biological data simultaneously, are streamlining the data acquisition process.

A Unified Approach to Deep Sea Exploration

The integration of geophysical, geochemical, and biological data represents a paradigm shift in deep sea exploration and mining. By combining these diverse datasets into cohesive models, operators can achieve a comprehensive understanding of the seafloor's mineral resources and ecological dynamics. This holistic approach not only enhances the precision and efficiency of exploration but also ensures that mining activities are conducted responsibly, with minimal impact on fragile marine ecosystems.

As the field of deep sea mining continues to evolve, the role of integrated data systems will only grow in importance. By leveraging cutting-edge technologies and interdisciplinary insights, the industry can navigate the challenges of the deep ocean while unlocking its immense potential. In doing so, it sets a standard for innovation and sustainability in one of the most challenging and promising frontiers of human enterprise.

3. AI for Exploration and Resource Characterization

3.1. Machine Learning for Mineral Prospectivity

Pattern recognition in geospatial data is one of the most transformative applications of machine learning (ML) in mineral prospectivity for deep sea mining. The vast scale of the seafloor, coupled with the high costs and logistical challenges of traditional exploration, demands a more efficient and accurate approach to identifying mineral-rich zones. Machine learning offers a solution by processing complex, multi-dimensional datasets to uncover patterns and relationships that indicate the presence of valuable mineral deposits. Through advanced algorithms and data integration, ML-driven pattern recognition enables operators to prioritize exploration areas, optimize resource allocation, and significantly reduce uncertainty.

The Challenges of Geospatial Data in Deep Sea Mining

Geospatial data in the context of deep sea mining encompasses a wide range of information, including bathymetric maps, sonar images, sediment composition, magnetic anomalies, and chemical gradients. These datasets are often collected at varying resolutions and scales, presenting a challenge for consistent analysis. Additionally, the seafloor's vastness and heterogeneity mean that mineral deposits are often scattered across remote and inaccessible regions, making exhaustive exploration impractical.

Traditional approaches to interpreting geospatial data rely heavily on manual analysis, where experts visually inspect maps and datasets for geological features associated with mineralization. While effective in certain contexts, this method is limited by human biases and the sheer volume of data generated by modern exploration technologies. Machine learning transcends these limitations, offering automated and scalable tools for extracting meaningful patterns from complex datasets.

Supervised Learning for Deposit Prediction

Supervised learning is a cornerstone of ML-based pattern recognition in geospatial data. In this approach, algorithms are trained on labeled datasets where the presence or absence of mineral deposits is already known. These training datasets typically include features such as topography, acoustic signatures, and geochemical anomalies, which are correlated with mineralization.

For example, a supervised learning model might be trained on data from regions of the Clarion-Clipperton Zone (CCZ) known to contain high densities of polymetallic nodules. The model learns to associate specific patterns, such as flat seafloor topography combined with certain sediment characteristics, with the likelihood of nodule presence. Once trained, the model can be applied to unexplored areas, generating prospectivity maps that highlight zones with similar patterns.

Algorithms commonly used for supervised learning in mineral prospectivity include random forests, support vector machines, and gradient boosting methods such as XGBoost. These algorithms are capable of handling large and heterogeneous datasets, identifying subtle correlations that might elude human analysts. The outputs of these models are probability maps that quantify the likelihood of mineralization at each location, enabling targeted exploration efforts.

Unsupervised Learning for Pattern Discovery

In regions where little is known about the geological characteristics of mineral deposits, unsupervised learning provides a powerful tool for discovering patterns without prior knowledge. Clustering algorithms, such as k-means and hierarchical clustering, group data points based on similarities in their features, revealing natural groupings that may correspond to different geological processes or mineralization styles.

For instance, unsupervised learning can analyze bathymetric data combined with magnetic and acoustic properties to identify distinct seafloor features, such as ridges, valleys, and vent fields. These features can then be cross-referenced with geological models to infer their potential association with mineral deposits. In the case of seafloor massive sulfides (SMS), clustering algorithms might identify regions with high acoustic reflectivity and magnetic anomalies, which are indicative of sulfide-rich zones.

Dimensionality reduction techniques, such as principal component analysis (PCA) and t-distributed stochastic neighbor embedding (t-SNE), further enhance pattern recognition by simplifying high-dimensional data into more manageable forms. These techniques help highlight key variables driving geological variation, enabling more focused analysis and interpretation.

Deep Learning for Complex Pattern Recognition

Deep learning, a subset of machine learning, excels at recognizing complex and non-linear patterns in geospatial data. Convolutional neural networks (CNNs), originally developed for image recognition, are particularly well-suited to processing gridded data such as sonar images and bathymetric maps. By learning hierarchical representations of the data, CNNs can detect subtle features that are indicative of mineralization, such as specific seafloor textures or geometric alignments.

For example, a CNN trained on high-resolution sonar images of hydrothermal vent fields could learn to identify the characteristic shapes and structures associated with active vent systems. These systems often host seafloor massive sulfides, making their detection a critical step in resource exploration. The model's ability to generalize across different datasets and geological settings ensures that it remains effective even in previously unexplored regions.

Recurrent neural networks (RNNs), another type of deep learning model, are used for analyzing temporal and sequential data. In the context of deep sea mining, RNNs can process time-series data from oceanographic sensors to identify patterns associated with changing environmental conditions or geological activity. For instance, fluctuations in temperature and chemical gradients near hydrothermal vents could signal active mineralization processes, guiding further exploration efforts.

Integrating Multimodal Data for Enhanced Prospectivity

One of the key strengths of machine learning is its ability to integrate diverse data types into a unified framework. Geospatial data in deep sea mining is inherently multimodal, encompassing acoustic signals, optical imagery, chemical analyses, and geophysical measurements. ML algorithms combine these disparate datasets to create richer and more accurate models of mineral prospectivity.

For instance, a combined dataset might include bathymetric maps from multibeam sonar, sediment composition data from coring samples, and magnetic field measurements from AUV surveys. Machine learning models can analyze these datasets simultaneously, identifying correlations and interactions that are indicative of mineral deposits. This integrated approach not only improves predictive accuracy but also provides a more holistic understanding of the geological processes at play.

Challenges and Future Directions

While machine learning has made significant strides in pattern recognition for mineral prospectivity, several challenges remain. The quality and consistency of training data are critical for model performance, yet deep sea exploration often suffers from sparse and heterogeneous datasets. Addressing this issue requires innovative approaches such as transfer learning, where models trained on well-studied regions are adapted to less-explored areas.

Another challenge lies in the interpretability of machine learning models. While advanced algorithms such as deep learning can achieve high predictive accuracy, their inner workings are often opaque, making it difficult to understand the reasoning behind their predictions. Developing explainable AI techniques is essential for gaining the trust of stakeholders and ensuring that ML outputs are actionable and reliable.

The future of ML-driven pattern recognition in geospatial data lies in its integration with real-time data acquisition systems. By deploying edge computing capabilities on AUVs and ROVs, machine learning models can process data in situ, providing immediate feedback to operators and enabling dynamic exploration strategies. This real-time capability will further enhance the efficiency and effectiveness of deep sea mineral exploration.

Unlocking the Patterns of the Deep

Pattern recognition in geospatial data, powered by machine learning, is transforming the search for mineral resources in the deep ocean. By automating the analysis of complex and heterogeneous datasets, ML enables the identification of mineral-rich zones with unprecedented accuracy and efficiency. Through techniques such as supervised and unsupervised learning, deep learning, and multimodal data integration, machine learning provides a powerful toolkit for navigating the vast and uncharted depths of the seafloor.

As this technology continues to evolve, its potential to revolutionize mineral prospectivity will only grow, enabling the deep sea mining industry to unlock the hidden treasures of the ocean while minimizing risks and maximizing returns. By harnessing the power of pattern recognition, the industry stands poised to achieve a new level of precision and sophistication in resource exploration.

Feature extraction is a critical step in leveraging seismic, electromagnetic, and multi-beam sonar data for mineral prospectivity. These datasets contain vast amounts of raw information, often complex and multi-dimensional, requiring sophisticated techniques to identify meaningful patterns associated with the presence of valuable mineral deposits. Machine learning (ML) enhances this process by automating the identification of key features, enabling the efficient and accurate interpretation of data to guide exploration efforts.

Seismic Data: Revealing Subsurface Structures

Seismic surveys are essential for understanding the subsurface geology of the seafloor. By analyzing the propagation of seismic waves through different geological layers, these surveys provide information about the structure, composition, and stratigraphy of the subsurface. Machine learning plays a transformative role in extracting and interpreting features from seismic data, enabling the detection of patterns indicative of mineralization.

In the context of deep sea mining, seismic data is used to identify subsurface formations associated with seafloor massive sulfides (SMS) and other mineral deposits. ML models, such as convolutional neural networks (CNNs), are particularly effective at analyzing seismic reflection profiles. These networks learn to recognize characteristic features, such as fault lines, hydrothermal conduits, and layered sedimentary structures, that are associated with mineral-rich zones.

Advanced seismic inversion techniques, enhanced by ML, further improve feature extraction by converting raw seismic data into quantitative models of subsurface properties, such as porosity, density, and acoustic impedance. These models provide a more detailed understanding of the geological conditions that influence mineral deposition. For example, regions with high porosity and low impedance may indicate the presence of porous hydrothermal vents, which are prime targets for SMS exploration.

Seismic attribute analysis, another application of ML, involves extracting specific features from seismic data, such as amplitude, frequency, and phase, which can highlight subtle variations in subsurface geology. ML algorithms automate the identification of these attributes, correlating them with known mineral deposits to generate predictive models for unexplored areas.

Electromagnetic Data: Detecting Conductive Minerals

Electromagnetic (EM) surveys are critical for identifying the electrical conductivity of subsurface materials, a property closely associated with certain types of mineral deposits. For example, sulfide minerals, such as those found in SMS deposits, exhibit high conductivity, making EM data a valuable tool for mineral prospectivity.

Machine learning enhances the analysis of EM data by automating the extraction of features related to conductivity anomalies. Supervised learning models, trained on datasets from known deposits, can identify patterns in EM responses that correlate with mineralization. These models are particularly effective in distinguishing between conductive minerals and other sources of electrical anomalies, such as saline pore fluids or geological noise.

One of the challenges in EM data interpretation is its sensitivity to depth and resolution. ML algorithms, such as gradient boosting machines and deep neural networks, address this issue by integrating data from multiple survey depths and frequencies. This multi-scale analysis enables the detection of both shallow and deep conductive anomalies, providing a comprehensive view of subsurface conditions.

In addition to identifying conductivity anomalies, ML algorithms can also infer the geometry and size of mineralized bodies. By analyzing the spatial distribution of EM signals, these models generate three-dimensional representations of potential deposits, guiding further exploration efforts.

Multi-Beam Sonar Data: Mapping the Seafloor in Detail

Multi-beam sonar systems are a cornerstone of deep sea exploration, providing high-resolution maps of the seafloor's topography and acoustic properties. These systems emit acoustic waves in a fan-shaped pattern and measure the time it takes for the waves to return, creating detailed bathymetric and backscatter maps.

Machine learning transforms the analysis of multi-beam sonar data by automating the extraction of features associated with mineral deposits. For example, polymetallic nodules often produce distinctive acoustic signatures in backscatter data due to their density and surface texture. ML models, such as random forests and support vector machines, are trained to recognize these signatures, enabling the rapid identification of nodule-rich areas.

Deep learning techniques, particularly convolutional neural networks, are increasingly used to analyze sonar imagery. These networks excel at recognizing complex patterns, such as the alignment of nodules along current-driven ridges or the acoustic shadows cast by hydrothermal vent structures. By learning these patterns from large datasets, CNNs can generalize to new survey areas, significantly improving the efficiency of prospectivity mapping.

Beyond detecting individual features, ML algorithms integrate multi-beam sonar data with other datasets, such as sediment composition and magnetic anomalies, to provide a holistic understanding of the seafloor. For instance, combining bathymetric maps with backscatter data and geochemical profiles can reveal areas where nodules are both abundant and compositionally rich, optimizing the prioritization of exploration targets.

Integrating Seismic, Electromagnetic, and Sonar Data

The true power of machine learning lies in its ability to integrate data from multiple geophysical sources, creating comprehensive models that capture the complexity of mineral prospectivity. By combining seismic, EM, and sonar datasets, ML algorithms can identify correlations and interactions that are not apparent when these data types are analyzed independently.

For example, regions with seismic signatures indicative of hydrothermal activity can be cross-referenced with EM conductivity anomalies to pinpoint sulfide-rich zones. Similarly, multi-beam sonar data can be used to identify surface features that align with subsurface structures revealed by seismic surveys, providing a more complete picture of mineral deposits.

Dimensionality reduction techniques, such as principal component analysis (PCA), play a key role in integrating these datasets by simplifying high-dimensional data into manageable forms without losing critical information. These reduced datasets are then used as inputs for predictive models, enabling the identification of mineralization patterns across different geological settings.

Challenges and Innovations in Feature Extraction

While ML has significantly advanced feature extraction from seismic, EM, and sonar data, challenges remain. Data quality and resolution are critical, as noise and inconsistencies can obscure meaningful patterns. Addressing this issue requires advanced preprocessing techniques, such as noise filtering and normalization, to ensure that datasets are suitable for ML analysis.

Another challenge is the interpretability of ML models, particularly deep learning algorithms, which often operate as "black boxes." Developing explainable AI techniques is essential for understanding how models identify features and ensuring that their predictions align with geological realities.

Emerging innovations, such as transfer learning and domain adaptation, are poised to further enhance ML-driven feature extraction. These techniques enable models trained on well-characterized datasets to be applied to new regions with limited data, expanding the reach of ML technologies in mineral prospectivity.

Unlocking Hidden Signals

Feature extraction from seismic, electromagnetic, and multi-beam sonar data is a cornerstone of modern mineral prospectivity, enabling the efficient and accurate identification of valuable deposits in the deep ocean. Machine learning has revolutionized this process, automating the analysis of complex datasets and uncovering patterns that would otherwise remain hidden.

By integrating these diverse geophysical datasets, ML-driven approaches provide a holistic understanding of the seafloor, guiding exploration efforts with unprecedented precision. As technology continues to evolve, the ability to extract and interpret features from seismic, EM, and sonar data will remain central to unlocking the mineral wealth of the deep ocean, advancing both the science and practice of deep sea mining.

Predictive modeling for resource occurrence and grade estimation is one of the most impactful applications of machine learning (ML) in deep sea mining. By combining vast and diverse datasets with advanced computational techniques, ML enables the prediction of mineral deposit locations and the estimation of their quality or "grade" with unprecedented precision. These predictive models are central to reducing the uncertainty and costs associated with exploration, optimizing resource allocation, and improving the economic feasibility of deep sea mining operations.

The Importance of Predictive Modeling

Accurately predicting the occurrence and grade of mineral resources is critical for planning and decision-making in deep sea mining. Traditional geological methods rely on field surveys, sampling, and manual interpretation of data, which are time-intensive and costly, especially in the remote and harsh environments of the deep ocean. Predictive modeling, powered by ML, offers a transformative alternative, using algorithms to identify patterns and relationships within geospatial, geochemical, and geophysical data to forecast resource distribution and quality.

These models go beyond simple detection, incorporating statistical probabilities and data-driven insights to quantify the likelihood of mineral occurrence and estimate the concentration of economically valuable elements. This allows operators to focus exploration efforts on the most promising areas, minimizing wasted effort and environmental impact.

Machine Learning Techniques for Predictive Modeling

Machine learning provides a suite of techniques for building predictive models tailored to the complexities of deep sea mining. These techniques range from traditional statistical approaches to state-of-the-art neural networks, each with its strengths depending on the nature of the data and the prediction goals.

Supervised Learning for Classification and Regression

Supervised learning is at the core of many predictive modeling efforts, using labeled training data to teach algorithms to recognize patterns associated with mineral occurrence and grade. Two key approaches are classification, which predicts whether a given area contains a mineral deposit, and regression, which estimates the grade or concentration of specific elements within a deposit.

For example, decision tree algorithms, such as random forests and gradient boosting machines, are widely used for supervised learning in mineral prospectivity. These models excel at handling heterogeneous data, such as combining bathymetric maps, geochemical assays, and acoustic backscatter data. By training on datasets from known mineralized areas, they learn to associate specific features, such as topographical ridges or geochemical anomalies, with the likelihood of resource occurrence.

Regression models, often based on techniques such as support vector regression (SVR) or neural networks, are used to estimate grade values, such as the concentration of cobalt in ferromanganese crusts or the copper content of seafloor massive sulfides. These models analyze continuous variables, incorporating a wide range of inputs to predict not only the presence of resources but their economic potential.

Probabilistic Models and Bayesian Methods

Uncertainty is a fundamental challenge in resource prediction, particularly in the data-scarce environments of the deep ocean. Probabilistic models, including Bayesian networks, offer a framework for addressing this uncertainty by incorporating prior knowledge and updating predictions as new data becomes available.

Bayesian methods are particularly useful for integrating diverse data sources with varying degrees of reliability. For instance, seismic data may provide strong evidence of geological structures, while geochemical assays offer more localized but highly specific information. A Bayesian network can combine these datasets to calculate the probability of resource occurrence, providing both a prediction and a measure of confidence.

Deep Learning for Complex Patterns

Deep learning, particularly convolutional neural networks (CNNs) and recurrent neural networks (RNNs), has become a powerful tool for predictive modeling in mineral prospectivity. CNNs are particularly effective at analyzing spatially structured data, such as sonar images or geophysical grids, to identify complex patterns indicative of mineralization. For example, a CNN trained on high-resolution sonar images could predict the likelihood of polymetallic nodule occurrence based on texture and reflectivity.

RNNs, on the other hand, are adept at processing sequential or temporal data, such as changes in geochemical signatures over a region or time-series data from oceanographic sensors. These models can identify dynamic patterns that might signal active mineral-forming processes, such as hydrothermal venting or sediment deposition.

Deep learning models are especially valuable for grade estimation, as they can handle high-dimensional datasets with intricate relationships between variables. For instance, a deep neural network could analyze the combined effects of sediment composition, acoustic properties, and magnetic anomalies to predict the concentration of economically valuable elements in a deposit.

Data Integration and Feature Engineering

Predictive modeling relies heavily on the integration of diverse datasets, including geophysical surveys, geochemical analyses, and biological observations. Feature engineering, the process of selecting and transforming raw data into meaningful inputs for ML models, is critical for maximizing predictive accuracy.

For example, in the case of polymetallic nodules, features might include bathymetric slope (derived from sonar data), sediment thickness (from sub-bottom profiling), and manganese concentration (from geochemical assays). Feature engineering also involves creating composite variables, such as combining acoustic reflectivity with sediment porosity to capture complex interactions that influence nodule formation.

The integration of these features into a single model allows for a holistic understanding of the factors driving mineral prospectivity. Machine learning algorithms can then analyze these inputs to generate predictions that are both precise and contextually informed.

Applications in Deep Sea Mining

Predictive models are being deployed in several key applications within the deep sea mining industry:

Targeting Exploration Efforts: Models highlight high-probability zones for further investigation, reducing the need for extensive and costly surveys.

Resource Estimation: Predictive outputs provide preliminary estimates of resource quantity and grade, supporting feasibility studies and economic evaluations.

Risk Management: By quantifying uncertainty, models help operators assess the likelihood of resource presence and potential challenges, such as geological hazards.

Environmental Planning: Predictive tools can identify areas where mineral resources overlap with ecologically sensitive habitats, enabling more responsible planning and mitigation strategies.

Challenges and Future Directions

Despite its promise, predictive modeling for resource occurrence and grade estimation faces challenges. The quality and availability of training data are critical, yet many deep sea regions are underexplored, leading to sparse datasets. Addressing this issue requires innovative approaches such as transfer learning, where models trained on well-characterized regions are adapted to less-explored areas.

Another challenge lies in model interpretability. While advanced models, such as deep learning algorithms, offer high predictive accuracy, their complexity can make it difficult to understand how predictions are generated. Developing explainable AI techniques is essential for building trust among stakeholders and ensuring that models are grounded in geological realities.

The future of predictive modeling lies in real-time applications and adaptive learning. By integrating ML models with autonomous underwater vehicles (AUVs) and real-time data acquisition systems, predictions can be continuously updated as new information is collected. This dynamic approach will further enhance the precision and efficiency of exploration efforts, transforming predictive modeling into a cornerstone of deep sea mining operations.

Machine learning-driven predictive modeling represents a paradigm shift in the exploration and assessment of deep sea mineral resources. By leveraging advanced algorithms to analyze complex datasets, these models enable accurate predictions of resource occurrence and grade, reducing uncertainty and optimizing decision-making. As technology continues to evolve, predictive modeling will play an increasingly central role in unlocking the potential of the deep ocean's mineral wealth, combining efficiency with environmental responsibility in one of humanity's most ambitious endeavors.

3.2. Remote Sensing and Big Data Analytics

Remote sensing serves as the critical first step in the exploration of deep sea mineral resources, providing vast quantities of data about the seafloor and its mineral wealth without the need for direct, time-consuming surveys. Both satellite-based and underwater remote sensing approaches are central to this effort, offering complementary methods for mapping the physical, chemical, and geological characteristics of the deep ocean. These technologies not only expand the scope of exploration but also feed into big data analytics frameworks, enabling the extraction of actionable insights from complex datasets.

Satellite-Based Remote Sensing: A Macro Perspective
Satellite-based remote sensing offers a broad, high-level view of the Earth's oceans, providing valuable information about the topography, geology, and physical properties of the seafloor. Although limited in resolution compared to underwater systems, satellites excel in delivering global coverage, making them indispensable for the initial identification of regions of interest.

One of the primary satellite technologies used in deep sea exploration is satellite altimetry. Altimeters measure variations in sea surface height caused by gravitational anomalies, which, in turn, reflect the topography of the underlying seafloor. For example, a seamount or ridge exerts a greater gravitational pull, creating a subtle bulge in the ocean surface. By analyzing these variations, scientists can infer the presence of seafloor features that may host cobalt-rich ferromanganese crusts or polymetallic nodules.

Another important satellite-based tool is optical remote sensing, which uses multispectral and hyperspectral imaging to analyze oceanographic parameters such as chlorophyll concentration, water clarity, and surface temperature. While these measurements are primarily useful for understanding biological productivity and environmental conditions, they can indirectly indicate regions of geological interest. For instance, high biological productivity often coincides with nutrient upwelling near underwater ridges and hydrothermal vent systems, which are also prime targets for seafloor massive sulfides (SMS).

Synthetic aperture radar (SAR), another satellite-based technology, provides data on ocean surface roughness and wave patterns, which can reveal underlying seafloor features. SAR is particularly effective in detecting shallow seamounts and ridges, offering a cost-effective means of identifying areas for further investigation.

Underwater Remote Sensing: A Closer Look

While satellites provide a macro perspective, underwater remote sensing offers the high-resolution detail necessary for identifying and characterizing specific mineral deposits. This approach employs a range of technologies deployed on autonomous underwater vehicles (AUVs), remotely operated vehicles (ROVs), and specialized survey platforms.

1. Acoustic Remote Sensing

Acoustic systems, including multibeam sonar and side-scan sonar, are the workhorses of underwater remote sensing. Multibeam sonar creates detailed bathymetric maps by emitting acoustic waves in a fan-shaped pattern and measuring their return times. These maps reveal the topography of the seafloor, highlighting features such as abyssal plains, ridges, and hydrothermal vent fields.

Side-scan sonar complements multibeam sonar by producing high-resolution images of seafloor textures. This technology is particularly effective in detecting polymetallic nodules and ferromanganese crusts, which exhibit distinct acoustic reflectivity due to their density and surface properties. Side-scan sonar also identifies geological structures, such as fault lines and volcanic formations, which are often associated with mineralization.

2. Electromagnetic and Magnetometric Sensing

Electromagnetic (EM) sensors detect variations in the electrical conductivity of the seafloor, enabling the identification of sulfide-rich deposits. These systems are highly effective in regions with active hydrothermal activity, where conductive minerals such as copper and zinc are abundant.

Magnetometers, which measure the magnetic field of the Earth, are used to detect ferromagnetic minerals like cobalt and nickel. These instruments are particularly valuable in exploring seamounts and other geologically complex areas where ferromanganese crusts are likely to occur.

3. Optical and Laser-Based Systems

Optical sensors, including high-resolution cameras and video systems, provide visual confirmation of seafloor features and mineral deposits. These systems are deployed on AUVs and ROVs to capture images of nodules, crusts, and vent structures, aiding in the interpretation of acoustic and geophysical data.

Laser-based systems, such as laser line scanners and LIDAR (Light Detection and Ranging), create three-dimensional models of the seafloor with millimeter-scale resolution. These models are invaluable for assessing the morphology and spatial distribution of mineral resources, particularly in rugged terrains.

4. In Situ Chemical Sensors

In situ chemical sensors, including mass spectrometers and spectrophotometers, analyze the chemical composition of seawater and pore fluids. These sensors detect geochemical anomalies, such as elevated concentrations of metals or sulfides, that indicate the presence of mineral deposits. For example, the detection of high manganese or cobalt levels in the water column can point to the presence of polymetallic nodules or crusts.

Integration of Satellite and Underwater Data

The integration of satellite-based and underwater remote sensing data is a critical step in building a comprehensive understanding of the seafloor. Satellites provide the broad spatial context needed to identify regions of interest, while underwater systems deliver the detailed information required for site-specific exploration and resource assessment.

For example, satellite altimetry might identify a seamount as a potential target for ferromanganese crust mining. An AUV equipped with multibeam sonar and chemical sensors could then conduct a detailed survey of the site, producing high-resolution maps and geochemical profiles. These datasets can be further integrated into machine learning models to refine predictions of resource occurrence and grade.

Big data analytics plays a central role in this integration, enabling the processing and analysis of massive datasets from diverse sources. By combining satellite imagery, acoustic maps, and geochemical measurements, advanced algorithms can identify correlations and patterns that inform exploration strategies.

Challenges and Future Directions

Despite its transformative potential, remote sensing for deep sea mining faces several challenges. Satellite-based methods, while cost-effective, lack the resolution needed for detailed resource characterization. Underwater systems, on the other hand, are limited by their operational range and high deployment costs.

Future advancements in sensor technology, autonomous platforms, and data integration frameworks are poised to address these challenges. For instance, the development of hybrid systems that combine acoustic, optical, and electromagnetic sensors on a single platform could provide more comprehensive datasets. Additionally, real-time data transmission and processing capabilities will enable dynamic exploration strategies, where remote sensing systems adapt their surveys based on incoming data.

The Foundation of Modern Exploration

Satellite-based and underwater remote sensing approaches form the foundation of modern deep sea exploration. Together, these technologies offer a powerful combination of global coverage and high-resolution detail, enabling the efficient and accurate identification of mineral-rich zones. As remote sensing technologies continue to advance, their integration with big data analytics and machine learning will further enhance their effectiveness, driving the future of sustainable and economically viable deep sea mining. Through these innovations, the deep ocean's mineral wealth can be unlocked with precision and responsibility, transforming one of the planet's last frontiers into a vital resource for global development.

Geospatial analytics and cluster analysis are indispensable tools in deep sea mining, providing the ability to interpret high-dimensional data generated by remote sensing technologies. These methods enable the identification of patterns, relationships, and clusters within complex datasets, guiding exploration efforts and enhancing the understanding of mineral prospectivity. By leveraging advanced computational techniques, geospatial analytics and cluster analysis transform raw data into actionable insights, streamlining the search for valuable resources beneath the ocean floor.

The Complexity of High-Dimensional Data in Deep Sea Mining

High-dimensional data in deep sea mining arises from the diverse and multifaceted nature of the seafloor environment. This data includes bathymetric maps, acoustic backscatter imagery, geochemical assays, electromagnetic signatures, and biological observations. Each dataset captures a unique aspect of the seafloor, such as its topography, composition, or ecological characteristics.

The integration of these datasets creates a high-dimensional space where each point represents a location on the seafloor, and each dimension corresponds to a specific variable or feature. For example, a single point might include values for depth, slope, sediment type, manganese concentration, and acoustic reflectivity. Navigating and interpreting such a data-rich environment is a daunting task, requiring sophisticated analytical techniques.

Traditional approaches to analyzing geospatial data often struggle with the complexity and scale of high-dimensional datasets. Manual interpretation is not only time-intensive but also prone to human biases and errors. Geospatial analytics and cluster analysis address these limitations by automating the identification of meaningful patterns and relationships, enabling more efficient and accurate exploration.

Geospatial Analytics: Mapping and Analyzing Spatial Relationships

Geospatial analytics involves the processing and analysis of spatially referenced data to understand patterns, relationships, and trends. In deep sea mining, this approach is critical for mapping the distribution of mineral resources and identifying regions with high prospectivity.

Spatial autocorrelation analysis is a key technique in geospatial analytics, measuring the degree to which similar values are clustered or dispersed across a spatial domain. For example, high acoustic reflectivity values might cluster around seamounts, indicating the presence of ferromanganese crusts. Understanding these spatial patterns allows operators to prioritize exploration efforts in areas with the highest potential.

Hot spot analysis, another geospatial technique, identifies statistically significant clusters of high or low values within a dataset. This method is particularly useful for detecting localized zones of mineral enrichment. For instance, hot spot analysis of geochemical data might reveal areas with elevated concentrations of cobalt or nickel, guiding targeted surveys and sampling.

Interpolation techniques, such as kriging and inverse distance weighting, are used to predict values at unsampled locations based on nearby measurements. These methods generate continuous geospatial surfaces, such as maps of manganese concentration or sediment thickness, providing a comprehensive view of the seafloor's mineral potential.

Cluster Analysis: Unveiling Hidden Patterns

Cluster analysis is a subset of machine learning that groups data points into clusters based on their similarity across multiple dimensions. This technique is particularly valuable in deep sea mining, where the relationships between geological, geochemical, and geophysical features are often complex and non-linear.

1. K-Means Clustering

K-means clustering is one of the most widely used algorithms in mineral prospectivity. It partitions data into a predefined number of clusters, minimizing the variance within each cluster while maximizing the differences between clusters. For example, k-means clustering could group locations on the seafloor based on their bathymetric slope, acoustic reflectivity, and geochemical signature. The resulting clusters might correspond to different geological environments, such as abyssal plains, hydrothermal vent fields, or ridge flanks.

2. Hierarchical Clustering

Hierarchical clustering builds a tree-like structure of nested clusters, providing a more flexible approach to identifying relationships within high-dimensional data. This technique is particularly useful when the number of clusters is unknown or when exploring data at multiple scales. For instance, hierarchical clustering might reveal that certain regions of the seafloor, while geochemically similar, differ in their acoustic properties, suggesting distinct mineralization processes.

3. Density-Based Clustering

Density-based clustering algorithms, such as DBSCAN (Density-Based Spatial Clustering of Applications with Noise), identify clusters based on the density of data points in a given region. This approach is effective for detecting irregularly shaped clusters, such as elongated mineralized zones along tectonic ridges. DBSCAN also distinguishes between core clusters and outliers, helping to identify anomalous data points that might indicate unique geological features.

4. Spectral Clustering

Spectral clustering leverages the eigenvalues of similarity matrices to group data points in non-linear spaces. This method is particularly useful for analyzing complex datasets where traditional clustering techniques struggle. In deep sea mining, spectral clustering might be applied to combine geophysical and geochemical data, revealing hidden relationships between subsurface structures and mineral deposits.

Integrating Geospatial Analytics and Cluster Analysis

The true power of geospatial analytics and cluster analysis lies in their integration, enabling a holistic understanding of the seafloor's mineral potential. By combining spatial analysis techniques with clustering algorithms, operators can uncover both the spatial distribution and the underlying relationships between geological, geochemical, and geophysical features.

For example, a geospatial analysis might reveal a region of interest based on its topography and acoustic properties. Cluster analysis could then be applied to geochemical data from this region, identifying sub-clusters that represent distinct mineralization styles or grades. This integrated approach provides a more nuanced understanding of the seafloor, guiding exploration efforts with greater precision.

Big data platforms and machine learning frameworks play a critical role in this integration, enabling the processing and analysis of large and complex datasets. Cloud computing and distributed systems allow operators to analyze geospatial and clustering data at scale, while visualization tools create intuitive maps and models for decision-making.

Challenges and Innovations in High-Dimensional Data Analysis

Despite its potential, analyzing high-dimensional data in deep sea mining presents several challenges. Data quality and consistency are critical, as noise and gaps in datasets can obscure meaningful patterns. Addressing this issue requires robust preprocessing techniques, such as data normalization and outlier detection.

Another challenge lies in the interpretability of clustering results. While algorithms can identify clusters, understanding their geological significance requires domain expertise and careful validation. Advanced visualization tools, such as three-dimensional mapping and interactive dashboards, help bridge this gap by presenting results in an accessible and actionable format.

Emerging innovations, such as explainable AI and hybrid models, are poised to further enhance geospatial analytics and cluster analysis. These approaches combine the predictive power of machine learning with the interpretability of statistical models, providing both accuracy and transparency in mineral prospectivity.

Unlocking Insights from Complexity

Geospatial analytics and cluster analysis are transforming the exploration of deep sea mineral resources, enabling operators to navigate the complexity of high-dimensional data with precision and efficiency. By uncovering patterns and relationships within diverse datasets, these techniques provide actionable insights that guide exploration efforts, optimize resource allocation, and reduce uncertainty.

As technology continues to evolve, the integration of geospatial and clustering methods with big data analytics and machine learning will further enhance their capabilities. Through these innovations, the deep ocean's mineral wealth can be explored and harnessed responsibly, paving the way for a new era of resource discovery and economic development.

The vast scale and complexity of data generated by deep sea exploration demand advanced systems for storage, processing, and analysis. Cloud-based data processing pipelines have emerged as an essential solution for handling these large-scale datasets, enabling real-time collaboration, efficient computation, and scalable storage. By leveraging the power of the cloud, deep sea mining operations can transform raw remote sensing data into actionable insights with unprecedented speed and precision.

The Challenges of Large-Scale Data in Deep Sea Exploration

Deep sea mining generates massive amounts of data from remote sensing technologies, including satellite altimetry, multibeam sonar, side-scan sonar, electromagnetic surveys, and geochemical assays. Each of these systems produces high-resolution, multi-dimensional datasets, often running into terabytes per mission. Managing and analyzing this data poses several challenges:

1. **Volume**: The sheer size of the datasets requires scalable storage solutions capable of accommodating exponential growth as exploration expands.
2. **Velocity**: Data from autonomous underwater vehicles (AUVs) and remotely operated vehicles (ROVs) often need to be processed in real time or near real time to guide dynamic exploration strategies.
3. **Variety**: Data comes in multiple formats, including raster images, time-series measurements, point clouds, and tabular geochemical profiles, complicating integration and analysis.
4. **Complexity**: High-dimensional datasets with intricate relationships between variables demand advanced analytical techniques and computational resources.

Cloud-based pipelines address these challenges by providing a flexible, scalable, and efficient infrastructure for data processing, enabling deep sea mining operations to harness the full potential of their data.

Key Components of Cloud-Based Data Processing Pipelines

A cloud-based data processing pipeline is a modular system designed to handle the entire data lifecycle, from ingestion to analysis and visualization. The key components of such pipelines include:

Data Ingestion and Storage

The first step in any cloud-based pipeline is data ingestion, where raw data from various sources is collected, uploaded, and organized for further processing.

Cloud platforms support multiple ingestion methods, including batch uploads, streaming data, and real-time feeds from remote sensors.

- **Scalable Storage**: Cloud storage systems, such as Amazon S3, Google Cloud Storage, and Microsoft Azure Blob Storage, offer virtually unlimited capacity, ensuring that even the largest datasets can be stored efficiently. These systems provide tiered storage options, allowing operators to balance cost and performance based on data access frequency.
- **Data Lakes**: Many pipelines utilize data lakes, which are centralized repositories that store raw data in its native format. Data lakes enable seamless integration of structured, semi-structured, and unstructured data, making them ideal for managing the diverse datasets generated by deep sea mining.

Data Preprocessing and Transformation

Once ingested, raw data must be cleaned, transformed, and standardized to prepare it for analysis. Preprocessing involves tasks such as noise filtering, outlier detection, normalization, and georeferencing. Cloud-based tools automate these processes, reducing manual effort and ensuring consistency.

- **Serverless Computing**: Cloud platforms offer serverless computing services, such as AWS Lambda and Google Cloud Functions, which execute preprocessing tasks on demand without requiring dedicated infrastructure. These services scale automatically based on workload, ensuring efficient use of resources.
- **Geospatial Data Transformation**: Specialized libraries and frameworks, such as GDAL (Geospatial Data Abstraction Library) and PySAL (Python Spatial Analysis Library), are integrated into cloud pipelines to handle geospatial data transformations, including coordinate system conversions and raster-to-vector operations.

Big Data Processing Frameworks

The core of a cloud-based pipeline lies in its ability to process large-scale datasets quickly and efficiently. Distributed computing frameworks, such as Apache Hadoop and Apache Spark, are commonly used for this purpose.

- **Distributed Processing**: These frameworks divide datasets into smaller chunks, processing them in parallel across multiple nodes. This approach significantly accelerates computation, making it feasible to analyze terabytes of data in a fraction of the time required by traditional methods.
- **Integration with Cloud Services**: Cloud providers offer managed versions of these frameworks, such as Amazon EMR (Elastic MapReduce) and Google Dataproc, which simplify deployment and reduce administrative overhead.

Machine Learning and Advanced Analytics

Once preprocessed, data is analyzed using machine learning models and advanced statistical techniques to generate insights. Cloud platforms provide powerful tools and environments for building, training, and deploying machine learning models.

- **AI and ML Platforms**: Services such as AWS SageMaker, Google AI Platform, and Azure Machine Learning offer pre-built algorithms and customizable environments for developing models tailored to deep sea mining applications. These platforms support tasks such as resource prospectivity mapping, anomaly detection, and grade estimation.
- **AutoML**: Automated machine learning tools streamline model development by selecting the best algorithms, tuning hyperparameters, and validating results. This democratizes access to machine learning, enabling geologists and engineers without extensive AI expertise to leverage these capabilities.

Data Visualization and Collaboration

The final step in the pipeline is transforming analytical results into actionable insights through visualization and collaboration tools. Cloud platforms support interactive dashboards and geospatial visualization frameworks that enable stakeholders to explore data intuitively.

- **Visualization Tools**: Services like Tableau, Power BI, and Google Data Studio integrate seamlessly with cloud pipelines, providing dynamic charts, heatmaps, and three-dimensional models of the seafloor.

- **Geospatial Mapping**: Tools such as ArcGIS Online and Google Earth Engine enable the creation of detailed maps that overlay bathymetric data, geochemical signatures, and predictive models, offering a comprehensive view of exploration targets.
- **Collaboration Platforms**: Cloud-native collaboration tools, such as Google Workspace and Microsoft Teams, facilitate real-time sharing and discussion of results among geographically dispersed teams.

Benefits of Cloud-Based Pipelines for Deep Sea Mining

The adoption of cloud-based data processing pipelines provides several key advantages for the deep sea mining industry:

1. **Scalability**: Cloud platforms can scale resources up or down based on workload, ensuring cost-effectiveness without compromising performance.
2. **Real-Time Processing**: Pipelines enable the real-time analysis of data from AUVs, ROVs, and other sensors, supporting dynamic decision-making during exploration missions.
3. **Interoperability**: The integration of diverse data sources into a single pipeline allows for holistic analysis, combining geospatial, geochemical, and geophysical data into unified models.
4. **Cost Efficiency**: Pay-as-you-go pricing models reduce upfront costs and align expenses with actual usage, making cloud pipelines accessible even for smaller operators.
5. **Collaboration and Accessibility**: Cloud platforms enable seamless collaboration among teams and provide secure, remote access to data and insights from anywhere in the world.

Challenges and Future Directions

Despite their benefits, cloud-based pipelines face challenges such as data security, bandwidth limitations, and the need for skilled personnel to design and manage complex workflows. Addressing these challenges requires investment in cybersecurity, optimization of data transfer protocols, and training programs to develop expertise in cloud technologies.

Future innovations, such as edge computing and federated learning, promise to further enhance cloud-based pipelines. Edge computing will enable data processing at or near the point of collection, reducing latency and bandwidth requirements. Federated learning will allow machine learning models to be trained across distributed datasets without centralizing data, improving privacy and efficiency.

Cloud-based data processing pipelines are transforming the exploration and assessment of deep sea mineral resources. By providing scalable, efficient, and integrated solutions for handling large-scale datasets, these pipelines empower operators to extract actionable insights from the complex and dynamic environments of the deep ocean. As technology continues to evolve, cloud-based systems will play an increasingly central role in the future of deep sea mining, enabling precision, collaboration, and innovation in one of humanity's most challenging and promising frontiers.

3.3. Uncertainty Quantification

Uncertainty quantification is a critical aspect of deep sea mining, addressing the inherent variability and unknowns in data collected from remote and inaccessible underwater environments. Among the most robust and widely used approaches for managing uncertainty are Bayesian methods and probabilistic modeling. These techniques allow operators to make informed decisions based on incomplete or uncertain information, balancing the need for exploration efficiency with the risks associated with high-stakes resource extraction.

The Nature of Uncertainty in Deep Sea Mining

Uncertainty in deep sea mining arises from several sources, including sparse datasets, measurement errors, and the inherent complexity of geological, geophysical, and geochemical processes. Despite advances in remote sensing and data acquisition, the deep ocean remains one of the least explored frontiers, with significant gaps in knowledge about resource distribution, grade, and environmental impact.

This uncertainty poses challenges for every stage of the mining value chain. Exploration efforts may target regions that turn out to be less productive than predicted, extraction methods could encounter unanticipated geological conditions, and environmental assessments might underestimate the potential impact of mining activities. Bayesian methods and probabilistic modeling address these challenges by quantifying uncertainty and incorporating it into predictive models and decision frameworks.

Bayesian Methods: A Framework for Updating Beliefs

Bayesian methods provide a systematic framework for updating probabilities as new information becomes available. Rooted in Bayes' theorem, these methods calculate the posterior probability of an event based on prior knowledge and observed data:

$$P(Hypothesis|Data) = \frac{P(Data|Hypothesis)P(Hypothesis)}{P(Data)}$$

In the context of deep sea mining, Bayesian methods are used to refine predictions of resource occurrence, grade, or environmental impact as new data is collected. For example:

1. **Resource Prospectivity**: Prior geological knowledge and historical data might suggest a high likelihood of polymetallic nodule occurrence in a given region. As multibeam sonar surveys and geochemical assays provide additional data, Bayesian methods update the probability of resource presence, incorporating the strength and reliability of the new evidence.
2. **Grade Estimation**: Bayesian models can estimate the concentration of valuable metals within a deposit. For instance, initial core samples might provide a rough estimate of cobalt content. As more samples are collected, the model refines this estimate, reducing uncertainty and providing a clearer picture of the deposit's economic potential.
3. **Environmental Impact**: Probabilistic assessments of sediment plume dispersion or biodiversity loss can be updated using Bayesian techniques as more data about local ocean currents and biological communities becomes available.

Bayesian networks, a specific application of Bayesian methods, represent complex relationships between variables in a graphical structure. In deep sea mining, these networks can model the interdependencies between geological, geophysical, and environmental factors, allowing for sophisticated risk assessments and scenario analysis.

Probabilistic Modeling: Quantifying and Managing Uncertainty

Probabilistic modeling extends the principles of Bayesian methods to create comprehensive frameworks for uncertainty quantification. These models represent variables as probability distributions rather than fixed values, capturing the range of possible outcomes and their associated likelihoods.

1. **Monte Carlo Simulations**: Monte Carlo methods are a cornerstone of probabilistic modeling, using repeated random sampling to simulate a wide range of scenarios. In deep sea mining, Monte Carlo simulations might estimate the total volume of recoverable resources within a target area, accounting for uncertainties in grade, spatial distribution, and recovery efficiency. By running thousands of simulations, operators can calculate probability distributions for key metrics, such as expected yield or economic return.
2. **Geostatistical Simulations**: Geostatistical techniques, such as sequential Gaussian simulation or co-simulation, are used to model the spatial variability of geological and geochemical properties. These methods generate multiple realizations of the seafloor, each representing a plausible distribution of resources based on available data. The resulting models provide probabilistic maps of resource occurrence, guiding exploration efforts and reducing the risk of unproductive surveys.
3. **Stochastic Differential Equations**: For dynamic systems, such as the evolution of hydrothermal vent activity or sediment plume dispersion, stochastic differential equations incorporate randomness into mathematical models. These equations simulate the behavior of systems over time, providing insights into the range of possible outcomes under different conditions.

Advantages of Bayesian and Probabilistic Approaches

The integration of Bayesian methods and probabilistic modeling offers several key advantages for deep sea mining:

1. **Data Integration**: These approaches seamlessly combine diverse data sources, such as geophysical surveys, core samples, and environmental measurements, into a unified framework. This integration allows operators to make decisions based on the most comprehensive and up-to-date information available.
2. **Dynamic Updating**: Bayesian methods enable real-time updates to models as new data becomes available, ensuring that predictions remain accurate and relevant throughout the exploration process.
3. **Risk Management**: Probabilistic models quantify the range of possible outcomes and their associated probabilities, helping operators assess and mitigate risks. For example, a probabilistic model might estimate the likelihood of encountering challenging geological conditions or exceeding environmental thresholds.
4. **Transparency and Communication**: By providing explicit measures of uncertainty, Bayesian and probabilistic approaches enhance transparency and facilitate communication with stakeholders, including regulators, investors, and environmental organizations.

Challenges and Innovations

Despite their strengths, Bayesian methods and probabilistic modeling face challenges, particularly in the deep sea mining context. The scarcity of high-quality training data can limit the reliability of initial models, while computational demands can be significant for large-scale simulations.

Emerging innovations are addressing these challenges. Transfer learning techniques enable Bayesian models to leverage knowledge from well-studied regions to improve predictions in less-explored areas. Advances in cloud computing and parallel processing are making Monte Carlo simulations and geostatistical modeling more accessible, even for data-intensive applications. Additionally, the integration of machine learning with probabilistic frameworks is enhancing the accuracy and scalability of these methods.

Navigating Uncertainty with Confidence

Bayesian methods and probabilistic modeling are transforming the way uncertainty is managed in deep sea mining. By quantifying and incorporating uncertainty into predictive models, these approaches enable operators to make more informed decisions, optimize resource allocation, and reduce risks. As exploration pushes into increasingly uncharted territories, the ability to navigate uncertainty with confidence will be a defining factor in the success and sustainability of deep sea mining operations. Through continuous innovation and refinement, Bayesian and probabilistic approaches will remain at the forefront of this transformative industry.

Sparse data environments are a hallmark of deep sea mining, where limited exploration and the vastness of the seafloor result in incomplete datasets. Active learning and semi-supervised approaches have emerged as innovative methods for addressing these challenges, enabling effective predictive modeling and decision-making even in the absence of comprehensive data. These machine learning techniques are designed to maximize the utility of available data while intelligently guiding the collection of new information, making them indispensable tools in the high-uncertainty domain of deep sea exploration.

The Challenge of Sparse Data in Deep Sea Mining

Deep sea mining operates in one of the most data-scarce environments on Earth. Unlike terrestrial mining, where extensive geological surveys and sampling campaigns provide relatively rich datasets, deep ocean exploration is constrained by logistical, financial, and technological limitations. Data from multibeam sonar, core samples, and geochemical assays are often sparse and unevenly distributed, leaving significant gaps in understanding resource distribution, grade, and environmental conditions.

This sparsity poses several challenges:

1. **Model Accuracy**: Machine learning models trained on sparse data may struggle to generalize, leading to inaccurate predictions.
2. **Exploration Efficiency**: Without sufficient data, exploration efforts risk targeting suboptimal areas, wasting resources and time.
3. **Uncertainty Management**: Sparse data increases uncertainty, complicating decision-making and risk assessment.

Active learning and semi-supervised approaches address these issues by optimizing the use of existing data and strategically acquiring new information to fill critical gaps.

Active Learning: Focusing on the Most Informative Data

Active learning is a machine learning technique that identifies the most informative data points to label or sample, thereby improving model performance with minimal additional data collection. In the context of deep sea mining, active learning guides exploration efforts by prioritizing regions or samples that provide the most value for reducing uncertainty and refining predictive models.

1. **Query Strategies**: Active learning relies on query strategies to identify the most informative data points. Common strategies include:
   - **Uncertainty Sampling**: The model selects data points where its predictions are most uncertain, focusing exploration on regions with the highest ambiguity.
   - **Diversity Sampling**: The model selects data points that represent diverse features, ensuring that the training dataset covers a broad range of conditions.
   - **Expected Model Change**: The model predicts which data points would lead to the greatest improvement in its performance, prioritizing those for sampling or labeling.

For example, an active learning model might identify specific locations within a survey area where geophysical or geochemical features are inconsistent with its predictions, suggesting the need for additional core samples or high-resolution mapping.

2. **Iterative Exploration**: Active learning operates in an iterative loop, where the model is trained on an initial dataset, identifies the most informative new data points, and updates its predictions based on the acquired information. This iterative process ensures that each new data point maximizes its contribution to the model, accelerating convergence to accurate predictions.
3. **Applications in Resource Prospectivity**: In polymetallic nodule

exploration, active learning might prioritize sampling locations where acoustic backscatter data is ambiguous, refining predictions of nodule density and distribution. Similarly, in seafloor massive sulfide (SMS) exploration, it might target areas with inconsistent geochemical signatures for further analysis.

Semi-Supervised Learning: Leveraging Unlabeled Data

Semi-supervised learning bridges the gap between supervised learning, which requires labeled data, and unsupervised learning, which works with unlabeled data. By combining a small labeled dataset with a larger unlabeled dataset, semi-supervised approaches extract useful patterns from the unlabeled data, significantly enhancing model performance in sparse data environments.

1. **Pseudo-Labeling**: One common semi-supervised technique is pseudo-labeling, where the model assigns labels to unlabeled data points based on its predictions. These pseudo-labels are then used to augment the training dataset, effectively increasing the amount of labeled data. In deep sea mining, pseudo-labeling might involve classifying acoustic backscatter signatures as likely indicative of polymetallic nodules or barren sediment.
2. **Self-Training**: Self-training is an iterative process where the model is trained on labeled data, generates pseudo-labels for the most confident predictions on unlabeled data, and retrains on the expanded dataset. This approach gradually improves the model's accuracy while leveraging the abundance of unlabeled data from large-scale surveys.
3. **Graph-Based Methods**: Semi-supervised learning also includes graph-based methods, where data points are represented as nodes in a graph connected by edges based on their similarity. Labeled and unlabeled data points are linked, enabling the model to propagate labels across the graph. For example, regions of the seafloor with similar geophysical or geochemical characteristics can be grouped, with knowledge from labeled points informing predictions for the entire cluster.
4. **Applications in Grade Estimation**: In estimating the grade of cobalt in ferromanganese crusts, semi-supervised learning can leverage unlabeled geochemical data collected during large-scale AUV surveys. By combining these data with a smaller set of labeled core

samples, the model can predict grade distributions across the region with greater accuracy.

Combining Active Learning and Semi-Supervised Approaches

Active learning and semi-supervised learning are highly complementary, and their integration offers a powerful framework for addressing sparse data challenges. Active learning identifies the most critical data points to label or sample, while semi-supervised learning leverages the remaining unlabeled data to enhance predictions. Together, these techniques optimize the use of limited resources, ensuring that every data point contributes to reducing uncertainty and improving model accuracy.

For instance, in a survey of polymetallic nodules:

- Active learning might prioritize areas with high uncertainty in acoustic reflectivity for additional sampling.
- Semi-supervised learning could use the broader dataset of acoustic readings to refine predictions for unexamined regions, improving overall coverage and reducing the need for extensive fieldwork.

Advantages of Active Learning and Semi-Supervised Approaches

1. **Efficiency**: These methods minimize the need for extensive data collection, focusing efforts on the most informative regions or samples.
2. **Scalability**: By leveraging unlabeled data, semi-supervised learning scales effectively with large datasets, accommodating the vast spatial extent of the seafloor.
3. **Dynamic Adaptation**: Active learning adapts to new data in real time, ensuring that exploration strategies remain responsive to emerging information.
4. **Cost Reduction**: By optimizing data acquisition and maximizing the utility of existing data, these approaches significantly reduce exploration costs.

Challenges and Innovations
Despite their potential, active learning and semi-supervised approaches face challenges, including:

- **Model Bias**: Initial models may introduce biases that propagate through subsequent iterations, requiring robust validation and correction mechanisms.
- **Data Quality**: Sparse or noisy datasets can limit the effectiveness of both techniques, underscoring the importance of high-quality preprocessing and feature engineering.

Emerging innovations, such as transfer learning and deep semi-supervised models, are addressing these challenges. Transfer learning enables models trained on well-characterized regions to inform predictions in less-explored areas, while deep learning architectures improve the scalability and robustness of semi-supervised methods.

Optimizing Exploration in Data-Scarce Environments
Active learning and semi-supervised approaches represent a paradigm shift in the exploration and assessment of deep sea mineral resources. By intelligently prioritizing data collection and leveraging unlabeled datasets, these techniques enable operators to overcome the challenges of sparse data environments, reducing uncertainty and enhancing decision-making. As deep sea mining continues to push into uncharted territories, the integration of these methods will be essential for unlocking the ocean's mineral wealth with precision, efficiency, and responsibility.

Decision-making under uncertainty is a critical challenge in deep sea exploration campaigns. The inherent variability of seafloor environments, the limited availability of data, and the high costs associated with marine operations necessitate strategies that can balance risks and rewards effectively. By integrating uncertainty quantification into decision-making frameworks, exploration campaigns can prioritize actions, optimize resource allocation, and maximize the probability of successful outcomes, even in the face of incomplete or ambiguous information.

The Nature of Uncertainty in Exploration Campaigns
Uncertainty in exploration campaigns stems from several factors:

1. **Sparse and Heterogeneous Data**: Despite advances in remote sensing and data acquisition, datasets in deep sea mining remain incomplete and unevenly distributed, leading to gaps in knowledge about resource occurrence and environmental conditions.
2. **Geological Variability**: The seafloor is highly heterogeneous, with mineral deposits influenced by complex and poorly understood geological processes.
3. **Technological and Operational Risks**: Equipment failures, harsh environmental conditions, and logistical challenges add layers of uncertainty to exploration operations.
4. **Economic and Regulatory Factors**: Fluctuations in metal prices, evolving regulatory frameworks, and geopolitical considerations further complicate decision-making.

Effective exploration campaigns must navigate these uncertainties, ensuring that decisions are data-driven and aligned with strategic objectives.

Decision-Making Frameworks for Exploration Under Uncertainty

A variety of frameworks and methodologies have been developed to support decision-making in uncertain environments. These approaches integrate uncertainty quantification, probabilistic modeling, and optimization techniques to guide exploration efforts.

1. **Bayesian Decision Theory** Bayesian decision theory provides a mathematical framework for making optimal decisions under uncertainty by combining prior knowledge with observed data to calculate expected outcomes. This approach involves:
   - **Defining Decision Alternatives**: Identifying the possible courses of action, such as targeting specific areas for surveys or deploying additional resources for sampling.
   - **Modeling Uncertainty**: Representing unknown variables, such as resource occurrence or grade, as probability distributions.
   - **Calculating Expected Utility**: Evaluating the expected value of each decision alternative, considering both potential rewards (e.g., resource discovery) and risks (e.g., operational costs or environmental impact).

For example, Bayesian decision theory might be used to evaluate whether to conduct a high-resolution survey of a promising seamount. The decision would weigh the potential economic benefits of discovering cobalt-rich ferromanganese crusts against the costs of the survey and the probability of success.

2. **Value of Information (VOI) Analysis** VOI analysis assesses the benefit of acquiring additional information before making a decision. This approach is particularly valuable in exploration campaigns, where data collection is expensive and time-consuming. VOI analysis helps determine whether the expected improvement in decision accuracy justifies the cost of additional surveys or samples.

For instance, VOI analysis could evaluate whether collecting more core samples in a region with ambiguous geophysical signatures would significantly reduce uncertainty about resource grade, enabling more confident investment decisions.

3. **Stochastic Optimization** Stochastic optimization incorporates uncertainty directly into optimization models, enabling the identification of robust solutions that perform well across a range of possible scenarios. This approach is particularly useful for planning multi-stage exploration campaigns, where decisions made at one stage influence subsequent actions.

In polymetallic nodule exploration, stochastic optimization might be used to design an optimal survey route for an autonomous underwater vehicle (AUV), balancing the need for comprehensive coverage with operational constraints such as battery life and data storage capacity.

4. **Decision Trees and Scenario Analysis** Decision trees provide a visual and quantitative representation of decision pathways, incorporating probabilities, costs, and rewards at each branch. This approach is often combined with scenario analysis, where multiple plausible futures are evaluated to identify the most resilient strategies.

For example, a decision tree for SMS exploration might include branches for different survey methods (e.g., sonar, electromagnetic sensing), potential outcomes (e.g., discovery, no discovery), and follow-up actions (e.g., additional sampling, abandonment). The tree helps identify the sequence of actions that maximizes expected value while accounting for uncertainties.

Key Tools and Technologies

Modern decision-making under uncertainty leverages advanced computational tools and technologies to enhance accuracy and scalability:

1. **Simulation Models**: Monte Carlo simulations generate thousands of scenarios, modeling the potential outcomes of different decisions. These simulations provide probability distributions for key metrics, such as expected resource yield or economic return.
2. **Geospatial Decision Support Systems**: Geospatial tools integrate spatial data with probabilistic models, enabling dynamic decision-making based on real-time information. For instance, these systems can prioritize survey areas by overlaying bathymetric maps, geochemical signatures, and uncertainty metrics.
3. **Machine Learning for Dynamic Adaptation**: Machine learning algorithms enhance decision-making by continuously updating predictions as new data becomes available. Active learning and reinforcement learning are particularly valuable, enabling exploration strategies to adapt dynamically to changing conditions.

Practical Applications in Exploration Campaigns

1. **Target Prioritization**: Decision frameworks guide the selection of high-priority exploration targets, balancing the potential value of resource discovery against the risks and costs of surveying specific areas.
2. **Resource Allocation**: Probabilistic models help allocate resources, such as AUVs, ROVs, and survey vessels, to maximize coverage and data quality while minimizing operational expenses.
3. **Environmental Risk Management**: Uncertainty quantification informs strategies to mitigate environmental impacts, such as selecting survey routes that avoid ecologically sensitive areas.

4. **Adaptive Exploration**: Dynamic decision-making frameworks enable real-time adjustments to exploration plans based on incoming data, ensuring that operations remain efficient and responsive.

Challenges and Future Directions

Despite its benefits, decision-making under uncertainty in deep sea mining faces several challenges:

- **Data Limitations**: Sparse and noisy datasets can reduce the reliability of probabilistic models, requiring robust validation and preprocessing techniques.
- **Computational Complexity**: Advanced models and simulations often demand significant computational resources, necessitating the use of cloud-based platforms and parallel processing.
- **Integration with Stakeholder Objectives**: Decisions must align with the priorities of diverse stakeholders, including investors, regulators, and environmental advocates, adding complexity to the decision-making process.

Future innovations, such as the integration of explainable AI and advanced visualization tools, promise to enhance transparency and stakeholder engagement. Additionally, advancements in edge computing and real-time data analytics will enable more dynamic and adaptive decision-making frameworks.

Decision-making under uncertainty is a cornerstone of successful exploration campaigns in deep sea mining. By incorporating Bayesian methods, stochastic optimization, and advanced simulation models, operators can navigate the complexities of sparse data and unpredictable conditions with confidence and precision. As exploration pushes into increasingly uncharted territories, these frameworks will play a critical role in optimizing outcomes, managing risks, and unlocking the immense potential of the deep ocean's mineral wealth. Through continuous innovation and refinement, decision-making under uncertainty will remain at the forefront of responsible and effective resource exploration.

4. Robotics, Autonomous Systems, and AI-Driven Extraction Technologies

4.1. AI-Enhanced Robotics in Extreme Environments

The exploration and exploitation of deep sea mineral resources present a series of challenges unparalleled in terrestrial environments. Among these, navigation and control of robotic systems in the hostile conditions of the deep ocean are among the most formidable. The crushing pressures, perpetual darkness, complex seafloor topography, and unpredictable currents demand a level of precision, adaptability, and autonomy that traditional control systems struggle to achieve. Enter deep reinforcement learning (DRL), a cutting-edge subset of artificial intelligence that equips robotic systems with the ability to learn, adapt, and optimize their performance in real-time. DRL has become a cornerstone of advancements in robotic navigation and control, enabling autonomous underwater vehicles (AUVs) and remotely operated vehicles (ROVs) to operate effectively in extreme environments.

At its core, deep reinforcement learning relies on an agent's ability to interact with its environment, learn from those interactions, and improve its decision-making over time. The process involves the agent receiving observations from its surroundings, taking actions, and then receiving feedback in the form of rewards or penalties. Over iterative cycles, the agent optimizes its behavior to maximize cumulative rewards, effectively learning to navigate complex scenarios with minimal human intervention. For deep sea robotics, this paradigm aligns perfectly with the challenges of navigating and performing tasks on the seafloor, where unpredictable terrain, sparse data, and dynamic conditions necessitate a level of autonomy that conventional algorithms cannot provide.

In the context of navigation, DRL enables robotic systems to chart optimal paths through intricate underwater landscapes, avoiding obstacles such as seamounts, hydrothermal vents, or dense sediment plumes while maximizing survey coverage. Unlike traditional path-planning algorithms, which rely on pre-programmed rules and static maps, DRL-equipped AUVs dynamically adapt to their environment in real time. For instance, if a vehicle encounters an unexpected obstruction or a strong ocean current, a DRL model can recalibrate its trajectory almost instantaneously, ensuring the mission continues with minimal disruption. This adaptability is particularly crucial in environments where pre-mapped data is either unavailable or insufficiently detailed, as is often the case in deep sea mining regions.

Moreover, DRL excels at optimizing multi-objective tasks, a frequent requirement in resource exploration. An AUV equipped with a suite of sensors might need to balance competing objectives such as maintaining a safe distance from fragile ecosystems, maximizing the resolution of bathymetric surveys, and conserving energy for extended missions. Through reinforcement learning, the vehicle learns to prioritize these goals dynamically, adapting its behavior based on situational demands and long-term mission objectives. This capability not only enhances operational efficiency but also aligns with sustainability goals, minimizing the ecological footprint of deep sea mining activities.

In addition to navigation, DRL is revolutionizing control systems for robotic manipulation. Tasks such as collecting polymetallic nodules, sampling hydrothermal vent deposits, or installing monitoring equipment on the seafloor require precise and adaptive control of robotic arms and end-effectors. Traditional control algorithms often struggle with the complexities of underwater manipulation, where hydrodynamic forces and uncertain interactions with the environment introduce significant variability. DRL addresses these challenges by enabling robotic manipulators to learn from trial and error, refining their movements to optimize efficiency and precision. For example, a DRL-trained robotic arm can adapt to the irregular shapes and sizes of polymetallic nodules, ensuring successful collection even in challenging conditions.

The application of DRL in extreme environments is not without its challenges. Training reinforcement learning models requires extensive data and computational resources, as the agent must simulate countless interactions to develop robust policies. In the deep sea context, collecting real-world data can be both costly and time-consuming, making simulation environments an essential component of the training process. Advanced simulation platforms replicate the physical and environmental conditions of the seafloor, allowing DRL agents to train safely and efficiently before deployment. These platforms incorporate high-fidelity models of underwater physics, including hydrodynamic forces, pressure effects, and sensor noise, ensuring that the learned policies translate effectively to real-world operations.

Another critical aspect of DRL is its reliance on feedback signals to guide learning. Designing reward structures that encourage desirable behaviors while avoiding unintended consequences is a nuanced process. In deep sea navigation, for instance, a simple reward function that prioritizes speed might lead to reckless behaviors, such as bypassing critical survey areas or damaging delicate ecosystems. To address this, researchers craft multi-faceted reward structures that balance efficiency with mission-specific priorities and ethical considerations, fostering behaviors that align with both operational and environmental goals.

The integration of DRL into deep sea robotics also raises questions of robustness and reliability. Unlike deterministic algorithms, which produce consistent outputs for identical inputs, reinforcement learning models exhibit stochastic behaviors, reflecting their adaptive nature. While this adaptability is a strength, it also necessitates rigorous testing and validation to ensure safety and reliability, particularly in high-stakes environments like the deep ocean. Robustness testing under diverse conditions, including simulated equipment failures and extreme environmental scenarios, is essential to building trust in these systems.

Despite these challenges, the transformative potential of DRL in deep sea mining is undeniable. The ability of DRL-equipped robots to learn, adapt, and optimize in real-time is reshaping what is possible in extreme environments. By enabling autonomous systems to navigate complex terrains, perform delicate manipulations, and respond dynamically to changing conditions, DRL is driving the next generation of innovation in underwater robotics. As the field continues to advance, the integration of DRL with complementary technologies such as sensor fusion, edge computing, and AI-driven decision support systems will further enhance the capabilities of deep sea robots, unlocking new possibilities in the exploration and sustainable utilization of the ocean's vast mineral wealth.

Through the lens of deep reinforcement learning, the future of underwater robotics comes into sharper focus, one where autonomous systems not only survive but thrive in the most challenging environments on Earth, pushing the boundaries of what humanity can achieve in the last great frontier.

The deep ocean is an environment characterized by constant flux and extreme variability. Its dynamic nature, shaped by factors such as shifting sediments, fluctuating currents, temperature gradients, and the presence of geological features, presents formidable challenges for robotic systems tasked with exploration and resource extraction. To thrive in such conditions, robots must be equipped with adaptive control systems that can respond intelligently and autonomously to the unexpected. These systems, powered by artificial intelligence (AI), transform rigid, pre-programmed robots into agile, self-adjusting machines capable of navigating, operating, and performing complex tasks with precision and resilience.

Adaptive control systems are the result of advances in AI and robotics that enable machines to modify their behavior in real time based on environmental inputs. Unlike traditional control systems, which rely on fixed algorithms designed for predictable conditions, adaptive systems continuously assess their surroundings and optimize their responses accordingly. In the deep sea, where the unexpected is the norm, this capability is essential for achieving operational success.

The seafloor presents a tapestry of challenges that demand adaptability. For instance, sediment plumes generated by mining or natural disturbances can obscure visual sensors, compromising navigation and task execution. Adaptive control systems address this by integrating data from multiple sensors—acoustic, optical, and inertial—to maintain situational awareness even in low-visibility conditions. By dynamically adjusting navigation strategies based on the reliability of each sensor, these systems ensure that robotic platforms can maintain their trajectories and objectives despite environmental disruptions.

Another critical application of adaptive control is in managing the hydrodynamic forces that affect robots operating near the seafloor. Currents, turbulence, and pressure gradients can destabilize vehicles and impede precision tasks such as sampling or resource collection. Adaptive control systems employ predictive algorithms that anticipate these forces and adjust propulsion, buoyancy, and manipulator movements in real time to counteract their effects. For example, an autonomous underwater vehicle (AUV) equipped with an adaptive control system might adjust its thrusters to maintain stability while navigating a steep slope or operating near a hydrothermal vent.

In addition to environmental variability, the seafloor's geological complexity demands adaptive behavior. Features such as ridges, seamounts, and vent fields create uneven and unpredictable terrains that challenge traditional robotic systems. Adaptive control systems use real-time mapping and terrain analysis to guide movement, enabling robots to traverse these landscapes with efficiency and safety. This capability is particularly valuable for resource extraction tasks, where precise positioning is required to access mineral deposits such as polymetallic nodules or cobalt-rich crusts.

The adaptability of these systems extends to task-specific functions as well. In resource collection, for example, robots equipped with adaptive control can modify their gripping or suction mechanisms based on the physical characteristics of the target material. A manipulator designed to harvest polymetallic nodules might adjust its force and angle of approach based on the size, shape, and surface texture of the nodules, minimizing damage to the resource and the surrounding environment.

AI-driven learning algorithms play a pivotal role in enabling such adaptability. Reinforcement learning, for instance, allows control systems to improve their performance over time by learning from successes and failures in the field. By analyzing patterns in environmental interactions, these systems refine their responses, becoming more efficient and effective with each deployment. Transfer learning further enhances adaptability by allowing robots to apply knowledge gained in one environment to similar but distinct scenarios, accelerating the learning process and reducing the need for extensive retraining.

Sensor fusion is another cornerstone of adaptive control systems. By combining inputs from multiple sensors, these systems create a comprehensive and coherent picture of the environment, even when individual sensors are compromised. For example, a robot operating in a region with high turbidity might rely on acoustic sensors to compensate for the limitations of optical cameras, ensuring continuous functionality. Advanced algorithms process and prioritize these inputs, enabling the system to make informed decisions based on the most reliable data available at any given moment.

The integration of edge computing further enhances the capabilities of adaptive control systems. By processing data locally on the robotic platform, edge computing minimizes latency and allows for real-time decision-making, even in the bandwidth-limited environments of the deep sea. This ensures that adaptive responses are immediate and precise, enabling robots to react swiftly to changes in their surroundings.

Despite their transformative potential, adaptive control systems face challenges that must be addressed to ensure their reliability and effectiveness. The robustness of these systems in the face of extreme environmental conditions is paramount. Extensive testing in simulated and real-world environments is essential to identify and mitigate vulnerabilities. Additionally, the complexity of adaptive algorithms necessitates rigorous validation to ensure that they perform as intended across a wide range of scenarios.

Another consideration is the ethical and environmental implications of adaptive control. While these systems enhance efficiency and reduce operational risks, they must also prioritize sustainability and ecological responsibility. For instance, adaptive algorithms designed for resource extraction must include safeguards to minimize habitat disruption and avoid contact with sensitive ecosystems.

The future of adaptive control systems in deep sea robotics lies in their continued evolution and integration with complementary technologies. Advances in AI, sensor technology, and computing power will enable even greater levels of autonomy and precision, allowing robots to tackle increasingly complex challenges. As these systems mature, their applications will extend beyond exploration and mining to include environmental monitoring, habitat restoration, and even scientific research, further expanding the role of robotics in the sustainable utilization of the ocean's resources.

Through adaptive control, robots transcend the limitations of static programming, transforming into intelligent agents capable of thriving in the unpredictable and unforgiving environments of the deep sea. As these systems become more sophisticated, they will redefine the possibilities of underwater exploration and resource management, ensuring that humanity's ventures into the depths are both successful and sustainable.

4.2. Real-Time Decision Systems and Autonomy

The deep ocean is an environment where uncertainty reigns, and every mission encounters a unique combination of challenges, from uncharted terrains to unpredictable obstacles. In this context, real-time decision systems, powered by advanced sensor fusion technologies, are indispensable for enabling autonomous robotic systems to navigate and operate effectively. By synthesizing data from a diverse array of sensors, these systems create a cohesive understanding of the environment, allowing for dynamic obstacle avoidance and optimized route planning. This capability is essential not only for ensuring the success of exploration and resource extraction missions but also for minimizing operational risks and environmental impacts.

Sensor fusion lies at the heart of real-time decision-making for autonomous underwater vehicles (AUVs) and remotely operated vehicles (ROVs). It involves the integration of data from multiple sensors, acoustic, optical, inertial, and environmental, into a unified framework that captures the complexities of the underwater environment. Each sensor type provides unique and complementary information. Acoustic systems, such as multibeam sonar and side-scan sonar, excel at detecting and mapping large-scale features, such as ridges, seamounts, and potential obstacles, even in low-visibility conditions. Optical cameras and laser scanners, on the other hand, deliver high-resolution imagery and fine-detail mapping, essential for tasks like identifying mineral deposits or inspecting equipment. Inertial sensors contribute critical information about the vehicle's orientation and movement, while environmental sensors monitor factors such as current strength, water temperature, and turbidity.

In isolation, each sensor has its limitations. Acoustic systems, for example, may struggle with ambiguity in regions where geological features produce similar sonar signatures, while optical sensors are rendered ineffective in areas of high turbidity or complete darkness. Sensor fusion overcomes these individual shortcomings by combining data streams to create a richer, more reliable understanding of the environment. For instance, an AUV navigating through a sediment plume might rely primarily on sonar data for obstacle detection while supplementing this with inertial measurements to maintain trajectory. As conditions improve, the system dynamically re-weights the contributions of optical sensors, incorporating visual cues for more precise navigation.

The integration of sensor data also enables real-time obstacle avoidance, a critical function in the unpredictable seafloor environment. Autonomous systems must contend with diverse hazards, from natural features like hydrothermal vents and steep escarpments to human-made structures such as subsea pipelines or abandoned equipment. Real-time decision systems process sensor inputs at high speeds, identifying potential obstacles and recalculating safe paths within milliseconds. For example, if an AUV encounters an unexpected ridge, the system instantly analyzes the obstacle's dimensions and position, then adjusts the vehicle's course to maintain a safe distance while continuing toward its destination.

Route planning in the deep ocean adds another layer of complexity. Unlike terrestrial environments, where pre-mapped routes can often be followed with minimal adjustments, underwater navigation requires continuous re-evaluation of paths due to the dynamic and unknown nature of the seafloor. Sensor fusion enables AUVs to perform this re-evaluation autonomously, integrating real-time data with prior maps and mission objectives. By continuously updating their internal models of the environment, these systems identify the most efficient and safe routes, balancing objectives such as minimizing energy consumption, maximizing data collection, and avoiding sensitive ecosystems.

One of the most significant advancements in this domain is the application of machine learning algorithms to sensor fusion and route planning. Neural networks and reinforcement learning models excel at recognizing patterns in high-dimensional sensor data, allowing systems to anticipate obstacles and optimize routes based on prior experience. For instance, an AUV trained on data from previous missions might learn to recognize the sonar signature of a vent field or the visual patterns associated with sediment disturbances. This predictive capability enhances the efficiency and safety of navigation, reducing the need for human intervention.

Edge computing plays a crucial role in enabling real-time sensor fusion and decision-making. By processing data locally on the vehicle, edge computing minimizes latency, ensuring that decisions are made quickly enough to respond to fast-changing conditions. This capability is particularly valuable in bandwidth-limited underwater environments, where transmitting data to a surface vessel for processing would introduce unacceptable delays. Advanced processors integrated into AUVs handle the computational demands of sensor fusion and decision-making, running algorithms that synthesize data, identify obstacles, and recalibrate routes in real time.

Despite its transformative potential, sensor fusion for real-time decision-making faces challenges that must be addressed to ensure reliability and robustness. Environmental noise, such as interference from thermal gradients or biological activity, can degrade sensor performance, introducing uncertainty into the fused data. To mitigate this, advanced filtering techniques, such as Kalman filters and Bayesian inference, are employed to smooth and refine sensor outputs, isolating meaningful signals from background noise.

Another challenge lies in the computational demands of processing and integrating high-volume data streams in real time. The development of optimized algorithms and hardware accelerators, such as GPUs and TPUs, is critical for meeting these demands. These advancements allow autonomous systems to handle increasingly complex missions, from multi-vehicle coordination to large-scale seafloor mapping.

The impact of sensor fusion extends beyond navigation and route planning. By enhancing situational awareness, it enables autonomous systems to operate with a higher degree of autonomy and safety, reducing the need for direct human control. This capability is particularly valuable for long-duration missions in remote or hazardous regions, where real-time decision-making is essential for mission success. Additionally, the insights gained from sensor fusion contribute to broader scientific and environmental objectives, providing high-quality data for studies of marine ecosystems, geological processes, and climate dynamics.

As sensor fusion technology continues to evolve, its integration with other advanced systems, such as swarm robotics and AI-driven decision support, will further enhance its capabilities. In the future, autonomous underwater systems will not only navigate and operate effectively in the most challenging environments but also collaborate seamlessly to achieve complex, multi-objective missions. Through the power of real-time sensor fusion, the promise of autonomous exploration and resource management in the deep ocean is being realized, driving humanity's understanding and utilization of this vast and mysterious frontier.

The concept of swarm intelligence, inspired by the collective behavior of social organisms such as ants, bees, and schools of fish, represents a transformative approach to autonomous systems in the deep sea. By enabling coordinated fleets of autonomous underwater vehicles (AUVs) and remotely operated vehicles (ROVs), swarm intelligence offers unprecedented efficiency, scalability, and adaptability in tackling the immense challenges of deep sea exploration and resource extraction. These systems, underpinned by real-time communication, distributed decision-making, and emergent collective behavior, redefine the possibilities of underwater robotics, enabling large-scale, multi-objective missions that are beyond the capabilities of individual vehicles.

The deep ocean is an environment of vast scope and complexity, making it an ideal domain for swarm robotics. The sheer scale of the seafloor, coupled with the need for high-resolution mapping, resource identification, and ecological monitoring, demands the simultaneous operation of multiple robotic platforms. A swarm approach addresses this need by leveraging the collective capabilities of dozens, or even hundreds, of AUVs and ROVs working together in harmony. Unlike traditional systems, where each vehicle operates independently or under direct human supervision, swarms function as decentralized networks, with each unit contributing to a shared goal while dynamically adapting to local conditions and inputs.

Swarm intelligence is built on the principles of decentralized control and local interaction. Each vehicle in the swarm operates autonomously, processing data from its immediate surroundings and making decisions based on predefined algorithms and real-time feedback from neighboring units. These local decisions aggregate into coordinated global behavior, allowing the swarm to accomplish complex tasks efficiently and flexibly. For example, a fleet of AUVs conducting a seafloor survey might distribute themselves to cover the maximum area without overlapping, dynamically adjusting their trajectories to avoid obstacles or compensate for sensor failures within the group. This emergent coordination eliminates the need for centralized control, reducing communication bottlenecks and enhancing system robustness.

The application of swarm intelligence to AUVs and ROVs requires a sophisticated framework of real-time communication and data sharing. Underwater acoustic communication systems, which transmit data through sound waves, are the primary medium for vehicle-to-vehicle interaction in swarms. While the bandwidth of acoustic communication is limited compared to terrestrial wireless systems, its range and reliability in water make it indispensable for coordinating swarm behavior. Advanced communication protocols ensure efficient data exchange, prioritizing critical information such as vehicle status, environmental hazards, and task updates. For instance, if an AUV detects a previously unmapped hydrothermal vent, it can immediately broadcast its location to nearby units, prompting the swarm to reconfigure and prioritize high-resolution mapping of the site.

In addition to communication, swarm intelligence relies on distributed algorithms that enable collective decision-making. Algorithms inspired by natural systems, such as ant colony optimization and particle swarm optimization, are commonly employed in this context. These algorithms simulate behaviors observed in nature, such as trail-following in ants or coordinated movement in bird flocks, adapting them to the challenges of underwater robotics. For example, particle swarm optimization might be used to determine the optimal distribution of AUVs for a resource survey, balancing coverage, energy efficiency, and data collection priorities.

Swarm intelligence also enhances the resilience and fault tolerance of autonomous operations. In a traditional system, the failure of a single vehicle or sensor could compromise an entire mission. In a swarm, however, the distributed nature of decision-making ensures that the system continues to function effectively even in the presence of individual failures. If an AUV in the swarm experiences a mechanical issue or loses communication, neighboring units can adapt by reassigning its tasks and compensating for its absence. This redundancy is particularly valuable in the deep ocean, where environmental conditions and equipment reliability often pose significant risks.

One of the most promising applications of swarm intelligence in deep sea mining is large-scale mapping and resource prospectivity. A coordinated fleet of AUVs can conduct simultaneous surveys over vast areas, collecting high-resolution data on seafloor topography, sediment composition, and mineral distribution. Each vehicle in the swarm operates autonomously within its designated sector, using sensor fusion to navigate and collect data while continuously sharing information with the rest of the fleet. This collaborative approach accelerates the mapping process, reducing the time and cost required to identify promising resource deposits.

Resource extraction is another area where swarm intelligence offers transformative potential. In polymetallic nodule collection, for example, swarms of small, agile ROVs equipped with specialized manipulator arms can work together to harvest nodules from the seafloor. Each ROV operates independently within its assigned zone, identifying and collecting nodules while avoiding obstacles and minimizing sediment disturbance. The swarm's decentralized control ensures that the workload is evenly distributed, optimizing efficiency and reducing wear and tear on individual units. Moreover, real-time communication allows the swarm to adapt dynamically to environmental changes, such as the onset of strong currents or the detection of sensitive habitats, ensuring that operations remain safe and sustainable.

Swarm intelligence also enables advanced environmental monitoring and impact assessment, which are critical components of responsible deep sea mining. A fleet of AUVs equipped with chemical, biological, and optical sensors can simultaneously monitor multiple parameters, such as turbidity, sediment plume dispersion, and biodiversity. By operating as a cohesive network, the swarm can provide comprehensive, real-time data on the environmental impact of mining activities, guiding mitigation measures and informing regulatory compliance. For instance, if a section of the swarm detects elevated turbidity levels indicative of a spreading sediment plume, the system can reallocate vehicles to track and map the plume's extent, providing actionable insights for minimizing ecological disruption.

The implementation of swarm intelligence in deep sea robotics is not without challenges. The harsh underwater environment imposes strict constraints on communication, energy consumption, and computational capacity. Acoustic communication systems, while effective, have limited bandwidth and are susceptible to interference, necessitating efficient data compression and prioritization algorithms. Energy efficiency is another critical consideration, as AUVs and ROVs must operate for extended periods without access to surface charging or resupply. Swarm algorithms must balance computational demands with the vehicles' power constraints, ensuring that the system remains functional throughout the mission.

Additionally, the complexity of swarm behavior requires rigorous testing and validation to ensure reliability and safety. Simulation environments play a vital role in this process, allowing researchers to model and refine swarm algorithms under realistic conditions before deploying them in the field. High-fidelity simulations replicate the physical and environmental dynamics of the deep ocean, providing a controlled setting for evaluating swarm performance and identifying potential vulnerabilities.

Despite these challenges, the potential of swarm intelligence to revolutionize deep sea operations is immense. Ongoing advancements in AI, sensor technology, and communication systems are expanding the capabilities of swarm robotics, enabling increasingly sophisticated and scalable applications. The integration of machine learning further enhances swarm performance, allowing vehicles to learn from experience and adapt their behavior to new scenarios. For example, reinforcement learning algorithms can optimize swarm strategies for specific tasks, such as maximizing resource recovery or minimizing energy consumption, by analyzing historical mission data and simulating future operations.

As swarm intelligence matures, its applications will extend beyond mining and exploration to include a wide range of scientific and industrial activities. Swarms of autonomous vehicles could be deployed for environmental restoration, such as reseeding damaged coral reefs or mitigating the effects of underwater pollution. They could also play a key role in advancing our understanding of the ocean's ecosystems, enabling large-scale studies of biodiversity, biogeochemical cycles, and climate dynamics. Furthermore, swarm robotics has the potential to enhance disaster response efforts, such as locating and recovering debris from maritime accidents or assessing damage to subsea infrastructure following natural disasters.

The deep ocean, with its vast expanses and formidable conditions, is a domain where the challenges of communication, data processing, and real-time decision-making are magnified. Autonomous underwater vehicles (AUVs) and remotely operated vehicles (ROVs) operate in environments where conventional communication infrastructure is unavailable, and data transmission to surface vessels or control centers is constrained by the limitations of underwater acoustic communication. In such bandwidth-limited environments, edge computing emerges as a transformative technology, enabling on-board processing of complex data and empowering robotic systems to make critical decisions autonomously, without reliance on constant human oversight or high-bandwidth communication.

Edge computing refers to the decentralized processing of data at or near the source of generation, in this case, the AUVs or ROVs operating on the seafloor. By equipping these vehicles with advanced computational capabilities, edge computing allows for real-time analysis of sensor inputs, execution of AI-driven algorithms, and adaptation to dynamic environmental conditions. This paradigm minimizes the latency associated with data transmission, reduces the burden on communication channels, and enhances the operational efficiency and autonomy of underwater robotic systems.

One of the primary drivers of edge computing in deep sea robotics is the sheer volume and complexity of data generated by modern sensors. Multibeam sonar systems, side-scan sonar, high-resolution cameras, laser scanners, and geochemical sensors produce terabytes of data during a single mission. Transmitting this data to surface vessels or onshore facilities for processing is not only impractical but also inefficient, particularly in bandwidth-limited underwater environments where acoustic communication offers a fraction of the bandwidth available in terrestrial networks. Edge computing addresses this bottleneck by processing raw data locally, extracting actionable insights and transmitting only the most relevant information to surface operators.

The implementation of edge computing begins with the integration of high-performance hardware into AUVs and ROVs. Compact yet powerful processors, such as GPUs (graphics processing units) and TPUs (tensor processing units), enable the execution of complex machine learning algorithms and real-time data analysis directly on-board. These processors are designed to handle the intensive computational demands of tasks such as image recognition, terrain mapping, and obstacle avoidance while operating within the energy constraints of underwater vehicles. Specialized edge computing devices are further optimized for the harsh conditions of the deep sea, ensuring reliability and durability in extreme pressure, temperature, and corrosive environments.

A critical application of edge computing in underwater robotics is real-time sensor fusion. By combining data streams from multiple sensors, edge systems create a coherent and comprehensive picture of the environment, enabling vehicles to navigate complex terrains and avoid obstacles autonomously. For instance, an AUV equipped with sonar, optical cameras, and inertial measurement units (IMUs) can integrate these inputs on-board to identify hazards, chart optimal routes, and adapt to unexpected changes in the environment, such as sediment plumes or strong currents. This capability is particularly valuable in regions where pre-mapped data is unavailable or insufficiently detailed, as is often the case in unexplored deep-sea mining zones.

In addition to navigation, edge computing enhances the precision and efficiency of resource exploration and extraction. For example, AUVs tasked with identifying polymetallic nodules or seafloor massive sulfides (SMS) rely on high-resolution sonar and optical imagery to detect and classify potential mineral deposits. Edge systems process this data in real time, using AI algorithms to differentiate between mineralized features and background noise. This localized processing reduces the need for post-mission analysis, accelerating decision-making and enabling operators to adapt exploration strategies dynamically. If an AUV identifies a high-potential area during its mission, it can immediately relay this information to the rest of the fleet or initiate a more detailed survey of the site.

One of the most advanced applications of edge computing is the use of deep learning models for in situ analysis. Convolutional neural networks (CNNs), for instance, are deployed on-board to analyze visual and acoustic data, identifying patterns and features indicative of mineralization or ecological significance. These models, trained on large datasets from prior missions, can recognize the shapes and textures of polymetallic nodules, the acoustic signatures of hydrothermal vents, or the presence of sensitive marine habitats. By processing data locally, edge systems eliminate the latency and bandwidth constraints associated with transmitting high-resolution images or sonar maps to surface vessels for analysis.

Edge computing also plays a pivotal role in enabling collaborative autonomy among fleets of AUVs and ROVs. In swarm operations, where multiple vehicles work together to survey vast areas or perform coordinated tasks, real-time data sharing and decision-making are essential. Edge systems facilitate this by processing and transmitting summarized data between vehicles, enabling the swarm to adapt its behavior dynamically based on collective inputs. For example, if one AUV encounters an obstacle or identifies a high-priority target, it can share this information with nearby units, prompting the swarm to adjust its configuration or objectives. This decentralized approach reduces reliance on surface operators and enhances the resilience and efficiency of multi-vehicle operations.

The advantages of edge computing extend beyond efficiency and autonomy. By processing data locally, these systems reduce the energy demands associated with long-distance communication, extending the operational endurance of underwater vehicles. This is particularly critical for long-duration missions in remote regions, where recharging or resupply opportunities are limited. Additionally, edge computing enhances data security by minimizing the transmission of sensitive information over potentially vulnerable communication channels. This is especially important in the context of deep-sea mining, where proprietary data on resource distribution and environmental conditions represents a valuable asset.

Despite its transformative potential, the implementation of edge computing in underwater robotics presents several challenges. The design and integration of edge systems must balance computational power with energy efficiency, ensuring that vehicles can perform complex tasks without compromising battery life. Thermal management is another critical consideration, as high-performance processors generate significant heat, which must be dissipated effectively in the thermally insulating environment of the deep ocean. Reliability and robustness are also paramount, as edge systems must operate flawlessly in extreme pressures, corrosive seawater, and variable temperatures.

The development of edge computing solutions for underwater robotics is an active area of research and innovation. Advances in hardware, such as the miniaturization of GPUs and the development of energy-efficient processors, are expanding the capabilities of edge systems. At the same time, software innovations, including lightweight machine learning models and real-time optimization algorithms, are enhancing the performance and versatility of these systems. The integration of edge computing with complementary technologies, such as sensor fusion, swarm intelligence, and cloud-based data analytics, promises to unlock new possibilities in autonomous underwater operations.

Looking to the future, edge computing will play a central role in the evolution of underwater robotics and the sustainable development of deep-sea resources. As these systems become more powerful and efficient, they will enable a new generation of AUVs and ROVs capable of performing increasingly sophisticated tasks, from large-scale seafloor mapping to precision mineral extraction and ecological monitoring. By empowering robotic systems to process data and make decisions autonomously, edge computing not only enhances the operational capabilities of these vehicles but also reduces their environmental footprint, contributing to the responsible and efficient exploration of the deep ocean.

Edge computing represents a paradigm shift in the design and operation of autonomous underwater systems. By enabling on-board processing in bandwidth-limited environments, this technology addresses the unique challenges of deep-sea robotics, enhancing efficiency, resilience, and autonomy. As the field continues to advance, edge computing will remain at the forefront of innovation, driving the exploration and sustainable utilization of the ocean's vast and mysterious frontier. Through its transformative potential, edge computing ensures that the deep ocean's resources can be harnessed with precision, responsibility, and respect for the delicate balance of marine ecosystems.

4.3. Advanced Drilling and Collection Mechanisms

Drilling and collection operations in the deep ocean represent one of the most technically demanding aspects of resource extraction. The seafloor, with its extreme pressures, low temperatures, and complex geological formations, imposes extraordinary demands on drilling rigs, requiring precision, resilience, and adaptability. Traditional drilling systems, while robust in terrestrial contexts, are ill-suited for the challenges of deep sea operations, where equipment failures or inefficiencies can result in significant delays, cost overruns, and environmental risks. To address these challenges, intelligent drilling rigs equipped with predictive maintenance capabilities are emerging as a transformative solution, redefining how drilling operations are managed and executed in the harsh conditions of the deep ocean.

Intelligent drilling rigs represent the convergence of advanced robotics, machine learning, and sensor technologies. These systems are designed to operate autonomously or semi-autonomously, using real-time data to optimize drilling performance and anticipate maintenance needs before failures occur. This proactive approach minimizes downtime, extends the operational lifespan of equipment, and ensures that drilling operations remain safe, efficient, and environmentally responsible.

At the heart of intelligent drilling rigs is a network of integrated sensors that continuously monitor the performance and condition of critical components. These sensors collect data on parameters such as vibration, temperature, pressure, and torque, providing a comprehensive view of the rig's operational status. For example, vibration sensors installed on drill bits and rotating shafts detect anomalies that could indicate wear or imbalance, while pressure sensors monitor hydraulic systems to ensure optimal performance. By analyzing this data in real time, the rig's onboard systems can identify potential issues long before they escalate into failures, allowing operators to schedule maintenance at optimal intervals.

Machine learning algorithms play a pivotal role in enabling predictive maintenance for intelligent drilling rigs. These algorithms analyze historical and real-time sensor data to identify patterns and trends associated with equipment degradation or failure. For instance, an algorithm might detect a gradual increase in torque over successive drilling cycles, signaling the onset of wear in a drill bit or motor. By correlating these trends with past maintenance records, the system can predict when the component is likely to fail and recommend preventive measures. This capability not only reduces unplanned downtime but also optimizes the use of maintenance resources, ensuring that interventions are targeted and cost-effective.

The use of digital twins further enhances the predictive maintenance capabilities of intelligent drilling rigs. A digital twin is a virtual replica of the physical rig, continuously updated with data from its sensors. This digital model allows operators to simulate various scenarios, such as the impact of prolonged drilling on specific components or the effects of extreme environmental conditions. By running these simulations, the digital twin provides insights into potential vulnerabilities and informs strategies for mitigating risks. For example, if a simulation predicts that a critical component will fail under increased load conditions, the system can recommend adjustments to the drilling parameters or initiate a preemptive replacement.

In addition to monitoring equipment health, intelligent drilling rigs leverage AI-driven optimization algorithms to enhance drilling performance. These algorithms analyze geological data, such as rock density, porosity, and fracture patterns, to determine the most efficient drilling strategy for a given location. By adjusting parameters such as drill speed, pressure, and bit type in real time, the system ensures that drilling operations are both effective and energy-efficient. This adaptability is particularly valuable in the heterogeneous and unpredictable environments of the seafloor, where geological conditions can vary significantly over short distances.

The integration of intelligent control systems also enables drilling rigs to respond dynamically to changing conditions. For instance, if sensors detect an increase in sedimentation or water ingress, the system can automatically adjust drilling parameters to maintain stability and prevent damage to the rig or surrounding environment. Similarly, if the rig encounters a sudden increase in rock hardness, it can switch to a more suitable drill bit or modify its approach to avoid excessive wear and tear.

Another key advantage of intelligent drilling rigs is their ability to operate autonomously in remote or hazardous environments. Equipped with advanced AI and real-time decision-making capabilities, these rigs can execute complex drilling operations with minimal human intervention. This autonomy not only reduces the risks associated with human presence in extreme environments but also enables continuous operation, maximizing productivity and reducing the time required to access valuable resources.

Predictive maintenance is particularly critical in the context of deep sea mining, where the consequences of equipment failure can be severe. A malfunctioning drill bit or hydraulic system, for example, could halt operations for days or weeks, incurring significant costs and jeopardizing project timelines. Moreover, the logistical challenges of deploying repair crews and spare parts to remote offshore locations further underscore the importance of preemptive maintenance strategies. By detecting and addressing issues proactively, intelligent drilling rigs minimize the likelihood of costly disruptions and ensure that operations remain on track.

The environmental benefits of predictive maintenance are equally significant. By preventing catastrophic failures, these systems reduce the risk of unintended environmental impacts, such as oil spills, sediment plumes, or damage to marine ecosystems. Furthermore, the ability to optimize drilling parameters in real time minimizes energy consumption and material waste, aligning with sustainability goals and regulatory requirements.

Despite their transformative potential, the implementation of intelligent drilling rigs with predictive maintenance capabilities presents several challenges. The development and deployment of these systems require significant investment in advanced technologies, including sensors, AI algorithms, and digital twin platforms. Ensuring the reliability and robustness of these technologies in the extreme conditions of the deep ocean is another critical consideration, as equipment must withstand high pressures, corrosive seawater, and temperature fluctuations.

Data security and communication also pose challenges, particularly in remote offshore environments where real-time connectivity is limited. To address this, intelligent rigs rely on edge computing systems that process data locally, reducing the need for constant communication with surface operators. These systems are designed to operate independently, ensuring that predictive maintenance and optimization algorithms function seamlessly even in bandwidth-limited conditions.

As intelligent drilling rigs continue to evolve, their integration with other advanced technologies promises to unlock new possibilities in deep sea mining. For example, the combination of predictive maintenance with swarm intelligence could enable fleets of autonomous drilling rigs to coordinate their activities, optimizing resource extraction across large areas. Similarly, the integration of machine learning with geophysical and geochemical data could enhance the accuracy of predictive models, further improving the efficiency and reliability of drilling operations.

In the broader context of deep sea exploration, intelligent drilling rigs with predictive maintenance capabilities represent a critical step toward achieving sustainable and economically viable resource extraction. By addressing the unique challenges of the deep ocean, these systems enable operators to access valuable minerals while minimizing risks and environmental impacts. As technology continues to advance, the principles of predictive maintenance and intelligent control will play an increasingly central role in shaping the future of underwater robotics and resource management.

Through their combination of advanced sensing, real-time data analysis, and proactive decision-making, intelligent drilling rigs exemplify the potential of AI-driven innovation in the most challenging and remote environments on Earth. These systems not only enhance the efficiency and reliability of deep sea mining operations but also set a new standard for sustainability and resilience, ensuring that the exploration and utilization of the ocean's resources are conducted with precision, care, and respect for the delicate balance of marine ecosystems. As they become more widespread, intelligent drilling rigs with predictive maintenance capabilities will redefine the possibilities of underwater resource extraction, unlocking new frontiers of discovery and development in the depths of the ocean.

In the realm of deep sea mining, the efficient separation and sorting of extracted materials is a critical challenge that determines both the economic viability and environmental impact of resource extraction operations. Once mineral-laden material is brought to the surface, or in some cases, processed subsea, it must be separated into valuable components and residual waste. Traditional approaches to material sorting rely on mechanical processes or rudimentary techniques that often lack the precision necessary for the complex compositions of deep sea deposits. The advent of artificial intelligence (AI), particularly computer vision and machine learning, has revolutionized material separation and sorting by enabling real-time, high-accuracy identification and classification of extracted resources. This innovation not only maximizes recovery rates but also minimizes waste, energy consumption, and ecological disruption, establishing new benchmarks for efficiency and sustainability in underwater mining.

AI-driven material separation systems are designed to process and analyze vast quantities of raw material with unmatched speed and precision. At their core, these systems rely on computer vision algorithms, which use cameras and sensors to capture high-resolution images or spectroscopic data of extracted materials. These inputs are then analyzed in real-time by AI models trained to distinguish between valuable minerals and waste based on visual, textural, and compositional features. For instance, a computer vision system might identify the specific shape, color, or surface texture of polymetallic nodules, cobalt-rich crust fragments, or sulfide minerals, separating them from unwanted sediment or debris with exceptional accuracy.

The training of these AI models begins with the creation of extensive datasets comprising labeled images or spectroscopic profiles of target materials. These datasets include examples of valuable resources in various conditions, pristine, partially coated in sediment, or fractured, as well as common forms of waste material. Using supervised learning techniques, the AI is trained to recognize the distinguishing features of each category, such as the distinctive metallic sheen of polymetallic nodules or the crystalline structure of seafloor massive sulfides (SMS). Once trained, the system applies this knowledge to classify and sort materials during active operations, continuously refining its performance through reinforcement learning and real-time feedback.

One of the key advantages of AI-driven sorting systems is their ability to operate in real-time, enabling dynamic adjustments to extraction and processing workflows. For example, as raw material is conveyed through a collection pipeline or sorting mechanism, the system analyzes each piece in milliseconds, determining its composition and directing it to the appropriate processing stream. High-value resources are directed to further refinement, while waste materials are either discarded responsibly or reused in other applications. This instantaneous decision-making eliminates bottlenecks, ensures consistent throughput, and reduces the reliance on manual sorting, which is both labor-intensive and error-prone.

A particularly transformative application of AI in material sorting is the integration of multi-sensor data. While computer vision excels at identifying surface-level features, many valuable resources exhibit characteristics that are not visible to the naked eye. To address this, advanced sorting systems incorporate additional sensors, such as X-ray fluorescence (XRF), hyperspectral imaging, and laser-induced breakdown spectroscopy (LIBS). These technologies provide detailed insights into the elemental composition and internal structure of materials, complementing the visual data captured by cameras. AI algorithms fuse these diverse data streams, creating a holistic understanding of each material's properties and enabling even more precise classification.

For instance, in polymetallic nodule collection, AI-driven systems analyze both the visual appearance and chemical composition of each nodule. Using hyperspectral imaging, the system detects subtle variations in manganese, cobalt, nickel, and copper concentrations, ensuring that only nodules with economically viable grades are retained. Similarly, in SMS extraction, LIBS can identify the presence of valuable metals such as gold, silver, and zinc, distinguishing them from less valuable or environmentally harmful materials. By integrating these capabilities, AI-driven sorting systems maximize resource recovery while minimizing the volume of non-target material processed, reducing energy consumption and environmental impact.

Another critical benefit of AI-driven material sorting is its adaptability to diverse and changing conditions. The composition of extracted material can vary significantly within a single mining site, reflecting the complex geology of the seafloor. Traditional sorting systems struggle to handle this variability, often requiring manual adjustments or recalibration. AI systems, however, excel at adapting to new scenarios. By continuously learning from incoming data, they refine their classification algorithms to account for variations in resource quality, mineral associations, or sediment coatings. This adaptability ensures consistent performance across heterogeneous environments, enhancing the efficiency and reliability of mining operations.

The environmental implications of AI-driven sorting systems are particularly noteworthy. By accurately separating valuable resources from waste, these systems reduce the amount of material that needs to be transported, processed, or discarded, lowering the carbon footprint of mining activities. Furthermore, by minimizing the processing of waste material, they reduce the generation of harmful byproducts, such as sediment plumes or chemical residues, which can have detrimental effects on marine ecosystems. Some advanced systems even incorporate mechanisms for returning non-target material to the seafloor in a controlled manner, mitigating disruption to benthic habitats and promoting ecosystem recovery.

The role of AI in material sorting also extends to the optimization of downstream processes. Once raw materials are classified and separated, they are typically subjected to further refinement, such as smelting, chemical extraction, or electrolysis. AI-driven sorting systems ensure that only high-quality feedstock enters these processes, improving their efficiency and reducing operational costs. For example, by pre-sorting nodules with high manganese content, the system reduces the energy and reagent consumption of subsequent hydrometallurgical processes, enhancing overall sustainability.

Despite its many advantages, the implementation of AI-driven sorting systems in deep sea mining presents challenges that must be addressed to fully realize their potential. One of the primary challenges is the harsh operating environment of the deep ocean, where high pressures, low temperatures, and corrosive seawater impose strict requirements on hardware durability and reliability. Cameras, sensors, and processing units must be designed to withstand these conditions while maintaining consistent performance over extended periods. Additionally, the integration of AI systems into existing mining infrastructure requires careful planning and investment, as well as collaboration between technology providers, mining operators, and regulators.

Another challenge lies in the development and maintenance of high-quality training datasets. The accuracy of AI-driven sorting systems depends on the breadth and diversity of their training data, which must encompass the full range of materials encountered in deep sea mining. Collecting and labeling such data is a resource-intensive process, requiring expertise in geology, mineralogy, and computer science. Furthermore, as new mining sites are explored, these datasets must be continuously updated to reflect the unique characteristics of each location, ensuring that the AI system remains effective in diverse contexts.

Looking to the future, the integration of AI-driven sorting systems with other advanced technologies promises to unlock new possibilities in deep sea mining. For example, the combination of AI with swarm robotics could enable distributed sorting, where fleets of small, mobile ROVs collect and pre-sort materials directly on the seafloor, reducing the volume of material transported to surface processing facilities. Similarly, the use of digital twins, virtual replicas of mining operations, could allow operators to simulate and optimize sorting workflows before deployment, enhancing efficiency and reducing risk.

AI-driven material separation and sorting systems represent a paradigm shift in the efficiency, precision, and sustainability of deep sea mining operations. By leveraging computer vision, multi-sensor data fusion, and machine learning, these systems enable real-time classification of extracted materials, maximizing resource recovery while minimizing waste and environmental impact. The capabilities of AI-driven sorting systems will expand, driving innovation and sustainability in one of the planet's most challenging and promising frontiers. Through these advancements, the vision of responsible and economically viable deep sea mining becomes increasingly attainable, unlocking the ocean's mineral wealth with care and precision.

The development and testing of advanced drilling and collection mechanisms for deep sea mining are essential steps toward the sustainable and efficient exploitation of seabed resources. Case studies and prototypes provide invaluable insights into the practical challenges and opportunities associated with subsea excavation, offering real-world validation for innovative technologies and concepts. These efforts not only demonstrate the feasibility of extracting valuable minerals from the ocean floor but also highlight the importance of balancing operational efficiency with environmental stewardship. This section explores key case studies and prototypes in subsea excavation, illustrating the progress made in this field and the lessons learned along the way.

Nautilus Minerals: Pioneering Seafloor Massive Sulfides Extraction

One of the most prominent case studies in subsea excavation is Nautilus Minerals' Solwara 1 project, which aimed to extract seafloor massive sulfides (SMS) from hydrothermal vent fields in the Bismarck Sea off Papua New Guinea. The project represented a groundbreaking attempt to commercialize deep sea mining, leveraging state-of-the-art technologies to access high-grade deposits of copper, gold, and zinc.

Nautilus Minerals developed a suite of specialized equipment for the project, including three large robotic seafloor production tools (SPTs): the auxiliary cutter, the bulk cutter, and the collecting machine. These tools were designed to operate in tandem, mimicking the processes of terrestrial mining in a subsea context. The auxiliary cutter prepared the seafloor by creating a stable working surface, while the bulk cutter broke apart the mineralized material using a massive rotating drum equipped with cutting teeth. The collecting machine then gathered the fragmented material and transferred it to a riser and lifting system for transportation to the surface.

The SPTs incorporated advanced features to address the unique challenges of the subsea environment. For example, the cutting drums were equipped with replaceable teeth made from super-hard materials to withstand the abrasive nature of SMS deposits. Hydraulic systems were optimized for deep-sea pressures, and onboard cameras and sensors provided real-time feedback to operators on the surface. Despite these technological innovations, the project encountered significant financial, logistical, and regulatory hurdles, ultimately halting progress. However, the Solwara 1 project remains a landmark in the history of subsea excavation, providing valuable insights into the technical and operational requirements of deep sea mining.

DeepGreen Metals: Polymetallic Nodule Collection Prototypes

DeepGreen Metals (now part of The Metals Company) has focused on the collection of polymetallic nodules from the Clarion-Clipperton Zone (CCZ) in the Pacific Ocean. These nodules, rich in critical minerals such as cobalt, nickel, and manganese, are viewed as a key resource for supporting the global transition to renewable energy and electric vehicles. DeepGreen's approach emphasizes environmental responsibility, seeking to minimize the ecological impact of nodule harvesting.

To achieve this, the company has developed and tested prototypes for nodule collection systems designed to operate on the abyssal plain at depths of 4,000 to 6,000 meters. These systems are engineered to gently lift nodules from the seafloor while minimizing sediment disturbance, which could harm nearby ecosystems. One prototype, a crawler-based collector vehicle, features a low-impact suction mechanism that separates nodules from sediment with minimal turbidity. The collected material is transferred to a riser system and transported to the surface for processing.

Field trials conducted in collaboration with independent scientific organizations have provided critical data on the performance and environmental impact of these prototypes. High-resolution sensors and cameras mounted on the vehicles have been used to monitor sediment plumes and assess the recovery of benthic habitats after disturbances. These studies have informed the iterative design of the collection systems, ensuring that they meet both operational and environmental objectives.

Japan's JAMSTEC: Advancing Seafloor Drilling Technology

The Japan Agency for Marine-Earth Science and Technology (JAMSTEC) has been at the forefront of developing advanced drilling technologies for deep sea exploration and mining. JAMSTEC's efforts have focused on extracting cobalt-rich ferromanganese crusts and investigating seafloor massive sulfides in Japan's exclusive economic zone.

One notable project is the development of the Benthic Multicoring System (BMCS), a versatile tool for collecting sediment and rock samples from the seafloor. The BMCS combines drilling, coring, and sampling capabilities, allowing researchers to assess the composition and distribution of mineral resources with high precision. The system is equipped with advanced navigation and positioning technology, enabling accurate deployment in challenging subsea environments.

JAMSTEC has also pioneered the use of autonomous drilling systems, such as the Deep-sea Boring Machine, which can operate without direct human intervention. These systems are designed to withstand the extreme pressures and temperatures of deep ocean drilling while providing high-quality core samples for geological analysis. By integrating AI and sensor technologies, JAMSTEC's drilling systems are capable of adapting to variable conditions, ensuring reliable performance in diverse settings.

Subsea Collection Technologies: The EU's Blue Nodules Project

The European Union's Blue Nodules project, part of the Horizon 2020 program, has focused on developing sustainable technologies for polymetallic nodule collection. The project brings together industry partners, research institutions, and environmental organizations to address the technical, economic, and ecological challenges of deep sea mining.

One of the project's key prototypes is a modular collector vehicle equipped with innovative material separation and sorting mechanisms. The vehicle uses advanced computer vision and sensor systems to identify and selectively harvest nodules while minimizing sediment disturbance. It also incorporates a modular design, allowing components to be easily replaced or upgraded, reducing maintenance costs and extending operational lifespan.

The Blue Nodules project has emphasized the importance of environmental monitoring and mitigation. Field tests have included extensive assessments of sediment plumes, noise pollution, and the recovery of benthic ecosystems. Data from these tests has been used to refine the design and operation of the collector systems, ensuring that they align with sustainability principles and regulatory standards.

Lessons Learned and the Path Forward

These case studies and prototypes highlight both the progress and challenges associated with subsea excavation technologies. They underscore the importance of innovation, collaboration, and environmental responsibility in advancing deep sea mining. Several key lessons emerge from these efforts:

First, the technical complexity of subsea excavation demands a multidisciplinary approach, integrating expertise from robotics, geology, engineering, and environmental science. Successful projects have leveraged this diversity of knowledge to develop systems that are both effective and adaptable.

Second, field trials and prototype testing are essential for validating technologies under real-world conditions. These trials provide invaluable data on system performance, operational challenges, and environmental impacts, enabling iterative improvements that enhance efficiency and sustainability.

Third, collaboration with independent scientific organizations and regulatory bodies is crucial for building trust and ensuring compliance with environmental standards. Transparent assessments of the ecological impacts of subsea excavation can inform best practices and support the development of policies that balance resource extraction with conservation.

Looking to the future, the continued development of subsea excavation technologies will benefit from advancements in AI, sensor systems, and materials science. AI-driven control systems, for example, will enable greater autonomy and precision in drilling and collection operations, while advanced sensors will provide real-time insights into geological and environmental conditions. Innovations in materials science will improve the durability and performance of equipment, reducing maintenance requirements and extending operational lifespans.

5. Environmental Impact Assessment and Monitoring

5.1. Biodiversity, Ecosystem Services, and AI

The integration of biological data into artificial intelligence (AI)-driven models represents a profound leap forward in our ability to understand and manage the complex ecosystems of the deep ocean. The ocean floor is not merely a repository of mineral wealth; it is a dynamic, interconnected environment that supports an astonishing array of life, from microbial communities thriving in hydrothermal vents to megafauna navigating vast abyssal plains. These organisms contribute to vital ecosystem services, including nutrient cycling, carbon sequestration, and biodiversity maintenance, all of which have far-reaching implications for global environmental health. Effective deep sea mining operations must therefore go beyond resource extraction, embracing advanced tools to monitor, predict, and mitigate their impact on these ecosystems. By integrating biological data, such as environmental DNA (eDNA) and remote sensing observations of megafauna, into AI-driven models, we gain a powerful mechanism for achieving these goals with precision and accountability.

The extraction of biological data from the deep ocean begins with the collection of diverse and often elusive information. Environmental DNA, or eDNA, is one of the most promising technologies for characterizing biological diversity in deep sea ecosystems. This approach involves collecting water, sediment, or other samples and extracting fragments of genetic material shed by organisms into their environment. These fragments provide a snapshot of the species present in a given area, including those that are difficult to observe through traditional means, such as small or cryptic organisms. By analyzing eDNA, scientists can construct a detailed inventory of biodiversity, identifying not only species composition but also variations in population size, genetic diversity, and community structure.

eDNA data, while invaluable, is inherently complex and voluminous. The sheer number of sequences extracted from a single sample can reach into the millions, reflecting the vast genetic diversity of marine life. AI-driven models play a crucial role in managing and interpreting this data, transforming it into actionable insights. Machine learning algorithms, particularly those designed for pattern recognition and classification, are used to process raw genetic sequences, match them to known species in genomic databases, and detect novel or uncharacterized organisms. These models not only accelerate the analysis process but also improve its accuracy, reducing the likelihood of false identifications and uncovering subtle patterns that might otherwise be missed.

In addition to eDNA, remote sensing technologies are revolutionizing our understanding of deep sea ecosystems, particularly at the level of megafauna, large, mobile species such as whales, sharks, and deep sea fish. High-resolution cameras and sonar systems mounted on autonomous underwater vehicles (AUVs) or remotely operated vehicles (ROVs) capture visual and acoustic data on the presence, behavior, and movements of these organisms. AI algorithms analyze this data in real-time, identifying species based on visual or acoustic signatures and tracking their interactions with the environment. For instance, convolutional neural networks (CNNs), widely used in image recognition, can distinguish between different fish species based on subtle differences in morphology or swimming behavior, while machine learning models process acoustic data to identify whale calls or other vocalizations.

The integration of eDNA and remote sensing data into AI-driven models creates a multidimensional understanding of deep sea biodiversity. These models synthesize genetic, visual, acoustic, and environmental data into cohesive frameworks that capture the complexity of marine ecosystems. For example, a predictive AI model might combine eDNA data on microbial diversity with remote sensing observations of megafauna and environmental parameters such as temperature, salinity, and oxygen levels. This integrated approach enables the identification of critical habitat zones, the mapping of species distributions, and the detection of changes in ecosystem health over time.

One of the most significant applications of these integrated models is in environmental impact assessments (EIAs) for deep sea mining. The extraction of polymetallic nodules, seafloor massive sulfides (SMS), and cobalt-rich crusts has the potential to disturb benthic habitats, alter sediment dynamics, and disrupt food webs. By incorporating biological data into AI-driven EIAs, operators can quantify these impacts with unprecedented precision. For instance, an AI model might predict how sediment plumes generated during mining will affect nearby coral gardens or vent communities, providing a basis for designing mitigation strategies that minimize harm to these ecosystems.

Moreover, these models facilitate the identification of ecologically sensitive areas that should be avoided or protected. By analyzing eDNA and remote sensing data, AI systems can flag regions with high biodiversity, unique species, or critical ecosystem services, guiding the placement of mining operations away from these zones. This capability aligns with international commitments to marine conservation, such as those outlined in the United Nations Sustainable Development Goal 14, which emphasizes the protection and sustainable use of ocean resources.

AI-driven models also support the monitoring and management of cumulative impacts in the deep ocean. Mining activities do not occur in isolation; they interact with other stressors, such as climate change, overfishing, and pollution. By integrating biological data with broader environmental datasets, AI systems can model the combined effects of these pressures on marine ecosystems, enabling more holistic and adaptive management approaches. For example, a model might predict how warming temperatures will exacerbate the impacts of mining on a deep sea community already stressed by oxygen depletion or acidification, informing strategies to mitigate these compounded risks.

The development and deployment of these AI-driven models present both opportunities and challenges. On the one hand, the integration of eDNA and remote sensing data into AI frameworks represents a significant advance in our ability to study and protect deep sea ecosystems. These models offer a level of precision, scalability, and adaptability that is unattainable through traditional methods, enabling operators and regulators to make informed decisions that balance resource extraction with conservation.

On the other hand, the complexity and variability of biological data pose challenges for model accuracy and reliability. eDNA, for example, is highly sensitive to environmental factors such as water currents, temperature, and degradation rates, which can influence the detectability and distribution of genetic material. Remote sensing data, while rich in detail, is subject to limitations in resolution, coverage, and signal noise. Addressing these challenges requires robust data preprocessing, rigorous validation, and the development of algorithms that can account for uncertainty and variability.

Another challenge lies in the ethical and regulatory dimensions of integrating biological data into deep sea mining. While these models provide valuable insights for minimizing environmental impact, they also raise questions about the trade-offs between resource extraction and ecosystem preservation. The use of AI in this context must be guided by clear ethical principles and governance frameworks that prioritize transparency, accountability, and the equitable sharing of benefits. For instance, stakeholders must consider how the data generated by these models will be used to inform decisions, who will have access to it, and how it will contribute to the broader goals of marine conservation and sustainability.

The future of integrating biological data into AI-driven models for deep sea mining holds great promise. Advances in machine learning, data collection technologies, and computational power will continue to enhance the accuracy, scalability, and usability of these models. Emerging techniques, such as federated learning, will enable the sharing and integration of data across multiple stakeholders while preserving privacy and data ownership. Innovations in sensor technology, including miniaturized eDNA samplers and next-generation remote sensing platforms, will expand our ability to capture biological data in real-time and at unprecedented scales.

Ultimately, the integration of eDNA, remote sensing, and AI represents a powerful paradigm shift in our approach to deep sea mining. By embedding a deep understanding of biodiversity and ecosystem services into the operational framework of resource extraction, these models enable a more sustainable and responsible interaction with the ocean's resources. They provide a roadmap for reconciling the competing demands of economic development and environmental stewardship, ensuring that the benefits of deep sea mining are realized without compromising the health and resilience of the marine ecosystems that underpin life on Earth. Through this synthesis of cutting-edge technology and ecological insight, humanity can navigate the complexities of the deep ocean with care, precision, and respect for its extraordinary biodiversity.

Predictive modeling of ecosystem impacts is a cornerstone of sustainable deep sea mining, allowing operators and regulators to anticipate and mitigate the effects of activities such as turbidity plume generation and seafloor disturbance. These impacts, while often localized, can have cascading consequences for marine ecosystems, altering habitats, disrupting food webs, and affecting biodiversity at multiple levels. Artificial intelligence (AI) has emerged as a transformative tool in this domain, enabling the development of sophisticated models that simulate and predict ecosystem responses to mining activities. By combining machine learning algorithms with vast datasets on oceanographic conditions, biological communities, and geological structures, AI-driven predictive models offer unprecedented insights into the interplay between human activities and the fragile environments of the deep ocean.

Turbidity plumes, one of the most significant environmental impacts of deep sea mining, occur when sediment is resuspended into the water column during activities such as nodule harvesting or drilling for seafloor massive sulfides (SMS). These plumes can spread over large areas, depending on ocean currents, sediment characteristics, and mining techniques, and may persist for extended periods. Their ecological consequences include reduced light penetration, which affects photosynthetic organisms in shallower waters, and the smothering of benthic communities when sediments eventually settle. Predictive modeling is essential for understanding these dynamics and designing strategies to minimize their impacts.

AI-driven models of turbidity plume behavior integrate data from multiple sources, including sediment composition, water column properties, and current patterns. These models use machine learning algorithms to identify relationships between these variables, enabling the prediction of plume spread, intensity, and duration under different scenarios. For example, a model might analyze data from prior mining operations to estimate the extent of sediment dispersion for a new project, accounting for site-specific factors such as current velocity and sediment grain size. By simulating the evolution of turbidity plumes over time, these models provide valuable insights into the spatial and temporal scales of their impacts.

One of the key strengths of AI in this context is its ability to handle high-dimensional and dynamic data. Oceanographic conditions, such as temperature, salinity, and turbulence, interact in complex ways to influence plume behavior. Traditional modeling approaches, which rely on simplified equations and assumptions, often struggle to capture this complexity. In contrast, AI-driven models can process vast amounts of data from sensors, satellite observations, and historical records, uncovering patterns and interactions that inform more accurate predictions. For instance, deep learning algorithms, such as convolutional neural networks (CNNs) and recurrent neural networks (RNNs), are used to analyze time-series data on current flows, predicting how plumes will evolve in response to changing conditions.

In addition to predicting plume behavior, AI-driven models assess their ecological impacts by integrating biological data. Sediment plumes can have varying effects on different species and communities, depending on factors such as their tolerance to suspended particles, their reliance on visual cues for feeding or navigation, and their ability to recover from sedimentation. By combining plume models with data on species distributions, population dynamics, and ecological interactions, AI systems can identify areas and organisms most at risk. For example, a model might predict that a plume generated near a hydrothermal vent field will disproportionately affect filter-feeding species adapted to clear water conditions, guiding the design of mitigation measures to protect these organisms.

Seafloor disturbance, another major impact of deep sea mining, involves the physical alteration of benthic habitats during resource extraction. Activities such as nodule harvesting, drilling, and cutting disrupt the structure of the seafloor, potentially destroying habitats, altering sediment properties, and displacing or killing resident organisms. Predictive modeling of seafloor disturbance is essential for understanding these impacts and identifying strategies to minimize habitat loss and promote recovery.

AI-driven models of seafloor disturbance incorporate data on geological structures, sediment characteristics, and benthic community composition. These models use machine learning techniques to simulate the effects of different mining techniques on seafloor habitats, providing insights into the spatial extent and severity of disturbance. For instance, a model might predict how a crawler vehicle used for nodule collection will compact sediments, affecting the burrowing behavior of infaunal species. By comparing different extraction methods, these models help operators select techniques that minimize ecological damage.

One of the most significant challenges in modeling seafloor disturbance is accounting for the variability and heterogeneity of benthic ecosystems. The deep ocean is not a uniform environment; it encompasses diverse habitats ranging from abyssal plains to hydrothermal vent fields, each with its own unique biological and physical characteristics. AI-driven models address this variability by using clustering algorithms and geospatial analysis to classify seafloor habitats based on their vulnerability to disturbance. For example, a clustering model might group areas of the seafloor into high-, medium-, and low-risk zones based on factors such as sediment stability, biodiversity, and proximity to sensitive habitats.

AI also plays a critical role in monitoring and validating predictions of seafloor disturbance. Autonomous underwater vehicles (AUVs) and remotely operated vehicles (ROVs) equipped with cameras, sensors, and sampling tools collect data on the immediate and long-term effects of mining activities. This data is fed back into predictive models, allowing them to refine their simulations and improve their accuracy over time. For instance, if post-mining surveys reveal that sediment compaction in a specific area was greater than predicted, the model can adjust its parameters to better account for similar conditions in future scenarios.

The integration of turbidity plume and seafloor disturbance models into a unified framework represents a significant advance in our ability to manage the environmental impacts of deep sea mining. These integrated models provide a holistic view of how mining activities interact with marine ecosystems, enabling more comprehensive impact assessments and mitigation strategies. For example, a unified model might simulate how sediment plumes generated by a nodule collector interact with disturbed seafloor habitats, predicting cumulative impacts on biodiversity and ecosystem services. By incorporating these insights into planning and decision-making, operators can design mining strategies that balance resource extraction with ecological protection.

Predictive modeling also supports the development of adaptive management strategies, which are essential for responding to the uncertainties and dynamic conditions of the deep ocean. AI-driven models enable real-time monitoring and adjustment of mining operations based on observed impacts, ensuring that mitigation measures remain effective under changing conditions. For example, if a turbidity plume spreads further than expected due to unexpected current patterns, operators can use model predictions to adjust extraction techniques or relocate activities to less sensitive areas.

Despite their transformative potential, AI-driven models for predicting ecosystem impacts face several challenges. The quality and availability of input data are critical determinants of model accuracy, yet deep sea environments are often underexplored and poorly understood. Addressing this challenge requires investment in data collection technologies, such as eDNA samplers, high-resolution sonar systems, and long-term monitoring platforms. Collaboration between industry, academia, and regulatory bodies is also essential for generating and sharing high-quality datasets that support model development and validation.

Another challenge lies in the interpretability and transparency of AI-driven models. While these models excel at processing complex data and making predictions, their inner workings can be opaque, leading to concerns about accountability and trust. Developing explainable AI techniques that make model outputs and decision-making processes more transparent is a critical priority for ensuring stakeholder confidence and regulatory compliance.

Looking to the future, the integration of predictive modeling with emerging technologies promises to further enhance our ability to manage ecosystem impacts. For instance, the use of digital twins, virtual replicas of mining operations and their surrounding environments, could allow operators to simulate and optimize impact mitigation strategies before deployment. Advances in edge computing and real-time data analytics will enable models to adapt dynamically to new information, ensuring that predictions remain accurate and actionable under rapidly changing conditions.

Predictive modeling of ecosystem impacts, such as turbidity plumes and seafloor disturbance, is a vital tool for achieving sustainable deep sea mining. By leveraging the power of AI, these models provide detailed and accurate simulations of how mining activities interact with marine ecosystems, informing strategies to minimize harm and promote recovery. As technology and data collection capabilities continue to evolve, these models will play an increasingly central role in guiding responsible resource extraction, ensuring that the benefits of deep sea mining are realized without compromising the health and resilience of the ocean's ecosystems. Through a commitment to innovation, collaboration, and stewardship, predictive modeling represents a pathway to balancing human ambition with environmental responsibility in one of the planet's most extraordinary frontiers.

5.2. Long-Term Monitoring and Mitigation

The sustainability of deep sea mining depends on a rigorous commitment to long-term environmental monitoring and the development of effective mitigation strategies. Among the most critical components of this framework is the ability to detect anomalies in environmental signals, deviations from baseline conditions that may indicate disturbances caused by mining activities. These anomalies could range from increased sediment plumes and changes in water chemistry to disruptions in the behavior of marine species. The integration of artificial intelligence (AI) into monitoring systems offers transformative capabilities in anomaly detection, enabling real-time identification of environmental changes and supporting rapid, data-driven responses to minimize ecological harm.

Environmental signals in the deep ocean are inherently complex and variable, reflecting the interplay of physical, chemical, and biological processes. The challenge lies in distinguishing natural fluctuations from mining-induced changes, especially when operating in ecosystems that are poorly understood and sparsely monitored. Traditional approaches to anomaly detection often rely on manual interpretation of data, which is not only time-intensive but also prone to bias and limited by the scale of human analysis. AI-based anomaly detection systems address these limitations by automating the identification of deviations, processing vast datasets with speed and precision, and uncovering patterns that may elude conventional methods.

At the core of AI-based anomaly detection is the use of machine learning algorithms trained to recognize baseline environmental conditions and identify departures from these norms. Baseline conditions are established through long-term data collection prior to the commencement of mining activities, capturing the natural variability of parameters such as turbidity, oxygen levels, temperature, and biodiversity. Once these baselines are defined, AI systems continuously analyze incoming data from sensors, comparing it to the reference state and flagging any significant deviations.

One of the most widely used techniques in anomaly detection is unsupervised learning, which identifies anomalies without requiring pre-labeled examples of normal or abnormal states. Algorithms such as clustering and density-based models group data points with similar characteristics, isolating outliers that deviate from these clusters. For instance, an unsupervised model might detect an unexpected increase in turbidity levels at a mining site, signaling the presence of a sediment plume that exceeds predicted parameters. By focusing on these outliers, AI systems can pinpoint potential impacts and guide targeted investigations.

Supervised learning, while requiring labeled training data, offers additional precision in detecting specific types of anomalies. For example, a supervised model trained on historical data of mining-induced sediment plumes might predict the likelihood and extent of plume formation under varying conditions. This approach is particularly valuable in cases where operators have access to detailed datasets from similar sites or past operations, allowing the model to learn from known patterns of disturbance.

A major advantage of AI-based systems is their ability to process multi-dimensional and heterogeneous data streams simultaneously. Environmental monitoring relies on a diverse array of sensors, including acoustic monitors, optical cameras, chemical analyzers, and biological samplers, each providing unique insights into the state of the ecosystem. AI algorithms integrate these data streams into a unified framework, enabling a holistic assessment of environmental health. For example, an integrated system might analyze acoustic data to detect changes in marine species' vocalizations, correlate this with water chemistry data showing a decline in dissolved oxygen levels, and identify the underlying cause as a mining-induced plume.

Time-series analysis is another critical component of AI-based anomaly detection, allowing systems to track changes over time and identify trends or sudden shifts. Recurrent neural networks (RNNs), a type of deep learning model, are particularly adept at analyzing temporal data, capturing dependencies between past and current states. This capability enables predictive insights, such as forecasting the spread of a sediment plume based on its current trajectory or anticipating the recovery time of a disturbed habitat. By predicting anomalies before they escalate, AI systems support proactive mitigation measures, reducing the likelihood of long-term ecological damage.

AI-based anomaly detection also extends to the biological impacts of mining activities, including changes in species behavior, population dynamics, and ecosystem interactions. Machine learning models analyze data from visual and acoustic surveys, identifying shifts that may indicate stress or displacement. For instance, an AI system might detect a reduction in the frequency of whale vocalizations near a mining site, suggesting avoidance behavior due to increased noise pollution. Similarly, computer vision algorithms applied to video footage can track the movements of benthic organisms, revealing disruptions in their foraging or reproductive patterns.

The integration of environmental DNA (eDNA) data further enhances the sensitivity and scope of anomaly detection. eDNA provides a genetic snapshot of the species present in a given area, capturing even those that are difficult to observe directly. AI algorithms process this data to detect changes in species composition or the appearance of invasive organisms, offering early warnings of ecological shifts. For example, if an eDNA analysis detects a sudden decline in key indicator species near a mining site, the AI system can flag this as a potential impact requiring immediate investigation.

One of the most significant benefits of AI-based anomaly detection is its capacity for real-time monitoring and response. Traditional monitoring methods often involve delays between data collection, analysis, and decision-making, limiting the ability to address impacts as they occur. AI systems, by contrast, process data in real-time, enabling near-instantaneous identification of anomalies and triggering automated alerts to operators. This capability is particularly valuable in dynamic environments where conditions can change rapidly, such as during nodule harvesting or drilling for seafloor massive sulfides. For instance, if a sediment plume begins to spread beyond the predicted zone, the AI system can alert operators to adjust mining techniques or temporarily halt operations to mitigate the impact.

The scalability and adaptability of AI-based systems are equally important in the context of long-term monitoring. Deep sea mining operations often span large geographic areas, requiring extensive coverage to capture the full scope of potential impacts. AI systems can analyze data from networks of sensors distributed across these areas, maintaining consistent performance regardless of scale. Moreover, as new data becomes available, these systems continuously refine their models, improving accuracy and adapting to the unique characteristics of each mining site.

Despite their transformative potential, the implementation of AI-based anomaly detection systems faces several challenges. The quality and availability of baseline data are critical determinants of model performance, yet baseline conditions in deep sea environments are often poorly characterized due to the logistical and financial constraints of pre-mining surveys. Addressing this requires sustained investment in data collection and the establishment of standardized protocols for defining baselines.

Another challenge lies in the interpretability of AI models, particularly those based on deep learning, which often operate as "black boxes." While these models excel at identifying anomalies, understanding the underlying reasons for their predictions can be difficult, complicating efforts to design appropriate mitigation measures. Developing explainable AI techniques is essential for bridging this gap, providing transparency and fostering trust among stakeholders.

The future of AI-based anomaly detection in environmental monitoring holds great promise, driven by advances in machine learning, sensor technologies, and computational power. Emerging techniques, such as transfer learning, will enable models trained on one mining site to be applied to others with minimal retraining, expanding their applicability and reducing development costs. Innovations in edge computing will further enhance real-time capabilities, allowing data processing to occur directly on autonomous underwater vehicles (AUVs) or remotely operated vehicles (ROVs), minimizing latency and enabling immediate responses to detected anomalies.

Effective long-term monitoring and mitigation of the environmental impacts of deep sea mining depend on the deployment of continuous, autonomous sensor networks capable of tracking critical variables such as water quality, sedimentation, and biological activity. These networks, distributed across mining sites and surrounding ecosystems, provide real-time data on changes induced by extraction activities, enabling operators and regulators to assess impacts and implement mitigation strategies promptly. By leveraging advanced technologies, including artificial intelligence (AI), robotics, and IoT (Internet of Things), continuous sensor networks transform environmental monitoring into a dynamic, adaptive process that aligns with the complexities and demands of the deep ocean.

The establishment of continuous sensor networks begins with the strategic deployment of autonomous, multifunctional sensors capable of withstanding the extreme conditions of the deep ocean. These sensors are designed to measure a wide range of parameters critical for understanding the ecological health of mining sites. Water quality sensors monitor variables such as dissolved oxygen levels, pH, salinity, and concentrations of heavy metals or other pollutants that may be released during mining activities. Sedimentation sensors track the resuspension, dispersion, and settling of particles generated by excavation, providing insights into the extent and intensity of sediment plumes. Biota monitoring devices, including cameras, hydrophones, and environmental DNA (eDNA) samplers, capture data on the presence, behavior, and diversity of marine organisms, highlighting changes in community structure and species interactions.

The integration of these sensors into a cohesive network is facilitated by advances in autonomous robotics and IoT. Autonomous underwater vehicles (AUVs) and stationary platforms serve as mobile and fixed nodes within the network, collecting and transmitting data across large geographic areas. These nodes are equipped with high-capacity batteries, low-power processors, and robust communication systems, enabling them to operate independently for extended periods. By maintaining continuous operation, the network provides a comprehensive picture of environmental conditions over time, capturing both short-term disturbances and long-term trends.

AI plays a critical role in processing and interpreting the vast amounts of data generated by sensor networks. Machine learning algorithms analyze real-time data streams, identifying patterns, correlations, and anomalies that signal ecological impacts. For example, an AI system might detect a gradual decline in dissolved oxygen levels in the vicinity of a mining site, indicating potential hypoxic conditions caused by sediment resuspension or organic matter release. Similarly, AI-driven models can track the spread of sediment plumes, predicting their trajectory and identifying areas at risk of sediment deposition. By automating these analyses, AI reduces the reliance on manual interpretation, ensuring that critical insights are delivered promptly and accurately.

The adaptability of continuous sensor networks is particularly valuable in the dynamic and heterogeneous environments of the deep ocean. Mining activities often occur in regions characterized by complex topography, variable currents, and diverse habitats, necessitating flexible monitoring strategies. Sensor nodes equipped with adaptive algorithms can adjust their measurement frequencies, spatial coverage, or operational parameters in response to changing conditions. For instance, if a sensor detects an increase in turbidity indicative of a sediment plume, it can prioritize measurements in the affected area, coordinating with nearby nodes to enhance data resolution. This dynamic allocation of resources ensures that the network remains focused on the most ecologically significant changes, optimizing the use of limited energy and computational resources.

One of the most significant advantages of autonomous sensor networks is their ability to integrate biological monitoring into environmental assessments. Traditional approaches to biota monitoring often rely on periodic surveys conducted by remotely operated vehicles (ROVs) or human divers, providing snapshots of ecosystem health at discrete points in time. In contrast, continuous sensor networks enable the collection of high-frequency data on marine organisms, capturing temporal variations and uncovering subtle, cumulative impacts. For example, hydrophones deployed as part of the network can record the vocalizations of marine mammals, tracking changes in their behavior or distribution in response to increased noise levels from mining operations. Cameras mounted on stationary platforms or AUVs provide visual data on benthic communities, revealing shifts in species composition, abundance, or activity patterns.

The use of eDNA further enhances the capabilities of biota monitoring within sensor networks. eDNA analysis involves the extraction and sequencing of genetic material shed by organisms into the environment, offering a non-invasive method for assessing biodiversity. Autonomous samplers integrated into sensor nodes periodically collect water or sediment samples, which are analyzed using onboard or remote AI systems. These systems identify species present in the area, detect invasive or rare organisms, and monitor population dynamics. By combining eDNA data with other environmental parameters, such as temperature or salinity, AI models can uncover complex relationships between mining activities and ecosystem responses, providing a deeper understanding of ecological impacts.

The scalability and interconnectivity of continuous sensor networks are critical for monitoring the cumulative effects of mining across large areas and multiple sites. The impacts of deep sea mining are not confined to individual locations; sediment plumes, pollutant dispersion, and species displacement can extend beyond the immediate vicinity of extraction activities, affecting regional ecosystems. Sensor networks address this challenge by creating a spatially distributed monitoring framework that captures interactions and feedbacks across different areas. Data from individual nodes is aggregated and analyzed within a centralized platform, allowing operators to assess the broader implications of their activities and identify opportunities for coordinated mitigation measures.

Despite their transformative potential, the deployment and operation of continuous sensor networks face several challenges. The harsh conditions of the deep ocean, including high pressures, corrosive seawater, and low temperatures, impose stringent requirements on sensor design and durability. Ensuring the reliability and longevity of sensor nodes requires the use of advanced materials, protective coatings, and energy-efficient technologies. Communication is another critical consideration, as the underwater environment limits the effectiveness of traditional wireless systems. Acoustic communication, the most commonly used method, offers reliable data transmission over long distances but is constrained by low bandwidth and susceptibility to interference. Addressing these challenges necessitates the development of hybrid communication systems that combine acoustic, optical, and satellite technologies to enhance data transfer efficiency.

The management and interpretation of data generated by sensor networks also present significant technical and logistical demands. The continuous operation of these networks produces massive datasets, requiring robust storage, processing, and visualization capabilities. Cloud computing platforms and distributed data architectures are essential for managing this influx of information, enabling real-time access and analysis. Visualization tools, such as interactive dashboards and three-dimensional models, facilitate the interpretation of complex datasets, providing stakeholders with intuitive insights into environmental trends and anomalies.

The integration of continuous sensor networks into regulatory frameworks and decision-making processes is a critical priority for ensuring their effectiveness. Data collected by these networks must be standardized, validated, and transparently shared among stakeholders, including mining operators, regulators, and conservation organizations. Collaborative platforms that promote data sharing and joint analysis can enhance the credibility and utility of monitoring efforts, supporting evidence-based policies and adaptive management strategies.

Looking to the future, the capabilities of continuous sensor networks will be further enhanced by emerging technologies and interdisciplinary innovation. The miniaturization of sensors and the development of low-power electronics will enable the deployment of denser and more energy-efficient networks, increasing spatial and temporal resolution. Advances in edge computing will allow sensor nodes to process data locally, reducing the need for high-bandwidth communication and enabling faster responses to detected changes. The integration of swarm robotics, where fleets of AUVs operate collaboratively as mobile sensor platforms, will expand the coverage and adaptability of monitoring systems, providing comprehensive assessments of dynamic and large-scale impacts.

Continuous, autonomous sensor networks represent a transformative approach to monitoring the environmental impacts of deep sea mining. By combining advanced sensing technologies, AI-driven analytics, and adaptive design, these networks provide real-time, high-resolution data on water quality, sedimentation, and biological activity. Their ability to operate continuously and independently in the challenging conditions of the deep ocean ensures that critical changes are detected and addressed promptly, supporting sustainable and responsible resource extraction. As these networks evolve, they will play an increasingly central role in safeguarding the health and resilience of marine ecosystems, ensuring that the benefits of deep sea mining are realized without compromising the integrity of the planet's last great frontier. Through innovation, collaboration, and a commitment to ecological stewardship, continuous sensor networks offer a pathway to a balanced and sustainable relationship with the deep ocean.

Adaptive management frameworks represent the pinnacle of responsible environmental governance, especially in complex and uncertain domains like deep sea mining. These frameworks are designed to integrate real-time data into decision-making processes, enabling dynamic adjustments to operations in response to observed impacts. By coupling advanced technologies, such as autonomous sensor networks, artificial intelligence (AI), and data analytics, adaptive management frameworks offer a proactive, iterative approach to mitigating the environmental consequences of mining activities while optimizing operational efficiency. This section explores the principles, technologies, and practical applications of adaptive management frameworks in the context of long-term monitoring and mitigation, highlighting their transformative potential for sustainable resource extraction in the deep ocean.

At its core, an adaptive management framework is built around a cycle of planning, monitoring, evaluating, and adjusting. Unlike static management strategies, which rely on predefined rules and assumptions, adaptive frameworks embrace uncertainty and variability, recognizing that deep ocean ecosystems are dynamic and that mining impacts can manifest in unpredictable ways. This flexibility is particularly critical in the context of deep sea mining, where scientific knowledge of ecosystems and their responses to disturbance is still evolving.

The foundation of an adaptive management framework is real-time environmental data, collected through an extensive network of sensors and monitoring platforms. These systems provide continuous streams of information on key parameters, such as water quality, sedimentation, and biological activity, enabling operators and regulators to maintain an up-to-date understanding of ecosystem conditions. By integrating this data into predictive models and decision-support systems, adaptive frameworks allow for the identification of emerging trends, anomalies, and risks, guiding timely interventions to mitigate impacts.

One of the most significant advantages of adaptive management frameworks is their ability to respond dynamically to unexpected events. In traditional approaches, decisions are often based on historical data and assumptions, which may not accurately reflect current conditions. For example, a mining operation might be planned based on an environmental impact assessment conducted years earlier, with limited ability to adjust to real-time changes. In contrast, an adaptive framework enables continuous evaluation of operations, ensuring that decisions are informed by the latest data and grounded in current realities. If a sensor network detects a sudden increase in turbidity or a decline in dissolved oxygen levels near a mining site, the framework can prompt immediate actions, such as modifying extraction techniques or suspending operations temporarily, to minimize harm.

AI plays a central role in enabling the real-time capabilities of adaptive management frameworks. Machine learning algorithms analyze incoming data streams to identify patterns, correlations, and anomalies that signal ecological impacts. For instance, an AI system might detect a correlation between sediment plume dispersion and changes in benthic community composition, providing early warnings of habitat disruption. These insights are then used to update predictive models, which simulate the potential consequences of different management actions and recommend optimal responses. By automating these analyses, AI reduces the reliance on manual interpretation, ensuring that decisions are data-driven, timely, and precise.

Predictive modeling within adaptive frameworks extends beyond immediate impacts, incorporating long-term and cumulative effects. Deep sea mining activities often interact with other stressors, such as climate change, overfishing, and pollution, creating complex feedback loops that amplify ecological risks. Adaptive frameworks address these challenges by integrating multi-dimensional data into comprehensive models that capture the interplay of multiple factors. For example, a model might simulate how warming ocean temperatures exacerbate the impacts of mining-induced sedimentation on coral communities, informing strategies to mitigate combined pressures.

The implementation of adaptive management frameworks also relies on advanced visualization and decision-support tools that facilitate collaboration and transparency among stakeholders. Interactive dashboards and geospatial platforms provide intuitive representations of environmental data, enabling operators, regulators, and conservation organizations to explore trends, assess risks, and evaluate potential responses. For example, a dashboard might display real-time maps of sediment plume trajectories, overlaid with data on sensitive habitats and species distributions, helping stakeholders identify areas of concern and prioritize mitigation efforts.

The iterative nature of adaptive management is particularly valuable for addressing the inherent uncertainties of deep ocean ecosystems. Despite advances in monitoring and modeling, gaps in scientific knowledge remain, particularly regarding the resilience and recovery of benthic communities. Adaptive frameworks accommodate this uncertainty by treating management decisions as experiments, with each action generating new data that informs subsequent decisions. This approach ensures continuous learning and improvement, enabling frameworks to evolve alongside our understanding of deep sea ecosystems.

Practical applications of adaptive management frameworks in deep sea mining span a wide range of activities, from resource extraction and waste management to habitat restoration and biodiversity conservation. For example, in polymetallic nodule collection, an adaptive framework might monitor sediment plumes in real time, adjusting suction pressures or collection speeds to minimize sediment resuspension and its impacts on nearby ecosystems. In seafloor massive sulfide (SMS) mining, the framework could track changes in hydrothermal vent communities, guiding the placement of exclusion zones to protect sensitive habitats.

Adaptive frameworks also play a critical role in mitigating long-term and cumulative impacts, which often extend beyond the immediate vicinity of mining activities. For instance, sediment plumes generated by nodule harvesting can travel significant distances, affecting ecosystems far from the extraction site. By modeling these dynamics and incorporating data from regional monitoring networks, adaptive frameworks provide a holistic view of impacts, supporting coordinated management strategies that address interconnected challenges. This regional perspective is essential for balancing the needs of resource extraction with the broader goals of marine conservation and sustainability.

The scalability of adaptive management frameworks is another key strength, enabling their application across diverse geographic and operational contexts. Whether managing a single mining site or coordinating activities across multiple locations, adaptive frameworks provide a consistent and flexible structure for integrating data, evaluating impacts, and implementing responses. Advances in cloud computing and IoT technologies further enhance scalability, allowing data from distributed sensor networks to be aggregated and analyzed in real time, regardless of the size or complexity of the monitoring system.

Despite their transformative potential, the implementation of adaptive management frameworks faces several challenges. The success of these frameworks depends on the availability and quality of real-time data, which requires sustained investment in monitoring technologies and infrastructure. Ensuring the reliability and durability of sensors, communication systems, and data processing platforms is critical, particularly in the harsh and remote conditions of the deep ocean. Data security and interoperability are additional considerations, as frameworks must balance the need for information sharing with the protection of sensitive data.

Stakeholder engagement and governance are equally important for the success of adaptive management frameworks. The integration of real-time data into decision-making processes requires collaboration among operators, regulators, scientists, and local communities, each of whom brings unique perspectives and priorities. Transparent and inclusive governance structures are essential for building trust, resolving conflicts, and aligning management actions with shared goals. For example, stakeholder consultations might be used to establish thresholds for ecological impacts, define mitigation priorities, or evaluate the trade-offs between resource extraction and conservation.

The future of adaptive management frameworks in deep sea mining will be shaped by ongoing technological and methodological advancements. Emerging innovations, such as digital twins and edge computing, promise to enhance the precision and responsiveness of these frameworks. Digital twins—virtual replicas of mining operations and their surrounding environments—enable operators to simulate and optimize management actions before implementation, reducing risks and improving outcomes. Edge computing, which processes data locally at the source, minimizes latency and ensures that real-time insights are available even in bandwidth-limited environments.

As adaptive management frameworks evolve, they will play an increasingly central role in ensuring the sustainability of deep sea mining. By integrating real-time data, predictive models, and collaborative decision-making, these frameworks provide a pathway for balancing resource extraction with the protection of marine ecosystems. They represent a dynamic, evidence-based approach to environmental stewardship, enabling humanity to navigate the complexities of the deep ocean with care, precision, and responsibility. Through innovation, collaboration, and a commitment to sustainability, adaptive management frameworks offer a powerful tool for harmonizing economic development with ecological resilience in one of the planet's most extraordinary and enigmatic frontiers.

5.3. Comparative Case Studies

The evolution of terrestrial mining and offshore oil and gas operations offers a wealth of lessons for the nascent deep sea mining industry. These established industries have navigated decades of technical, economic, and environmental challenges, making them invaluable reference points for understanding the opportunities and risks associated with resource extraction in extreme environments. While the deep ocean presents unique conditions, such as immense pressure, perpetual darkness, and poorly understood ecosystems, many of the principles and practices honed in terrestrial and offshore contexts are directly applicable. By examining the successes and failures of these industries, deep sea mining can leverage proven strategies, avoid costly mistakes, and chart a path toward sustainable and responsible operations.

One of the most significant lessons from terrestrial mining is the importance of environmental impact assessments (EIAs) as a cornerstone of responsible resource extraction. Terrestrial mining operations, particularly in the 20th century, often prioritized economic gains over ecological concerns, leading to widespread deforestation, habitat destruction, and pollution. These failures spurred the development of comprehensive EIAs, which assess the potential environmental and social impacts of mining activities before operations begin. The deep sea mining industry can draw on this legacy by implementing rigorous EIAs tailored to the unique challenges of marine environments. These assessments must account for the slow recovery rates of deep sea ecosystems, the dispersal of sediment plumes, and the cumulative impacts of multiple mining operations within a region. By adopting a precautionary approach and incorporating adaptive management frameworks, as seen in progressive terrestrial mining projects, the industry can balance economic development with ecological stewardship.

The role of technology in minimizing environmental impacts is another critical lesson from terrestrial mining. Over the past several decades, advancements in extraction and processing technologies have significantly reduced the footprint of mining activities on land. For example, the adoption of in situ leaching for certain types of mineral deposits has eliminated the need for large-scale open-pit mining, preserving surface ecosystems. Similarly, precision drilling and automated machinery have reduced waste and energy consumption. These innovations underscore the importance of investing in technology to enhance efficiency and sustainability. In the context of deep sea mining, this means developing low-impact collection systems, such as suction-based nodule harvesters and AI-driven sorting mechanisms, that minimize sediment disturbance and ecological disruption.

The offshore oil and gas industry provides equally valuable insights, particularly in the areas of operational safety and risk management. Extracting hydrocarbons from beneath the seafloor involves significant technical challenges, including high pressures, volatile materials, and extreme weather conditions. Over decades of operations, the industry has developed robust safety protocols, engineering standards, and contingency plans to address these risks. The Deepwater Horizon disaster of 2010, however, remains a stark reminder of the catastrophic consequences of neglecting safety and environmental safeguards. The spill, which released millions of barrels of oil into the Gulf of Mexico, caused extensive ecological damage and underscored the need for stringent oversight and accountability.

Deep sea mining can learn from both the successes and failures of the offshore oil and gas sector. Robust risk assessment and management systems are essential to prevent accidents and mitigate their impacts. For instance, the development of pressure-resistant equipment, redundancy in critical systems, and real-time monitoring of operations can enhance safety and reliability. Furthermore, the adoption of spill-response strategies from the oil and gas industry, such as containment booms and dispersants, could inform contingency plans for sediment plumes or chemical leaks in mining operations.

One of the most notable achievements of the offshore oil and gas industry is its ability to operate in remote and extreme environments. Technological innovations, such as dynamic positioning systems, remotely operated vehicles (ROVs), and subsea processing facilities, have enabled the efficient extraction of resources in ultra-deepwater locations. These technologies provide a blueprint for overcoming the logistical challenges of deep sea mining, where autonomous systems and remote monitoring will play a central role. For example,

ROVs equipped with AI and advanced sensors can perform tasks such as sediment sampling, habitat mapping, and equipment maintenance without direct human intervention, reducing costs and enhancing safety.

The importance of stakeholder engagement is another critical lesson from both terrestrial mining and offshore oil and gas. Historically, a lack of consultation with local communities and environmental organizations has led to conflicts, delays, and reputational damage for resource extraction projects. Modern best practices emphasize the need for transparent communication, collaborative planning, and equitable benefit-sharing. In the deep sea mining context, this translates to engaging with a broad range of stakeholders, including coastal states, international regulators, scientific researchers, and conservation groups. By fostering dialogue and building consensus, the industry can align its activities with societal expectations and global sustainability goals.

A related insight is the value of public-private partnerships in advancing technological and environmental objectives. The offshore oil and gas industry has a long history of collaborating with academic institutions, government agencies, and private companies to develop innovative solutions to shared challenges. For example, partnerships between oil companies and research institutions have led to breakthroughs in seismic imaging, subsea robotics, and environmental monitoring. Deep sea mining can adopt a similar approach, leveraging partnerships to accelerate the development of cutting-edge technologies and expand the scientific knowledge base. Collaborative initiatives, such as the sharing of environmental data and the funding of independent impact assessments, can also enhance transparency and trust.

One of the most significant challenges shared by terrestrial mining, offshore oil and gas, and deep sea mining is the management of cumulative impacts. In terrestrial settings, the expansion of mining activities into previously untouched regions has often led to the fragmentation of habitats, loss of biodiversity, and degradation of ecosystem services. Similarly, offshore oil and gas operations have contributed to localized pollution, habitat alteration, and conflicts with other ocean users, such as fisheries. In the deep sea, the cumulative impacts of multiple mining operations, coupled with other stressors such as climate change, pose a serious threat to marine ecosystems. Addressing these challenges requires a regional perspective, where cumulative impacts are assessed and managed at the scale of entire ecosystems rather than individual sites. This approach aligns with the principles of ecosystem-based management, which seeks to balance resource use with the preservation of ecological integrity.

Another critical lesson is the role of innovation in waste management and reclamation. Terrestrial mining operations have historically struggled with the disposal of tailings and the restoration of disturbed lands. However, advances in waste management techniques, such as dry stacking and the use of tailings as construction materials, have reduced environmental impacts and supported post-mining land use. Similarly, the offshore oil and gas industry has developed techniques for decommissioning platforms and restoring marine habitats, such as converting retired rigs into artificial reefs. These practices highlight the importance of incorporating reclamation and waste management strategies into the design of deep sea mining projects. For example, the careful placement of returned sediment or the creation of artificial substrates could promote the recovery of benthic ecosystems after mining.

The integration of monitoring and adaptive management systems is another key takeaway from both terrestrial and offshore contexts. In terrestrial mining, real-time monitoring of air and water quality has become standard practice, enabling operators to detect and respond to environmental impacts quickly. Offshore oil and gas platforms use advanced monitoring systems to track emissions, leaks, and marine life interactions, ensuring compliance with environmental regulations. For deep sea mining, continuous monitoring networks equipped with AI-driven analytics can provide similar benefits, offering a real-time view of ecological conditions and supporting dynamic adjustments to operations. These systems are essential for building resilience into mining activities, ensuring that they can adapt to new information and unforeseen challenges.

The rapid advancement of technology in deep sea mining brings with it a profound responsibility to ensure that innovation serves not only economic objectives but also ethical and environmental imperatives. As the industry moves closer to commercial operations, it must grapple with fundamental questions about the long-term implications of its activities for marine ecosystems, global equity, and future generations. Ethical frameworks for responsible innovation provide a structured approach to addressing these challenges, offering principles and guidelines that balance the pursuit of resource extraction with the preservation of ecological integrity and social justice.

The development of ethical frameworks for responsible innovation is not a new concept. Industries such as biotechnology, artificial intelligence, and renewable energy have long recognized the need to align technological progress with societal values. These frameworks are particularly relevant to deep sea mining, which operates at the intersection of environmental science, economic development, and global governance. Drawing on comparative case studies from other sectors, we can identify key elements of effective ethical frameworks and explore how they can be adapted to the unique context of deep sea mining.

At the heart of any ethical framework is the principle of *do no harm*, which prioritizes the prevention of adverse impacts on ecosystems and communities. In the context of deep sea mining, this principle underscores the importance of protecting fragile marine environments from irreversible damage. The deep ocean is one of the least understood and most vulnerable ecosystems on Earth, hosting a diversity of life that has evolved over millions of years in extreme and isolated conditions. Mining activities, such as nodule harvesting or seafloor massive sulfide (SMS) extraction, have the potential to disrupt these ecosystems through habitat destruction, sediment plumes, and noise pollution.

To operationalize the principle of *do no harm*, ethical frameworks emphasize the need for precautionary measures and adaptive management. The precautionary approach, widely adopted in environmental law and policy, dictates that the absence of complete scientific certainty should not be used as a justification for proceeding with potentially harmful activities. For deep sea mining, this means conducting rigorous environmental impact assessments (EIAs) and baseline studies before operations begin, ensuring that potential risks are thoroughly understood and mitigated. Adaptive management complements this by providing a mechanism for continuous learning and adjustment, enabling operators to respond dynamically to new information and unexpected impacts.

Another cornerstone of ethical frameworks is the principle of *equity*, which seeks to ensure that the benefits and burdens of innovation are distributed fairly across society. Deep sea mining has the potential to generate significant economic value, providing critical minerals for renewable energy technologies, electric vehicles, and other applications essential to the global transition to a low-carbon economy. However, the distribution of these benefits raises important ethical questions. Who will profit from the exploitation of marine resources? How will revenues be shared between mining companies, coastal states, and the international community? And what measures will be taken to address the needs of communities that may be disproportionately affected by mining activities, such as small island developing states (SIDS) that depend on healthy marine ecosystems for fisheries and tourism?

To address these questions, ethical frameworks advocate for benefit-sharing mechanisms that promote global equity. The United Nations Convention on the Law of the Sea (UNCLOS) and its implementing body, the International Seabed Authority (ISA), provide a foundational framework for ensuring that the resources of the deep seabed are managed as the "common heritage of mankind." This principle mandates that the benefits derived from deep sea mining in international waters should be shared equitably among all nations, particularly developing countries. Operationalizing this principle requires transparent and accountable governance structures, such as royalty systems, environmental funds, and capacity-building initiatives that support sustainable development.

Transparency is a third pillar of ethical frameworks for responsible innovation. Transparent processes foster accountability, build trust, and enable informed decision-making by stakeholders. In deep sea mining, transparency involves the open sharing of scientific data, environmental assessments, and operational plans with regulators, researchers, and the public. This transparency is particularly important for addressing the uncertainties and risks associated with mining activities, as it allows stakeholders to scrutinize claims, evaluate trade-offs, and advocate for necessary safeguards.

Comparative case studies from industries such as biotechnology and AI demonstrate the value of participatory approaches to transparency. For example, public consultations and stakeholder forums have become standard practices in the approval of genetically modified organisms (GMOs) and the deployment of facial recognition technologies. These mechanisms provide a platform for diverse perspectives, enabling communities, civil society organizations, and experts to contribute to decision-making processes. In the context of deep sea mining, participatory approaches can help reconcile competing interests, such as the economic aspirations of mining operators and the conservation priorities of environmental organizations.

A related principle is *accountability*, which ensures that those responsible for mining activities are held to high standards of environmental and social performance. Ethical frameworks advocate for clear, enforceable regulations that define the responsibilities of operators, regulators, and other stakeholders. These regulations must be backed by robust monitoring and enforcement mechanisms, ensuring that violations are identified and addressed promptly. For example, if a mining operation exceeds its permitted levels of sediment discharge or fails to implement agreed-upon mitigation measures, accountability mechanisms should include penalties, remediation requirements, and, if necessary, the suspension of operations.

Accountability is further supported by independent oversight, which provides an objective evaluation of compliance and performance. Comparative case studies highlight the role of third-party audits and certification schemes in promoting responsible practices. In forestry, for instance, certification programs such as the Forest Stewardship Council (FSC) provide assurance that timber is sourced sustainably. Similarly, the offshore oil and gas industry has adopted third-party verification for environmental management systems. For deep sea mining, independent oversight could take the form of environmental certification for operators, periodic audits of monitoring data, and public reporting of compliance metrics.

Ethical frameworks also emphasize the importance of *intergenerational equity*, which recognizes the rights and responsibilities of future generations. This principle is particularly relevant to deep sea mining, as the long-term impacts of resource extraction on marine ecosystems remain uncertain. Protecting the interests of future generations requires a commitment to sustainable resource management, where extraction rates are aligned with ecosystem recovery times, and biodiversity is conserved for the benefit of all. Ethical frameworks advocate for the adoption of science-based targets and thresholds, such as biodiversity offset requirements or no-net-loss policies, to ensure that mining activities do not compromise the ecological integrity of the deep ocean.

Finally, ethical frameworks call for a *holistic perspective* that considers the interconnectedness of environmental, social, and economic factors. Deep sea mining does not occur in isolation; it interacts with other ocean uses, such as shipping, fisheries, and conservation, as well as global challenges such as climate change. Ethical frameworks encourage a systems-based approach to decision-making, where the cumulative and synergistic impacts of activities are assessed and managed. For example, mining-induced sediment plumes may affect fisheries beyond the immediate vicinity of the extraction site, requiring coordination between the mining and fisheries sectors to mitigate conflicts and ensure sustainable outcomes.

The implementation of ethical frameworks for responsible innovation in deep sea mining requires collaboration and commitment from all stakeholders, including industry leaders, governments, regulators, scientists, and civil society organizations. It also demands a willingness to innovate, not only in technology but also in governance, policy, and public engagement. By integrating ethical considerations into every stage of mining operations, from exploration and feasibility studies to extraction and decommissioning, the industry can demonstrate its commitment to sustainability and social responsibility.

Ethical frameworks for responsible innovation provide a vital foundation for balancing the economic potential of deep sea mining with the need to protect marine ecosystems and promote global equity. By adhering to principles such as *do no harm*, *equity*, *transparency*, *accountability*, and *intergenerational equity*, the industry can navigate the complex challenges of resource extraction in one of the planet's most vulnerable and enigmatic environments. Through collaboration, innovation, and a commitment to ethical stewardship, deep sea mining has the opportunity to become a model for sustainable and responsible practices, ensuring that the treasures of the deep ocean are harnessed for the benefit of all humanity, now and in the future.

6. Economic Feasibility and Value Chain Optimization

6.1. Economic Modeling and AI

The integration of predictive analytics into economic modeling has transformed the way industries forecast prices and anticipate demand-supply trends, offering a level of precision and adaptability unattainable through traditional methods. In the context of deep sea mining, predictive analytics serves as a critical tool for navigating the complexities of global commodity markets. The minerals targeted by this nascent industry, cobalt, nickel, manganese, and rare earth elements, are foundational to emerging technologies such as electric vehicles, renewable energy systems, and advanced electronics. These markets are characterized by high volatility, geopolitical influences, and rapidly evolving demand patterns, necessitating sophisticated tools to guide investment decisions, operational strategies, and long-term planning.

At the core of predictive analytics lies the ability to process vast amounts of structured and unstructured data, extracting patterns and insights that inform future trends. For deep sea mining, this data encompasses a wide array of variables, including historical price movements, production volumes, technological advancements, and policy changes. Advanced machine learning algorithms analyze these inputs, identifying correlations and drivers that influence market behavior. Unlike traditional econometric models, which often rely on linear assumptions and limited datasets, machine learning approaches excel at capturing the non-linear and dynamic relationships that define commodity markets.

One of the most critical applications of predictive analytics in deep sea mining is price forecasting. Commodity prices are influenced by a complex interplay of factors, including macroeconomic conditions, supply chain dynamics, and shifts in consumer behavior. For example, the price of cobalt, a key component in lithium-ion batteries, is shaped not only by mining output but also by the adoption rates of electric vehicles and advancements in battery technology. Predictive analytics enables mining companies and investors to anticipate price fluctuations with greater accuracy, providing a competitive edge in a market where timing and precision are paramount.

The process of price forecasting begins with the aggregation of historical data, capturing trends and cycles that offer insights into future behavior. Machine learning models, such as recurrent neural networks (RNNs) and long short-term memory (LSTM) networks, are particularly adept at analyzing time-series data, recognizing patterns that signal upward or downward movements. For instance, an LSTM model trained on ten years of cobalt price data might identify seasonal trends, such as increased demand in the fourth quarter due to manufacturing cycles, or longer-term cycles driven by technological transitions. By integrating these insights with real-time market data, the model generates forecasts that account for both historical trends and current conditions.

One of the defining features of predictive analytics is its ability to incorporate external variables that influence commodity markets. For deep sea mining, these variables might include changes in geopolitical relations, environmental regulations, or global economic growth rates. Natural language processing (NLP) algorithms are used to analyze news articles, policy announcements, and social media posts, identifying signals that could impact demand or supply. For example, an NLP model might detect an increase in discussions about green energy policies in major economies, predicting a corresponding rise in demand for critical minerals such as cobalt and nickel. By integrating this qualitative data with quantitative models, predictive analytics provides a holistic view of market dynamics.

The role of predictive analytics extends beyond price forecasting to encompass demand-supply modeling, a critical component of strategic planning for deep sea mining. The demand for minerals targeted by this industry is heavily influenced by technological innovation and policy shifts. The transition to renewable energy, for instance, has created unprecedented demand for metals used in wind turbines, solar panels, and energy storage systems. Similarly, the global push toward electric vehicles has driven demand for lithium, cobalt, and nickel, while the rollout of 5G networks has increased the need for rare earth elements.

Predictive analytics models analyze these demand drivers, forecasting consumption patterns across different sectors and regions. By combining historical consumption data with projections for technological adoption, these models estimate future demand with a high degree of granularity. For example, a model might predict that cobalt demand will grow at a compound annual rate of 6% over the next decade, driven by increasing electric vehicle production in China and Europe. These forecasts inform investment decisions, guiding mining companies toward the most lucrative opportunities and helping policymakers anticipate resource bottlenecks.

On the supply side, predictive analytics provides insights into production trends, inventory levels, and potential disruptions. Machine learning algorithms analyze data from mining operations, trade flows, and infrastructure developments, identifying risks and opportunities that influence supply dynamics. For instance, a model might detect that a new mining project in the Democratic Republic of Congo, the world's largest cobalt producer, is likely to increase global supply by 10% over the next five years. Alternatively, the model could flag potential disruptions, such as labor strikes, regulatory changes, or geopolitical conflicts, that could constrain supply and drive up prices.

The integration of demand-supply models with price forecasting creates a powerful framework for scenario analysis, enabling stakeholders to evaluate the implications of different market conditions. For example, a predictive analytics platform might simulate the impact of a hypothetical scenario where cobalt demand surges by 20% due to accelerated electric vehicle adoption, while supply remains constrained by delays in new mining projects. The model would estimate the resulting price trajectory, identifying potential profit margins and guiding investment strategies.

One of the most transformative aspects of predictive analytics is its ability to provide real-time insights, enabling dynamic decision-making in rapidly changing markets. In traditional commodity forecasting, updates to projections are often infrequent, limiting their relevance in volatile conditions. Predictive analytics, by contrast, continuously updates forecasts based on the latest data, ensuring that stakeholders have access to timely and accurate information. For deep sea mining, this capability is particularly valuable given the nascent and fast-evolving nature of the industry. Real-time analytics allows operators to adjust production schedules, optimize inventory levels, and respond proactively to market fluctuations.

The integration of AI-driven predictive analytics into deep sea mining also supports long-term strategic planning, aligning resource extraction with global sustainability goals. For example, models can forecast the future availability of critical minerals under different scenarios, such as increased recycling rates, advances in material substitution, or changes in consumer behavior. These insights inform decisions about exploration priorities, ensuring that mining activities are aligned with anticipated demand while minimizing environmental impacts. Predictive analytics can also identify opportunities for collaboration across industries, such as partnerships between mining companies and battery manufacturers to secure stable supply chains and promote circular economy initiatives.

Despite its transformative potential, the application of predictive analytics in deep sea mining faces challenges that must be addressed to fully realize its benefits. The accuracy of forecasts depends on the quality and comprehensiveness of the underlying data, which can be limited by gaps in historical records, inconsistent reporting, or the opacity of global supply chains. Addressing these challenges requires investment in data collection and standardization, as well as collaboration among industry stakeholders to share information and build robust datasets.

Another challenge lies in the interpretability of machine learning models, particularly those based on deep learning, which often operate as "black boxes." While these models excel at identifying patterns and generating predictions, understanding the rationale behind their outputs can be difficult. Developing explainable AI techniques is essential for building trust in predictive analytics, ensuring that stakeholders can evaluate the validity of forecasts and make informed decisions.

The economic viability of deep sea mining hinges on the careful coordination of resource extraction schedules, where timing, efficiency, and adaptability are paramount. Unlike terrestrial mining, deep sea operations face a unique set of uncertainties, including volatile commodity markets, unpredictable seafloor conditions, and dynamic environmental regulations. Stochastic optimization, an advanced mathematical and computational approach, provides a robust framework for managing these uncertainties. By integrating probabilistic modeling with resource extraction strategies, stochastic optimization enables decision-makers to design schedules that maximize profitability, minimize risks, and ensure compliance with environmental and operational constraints.

Stochastic optimization differs from traditional optimization techniques by explicitly incorporating uncertainty into the decision-making process. While deterministic models assume that all parameters are known and fixed, stochastic models account for variability in key factors such as resource availability, market demand, and operational costs. This flexibility allows for the development of extraction schedules that are resilient to unforeseen changes, ensuring that mining operations remain viable under a range of scenarios.

In the context of deep sea mining, resource extraction scheduling involves determining the optimal sequence, timing, and intensity of activities such as seafloor mapping, nodule collection, and material processing. These decisions are influenced by a variety of factors, including the spatial distribution of resources, the capacity and reliability of mining equipment, and the economic value of extracted materials. Stochastic optimization addresses these complexities by generating solutions that balance multiple objectives, such as maximizing net present value (NPV), minimizing environmental impact, and adhering to regulatory requirements.

A key component of stochastic optimization is the use of probabilistic models to represent uncertainty in input parameters. For example, the availability and quality of mineral deposits may vary across different regions of the seafloor, influenced by geological factors and the accuracy of exploration data. Similarly, commodity prices may fluctuate due to market dynamics, geopolitical events, or technological advancements. Stochastic models capture these uncertainties by representing input parameters as probability distributions rather than fixed values. This probabilistic representation allows decision-makers to evaluate the expected outcomes of different scheduling strategies and identify options that perform well across a range of scenarios.

The process of developing a stochastic optimization model begins with the definition of decision variables, constraints, and objectives. Decision variables represent the choices to be made, such as which mining sites to prioritize, how much material to extract, and when to deploy equipment. Constraints ensure that the schedule adheres to operational and environmental limits, such as equipment capacity, safety standards, and ecological thresholds. Objectives define the desired outcomes, such as maximizing revenue, minimizing costs, or achieving specific sustainability targets. These elements are combined into a mathematical model that describes the optimization problem in terms of probabilistic relationships.

One of the most widely used techniques in stochastic optimization is scenario analysis, where the model evaluates multiple possible futures, or scenarios, to identify robust solutions. Each scenario represents a specific realization of uncertain parameters, such as a high-demand, low-price market condition or a scenario where equipment failures reduce extraction capacity. The optimization process identifies schedules that perform well across these scenarios, ensuring that the chosen strategy is resilient to variability. For example, a stochastic optimization model for polymetallic nodule collection might evaluate scenarios where sediment plumes spread further than expected, guiding the selection of routes and collection techniques that minimize environmental risks.

Another powerful approach within stochastic optimization is Monte Carlo simulation, which generates thousands of random samples from the probability distributions of input parameters. These samples are used to simulate the outcomes of different extraction schedules, providing a detailed understanding of the risks and trade-offs associated with each option. For instance, a Monte Carlo simulation might estimate the expected revenue, costs, and environmental impacts of mining operations under varying market conditions, enabling decision-makers to prioritize strategies that offer the best balance of profitability and sustainability.

The integration of machine learning with stochastic optimization further enhances its capabilities, enabling the analysis of complex, high-dimensional datasets and the identification of patterns that inform scheduling decisions. Machine learning algorithms process data from exploration surveys, historical market trends, and operational performance to improve the accuracy of probabilistic models and generate more effective solutions. For example, a machine learning model might analyze sonar and geochemical data to predict the spatial distribution of high-grade nodules, guiding the selection of target areas for extraction. Similarly, neural networks can be used to model commodity price dynamics, providing more precise inputs for stochastic optimization models.

Stochastic optimization also supports adaptive scheduling, where decisions are updated in real-time based on new information. In deep sea mining, this capability is critical given the dynamic nature of the operating environment. For example, if a sediment plume spreads beyond predicted boundaries or if a surge in market prices increases the value of certain minerals, an adaptive scheduling model can adjust the extraction plan to reflect these changes. This real-time adaptability ensures that operations remain efficient and responsive, maximizing the value of extracted resources while minimizing risks.

The environmental implications of resource extraction scheduling are a central consideration in stochastic optimization models. Deep sea mining has the potential to impact fragile marine ecosystems, including benthic habitats and the broader water column. Stochastic models incorporate environmental constraints and objectives to ensure that schedules align with sustainability goals.

For example, a model might limit the intensity or duration of extraction activities in ecologically sensitive areas, balance resource extraction with habitat recovery times, or prioritize low-impact techniques in regions with high biodiversity. By integrating these considerations, stochastic optimization supports the development of schedules that minimize environmental harm while achieving economic objectives.

The implementation of stochastic optimization in deep sea mining faces several challenges, including the availability and quality of input data. Accurate and comprehensive datasets are essential for developing reliable probabilistic models, yet deep ocean environments are often poorly characterized due to the high costs and logistical difficulties of exploration. Addressing this challenge requires investment in data collection technologies, such as high-resolution sonar, environmental DNA (eDNA) analysis, and long-term monitoring platforms. Collaborative efforts among mining operators, research institutions, and regulatory bodies can further enhance data quality and availability, supporting the development of robust optimization models.

Computational complexity is another challenge, as stochastic optimization models often require significant processing power to evaluate multiple scenarios and generate solutions. Advances in cloud computing and parallel processing are addressing this issue, enabling the rapid analysis of large datasets and the exploration of high-dimensional solution spaces. The integration of edge computing technologies further enhances real-time adaptability, allowing optimization models to process and respond to data locally at the source.

The benefits of stochastic optimization extend beyond individual mining operations, offering insights for regional and global resource management. By analyzing the cumulative impacts of multiple extraction activities and evaluating the trade-offs between economic and environmental objectives, stochastic models support coordinated decision-making across stakeholders. For example, a regional optimization model might allocate extraction schedules among multiple operators to balance resource utilization with ecosystem preservation, ensuring that the benefits of deep sea mining are distributed equitably and sustainably.

Stochastic optimization represents a powerful tool for resource extraction scheduling in the deep sea mining industry. By integrating probabilistic modeling, machine learning, and scenario analysis, these models provide a robust framework for managing uncertainty and complexity. They enable the development of schedules that maximize profitability, minimize risks, and align with environmental and operational constraints. As the industry continues to evolve, the adoption of stochastic optimization will be essential for navigating the challenges and opportunities of deep sea mining, ensuring that resource extraction contributes to sustainable economic development while safeguarding the health and resilience of marine ecosystems. Through innovation and collaboration, stochastic optimization offers a pathway to a more efficient, adaptable, and responsible approach to deep sea resource management.

Techno-economic assessments (TEAs) are critical tools for evaluating the feasibility, profitability, and sustainability of deep sea mining projects. These assessments integrate technical parameters, such as resource availability and extraction methods, with economic factors, including capital costs, operating expenses, and market dynamics. In the context of global market uncertainty, TEAs play a pivotal role in guiding investment decisions and operational strategies, enabling stakeholders to navigate the complex interplay of economic risks and opportunities. By incorporating artificial intelligence (AI) and advanced analytics, TEAs have evolved into dynamic, data-driven frameworks that offer unparalleled precision and adaptability, ensuring that deep sea projects are resilient to the volatility and unpredictability of global markets.

At the heart of a techno-economic assessment is the evaluation of a project's cost structure, which encompasses capital expenditures (CAPEX), operating expenditures (OPEX), and financial risks. Deep sea mining projects involve substantial upfront investments in exploration, equipment, and infrastructure, as well as ongoing costs for maintenance, personnel, and energy consumption. These costs are influenced by a range of technical variables, such as the depth and distribution of resources, the efficiency of extraction technologies, and the logistical challenges of operating in remote and extreme environments. TEAs quantify these factors, providing a comprehensive view of the economic viability of a project.

The revenue side of a TEA is equally complex, shaped by the quantity and quality of recoverable resources and the prices these materials can command in the market. For deep sea mining, the targeted resources, cobalt, nickel, manganese, rare earth elements, and others, are integral to industries such as renewable energy, electric vehicles, and advanced electronics. However, their prices are subject to significant fluctuations driven by technological advancements, policy changes, and shifts in global supply and demand. AI-driven models enhance the accuracy of revenue projections by analyzing historical data, market trends, and external drivers, such as geopolitical events or environmental regulations. For instance, an AI system might predict an increase in cobalt demand due to accelerated electric vehicle adoption, providing a basis for projecting future revenue streams.

One of the most significant challenges in conducting TEAs for deep sea projects is accounting for global market uncertainty. Commodity markets are inherently volatile, influenced by factors such as trade policies, supply chain disruptions, and macroeconomic trends. Traditional TEAs often rely on static assumptions about prices and demand, which can lead to inaccurate projections and suboptimal decisions. By contrast, AI-powered TEAs incorporate probabilistic modeling and scenario analysis, capturing the variability and uncertainty of market conditions. These models simulate a range of potential outcomes, allowing stakeholders to evaluate the risks and rewards associated with different project scenarios.

For example, a probabilistic TEA for a polymetallic nodule mining project might analyze the impact of fluctuating nickel prices on the project's net present value (NPV) and internal rate of return (IRR). Using machine learning algorithms, the model could generate thousands of price trajectories based on historical volatility, macroeconomic indicators, and sector-specific trends. The results would provide a distribution of possible financial outcomes, highlighting the probability of achieving key economic milestones under various market conditions. This probabilistic approach enables decision-makers to identify strategies that are resilient to uncertainty, such as diversifying resource portfolios or securing long-term offtake agreements.

The integration of machine learning with TEAs extends beyond market analysis to include technical and operational parameters. AI algorithms process data from exploration surveys, operational performance, and environmental monitoring to refine cost estimates and optimize resource extraction strategies. For instance, a machine learning model might analyze sonar and geochemical data to estimate the spatial distribution and grade of mineral deposits, guiding the selection of high-value target areas. Similarly, predictive maintenance algorithms can estimate the likelihood of equipment failures and their associated costs, ensuring that these risks are accounted for in the TEA.

Another critical component of TEAs for deep sea projects is the assessment of environmental and social impacts, which are increasingly recognized as key determinants of project viability. Regulatory frameworks, public perception, and stakeholder expectations place growing emphasis on sustainability, requiring mining companies to demonstrate their commitment to minimizing ecological disruption and promoting social equity. AI-driven TEAs incorporate environmental data and sustainability metrics, providing a holistic view of a project's impacts and trade-offs. For example, a TEA might evaluate the costs and benefits of using low-impact nodule collection technologies or investing in habitat restoration initiatives, ensuring that these considerations are integrated into financial projections.

The role of TEAs in decision-making extends to the evaluation of financing options and risk mitigation strategies. Deep sea mining projects often require substantial capital investment, which may be sourced from a combination of equity, debt, and government incentives. TEAs analyze the financial implications of different funding structures, assessing their impact on cash flow, profitability, and risk exposure. For instance, a TEA might compare the benefits of securing a government-backed loan with the costs of issuing equity, providing stakeholders with insights into the optimal financing strategy.

Risk mitigation is another critical aspect of TEAs, particularly in the face of global market uncertainty. AI-powered TEAs identify and quantify key risks, such as price volatility, supply chain disruptions, or regulatory changes, and evaluate the effectiveness of mitigation measures. For example, a TEA might assess the impact of hedging strategies, such as entering into forward contracts or using derivatives to lock in favorable commodity prices. Alternatively, the model could evaluate the benefits of forming strategic partnerships with downstream industries, such as battery manufacturers or renewable energy companies, to secure stable demand and reduce exposure to market fluctuations.

The scalability and adaptability of AI-driven TEAs are particularly valuable in the dynamic and interconnected context of global markets. Traditional TEAs are often tailored to specific projects, limiting their applicability across different scenarios or regions. By contrast, AI-powered models can be customized and scaled to address a wide range of questions, from the economic feasibility of individual mining sites to the cumulative impacts of multiple operations within a region. These models are also inherently adaptive, updating their projections in real-time as new data becomes available. This capability ensures that TEAs remain relevant and actionable in rapidly changing environments, supporting informed decision-making at every stage of the project lifecycle.

Despite their transformative potential, the implementation of AI-driven TEAs faces challenges that must be addressed to fully realize their benefits. The accuracy and reliability of TEAs depend on the availability and quality of input data, which can be limited by gaps in exploration surveys, market intelligence, or environmental monitoring. Addressing these challenges requires investment in data collection and sharing, as well as collaboration among industry stakeholders to build comprehensive and standardized datasets. Transparency and explainability are also critical, as TEAs must provide stakeholders with a clear understanding of their assumptions, methodologies, and results. Developing explainable AI techniques is essential for building trust and ensuring that TEAs are used effectively in decision-making processes.

Techno-economic assessments are indispensable for evaluating the feasibility and sustainability of deep sea mining projects, particularly in the face of global market uncertainty. By integrating AI and advanced analytics, TEAs provide a dynamic, data-driven framework for navigating the complexities of resource extraction, balancing economic objectives with environmental and social considerations. These assessments enable stakeholders to make informed decisions, optimize resource allocation, and mitigate risks, ensuring that deep sea mining contributes to sustainable economic development while safeguarding the health and resilience of marine ecosystems. The adoption of AI-powered TEAs will be essential for harnessing the potential of deep sea resources responsibly and effectively, setting a new standard for innovation and accountability in resource management. Through collaboration, transparency, and a commitment to sustainability, techno-economic assessments offer a pathway to a more resilient and equitable future for deep sea mining.

6.2. Supply Chain and Logistics Optimization

The efficient transportation of extracted resources from the depths of the ocean to surface processing facilities represents a critical aspect of the deep sea mining value chain. With operations often located in remote and challenging environments, the optimization of supply chain logistics is vital to ensure economic viability and environmental sustainability. Artificial intelligence (AI) has emerged as a transformative tool in this domain, offering sophisticated solutions for route planning and fleet management that enhance efficiency, reduce costs, and minimize ecological impact. By leveraging machine learning, predictive analytics, and real-time decision-making, AI enables deep sea mining operations to address the logistical complexities of resource transportation with precision and adaptability.

The logistics of deep sea mining are inherently complex, involving the movement of materials across multiple phases of extraction, collection, and delivery. Resources such as polymetallic nodules or seafloor massive sulfides (SMS) must be transported from the seafloor to surface vessels or platforms, and subsequently transferred to processing facilities, often located far from the extraction site. Each phase of this process presents unique challenges, including variable seafloor topography, dynamic ocean conditions, and the need to balance operational efficiency with environmental considerations. AI-based route planning and fleet management systems address these challenges by optimizing the allocation of resources, coordinating fleet activities, and ensuring seamless integration across the supply chain.

At the heart of AI-driven logistics optimization is the ability to process and analyze vast amounts of data from diverse sources, including sensor networks, satellite imagery, and operational records. For route planning, machine learning algorithms integrate real-time data on ocean currents, wave heights, and weather patterns to identify the most efficient and safe paths for transport vessels and subsea vehicles. By accounting for these dynamic conditions, AI systems reduce transit times, fuel consumption, and the risks associated with navigating complex marine environments. For instance, an AI-powered route planning system might suggest alternative paths to avoid strong currents or adverse weather, ensuring that operations remain on schedule while minimizing disruptions.

In addition to route optimization, AI plays a central role in fleet management, enabling the coordinated operation of multiple vehicles and vessels involved in resource transportation. Fleet management systems use AI to monitor the status and performance of each asset in real time, ensuring that resources are allocated efficiently and that potential bottlenecks or delays are addressed proactively. For example, if a transport vessel encounters mechanical issues or unexpected delays, the AI system can reroute other vessels to compensate, maintaining the flow of materials and avoiding downtime. Similarly, machine learning models can predict maintenance needs for fleet assets, scheduling repairs or replacements to minimize disruptions and extend the operational lifespan of equipment.

One of the most significant advantages of AI-based logistics optimization is its ability to adapt dynamically to changing conditions. Unlike traditional planning approaches, which rely on static assumptions and pre-defined schedules, AI systems continuously update their models based on incoming data, enabling real-time adjustments to operations. This adaptability is particularly valuable in the unpredictable environments of deep sea mining, where conditions such as sediment plumes, equipment failures, or regulatory constraints can necessitate rapid changes to logistics plans. By providing operators with actionable insights and recommendations, AI systems ensure that operations remain resilient and responsive, even in the face of unforeseen challenges.

AI-driven logistics optimization also supports the integration of environmental considerations into route planning and fleet management. The transportation of resources can generate significant environmental impacts, including greenhouse gas emissions from fuel consumption and disturbances to marine ecosystems from vessel traffic. AI systems address these impacts by identifying routes and strategies that minimize environmental footprints. For example, machine learning algorithms might prioritize paths that avoid ecologically sensitive areas, such as coral reefs or marine protected zones, while also optimizing fuel efficiency to reduce carbon emissions. These environmentally informed decisions align with the broader goals of sustainable resource extraction, demonstrating the industry's commitment to responsible practices.

Another critical application of AI in logistics optimization is the management of multi-modal transportation networks, where resources are moved across different types of vehicles and infrastructure. In deep sea mining, this often involves the transfer of materials from subsea collection systems to surface vessels, and subsequently to onshore processing facilities via bulk carriers or pipelines. AI systems coordinate these transitions, ensuring that materials flow seamlessly through the supply chain and that each mode of transport operates at maximum efficiency. For instance, an AI-based fleet management platform might synchronize the arrival of surface vessels with the loading capacity of onshore facilities, reducing waiting times and optimizing throughput.

The scalability of AI-based logistics systems is particularly valuable for deep sea mining operations that span large geographic areas or involve multiple extraction sites. Traditional logistics planning can become unwieldy as the scale and complexity of operations increase, leading to inefficiencies and higher costs. AI systems, by contrast, excel at managing large-scale and high-dimensional data, providing a centralized platform for coordinating activities across diverse locations. This capability ensures that resources are allocated effectively, bottlenecks are minimized, and overall performance is optimized, regardless of the scope of operations.

The integration of predictive analytics further enhances the capabilities of AI-based logistics systems, enabling operators to anticipate and prepare for future conditions. By analyzing historical data and identifying trends, predictive models provide insights into potential challenges and opportunities, such as seasonal variations in weather, fluctuations in fuel prices, or changes in regulatory requirements. These forecasts inform strategic planning, allowing operators to adjust their logistics strategies proactively and avoid costly disruptions. For example, a predictive model might suggest stockpiling critical supplies in advance of a predicted storm season, ensuring that operations can continue uninterrupted despite adverse conditions.

Despite their transformative potential, the implementation of AI-based logistics systems in deep sea mining faces challenges that must be addressed to fully realize their benefits. The reliability of these systems depends on the quality and availability of input data, which can be limited by gaps in monitoring coverage, inconsistencies in data formats, or the inherent difficulties of collecting information in remote and extreme environments. Addressing these challenges requires investment in data collection technologies, such as advanced sensors, satellite systems, and autonomous monitoring platforms, as well as collaboration among industry stakeholders to establish standardized data-sharing protocols.

Another challenge lies in the complexity of integrating AI systems with existing logistics infrastructure and processes. Many mining operations rely on legacy systems that may not be compatible with modern AI platforms, necessitating upgrades or the development of custom solutions. Ensuring the interoperability of these systems requires careful planning and the involvement of cross-disciplinary teams, including logistics experts, data scientists, and IT specialists.

The future of AI-based route planning and fleet management in deep sea mining is closely tied to ongoing advancements in machine learning, edge computing, and IoT. Emerging technologies, such as swarm robotics and real-time digital twins, promise to further enhance the efficiency and adaptability of logistics systems. Swarm robotics, where fleets of autonomous vehicles operate collaboratively, could enable decentralized decision-making and dynamic reconfiguration of logistics plans, reducing the reliance on centralized control. Digital twins, which create virtual replicas of physical assets and operations, provide a powerful tool for simulating and optimizing logistics strategies before implementation, minimizing risks and maximizing efficiency.

AI-based route planning and fleet management represent a paradigm shift in the logistics of deep sea mining, enabling operators to address the challenges of resource transportation with precision, adaptability, and environmental responsibility. By leveraging advanced analytics, real-time decision-making, and predictive capabilities, these systems optimize the flow of materials across the supply chain, ensuring that operations are efficient, resilient, and aligned with sustainability goals. The adoption of AI-driven logistics systems will be essential for achieving the economic and environmental objectives of deep sea mining, setting a new standard for innovation and accountability in resource extraction. Through collaboration, investment, and a commitment to excellence, AI-based logistics optimization offers a pathway to a more efficient and sustainable future for the deep ocean's mineral wealth.

As global demand for critical minerals intensifies, transparency and accountability in supply chains have become essential priorities for industries reliant on these resources. Deep sea mining, with its potential to provide significant quantities of valuable metals like cobalt, nickel, and manganese, faces increasing scrutiny over the environmental and ethical implications of its operations. Consumers, regulators, and stakeholders demand assurance that the materials extracted from the ocean floor are sourced responsibly, processed ethically, and transported efficiently. Blockchain and Internet of Things (IoT)-driven solutions have emerged as transformative technologies for enabling traceability and provenance, creating a transparent, secure, and verifiable supply chain for deep sea resources.

Traceability refers to the ability to track the movement of materials through each stage of the supply chain, from extraction and transportation to processing and final use. Provenance adds an additional layer of detail, documenting the origin, composition, and ethical credentials of materials. Together, these concepts ensure that stakeholders can verify the journey of resources, addressing concerns about environmental impact, regulatory compliance, and ethical sourcing. Blockchain and IoT technologies offer complementary solutions for achieving these goals, combining immutable data records with real-time monitoring to create an integrated system of trust and accountability.

Blockchain technology, a decentralized and distributed ledger system, is ideally suited for enhancing transparency in supply chains. Each transaction or event in the supply chain is recorded as a block of data, cryptographically secured and linked to the previous block, creating a tamper-proof chain of information. For deep sea mining, blockchain can be used to document every step of the process, from the location and method of resource extraction to the transfer of materials between operators, transporters, and processors. Each transaction is verified by consensus among network participants, ensuring the accuracy and integrity of the data.

The implementation of blockchain in deep sea mining begins with the creation of digital tokens or identifiers for extracted materials. For instance, polymetallic nodules harvested from the seafloor could be assigned unique digital tags containing information about their origin, composition, and extraction method. These tags are recorded on the blockchain, along with supporting documentation such as environmental impact assessments, compliance certifications, and chain-of-custody records. As the materials move through the supply chain, each transaction or transfer is updated on the blockchain, creating a complete and transparent record of their journey.

IoT devices play a critical role in enhancing the functionality and granularity of blockchain-based traceability systems. IoT refers to the network of interconnected sensors, devices, and systems that collect and exchange data in real time. In the context of deep sea mining, IoT devices can monitor key parameters such as the location, temperature, and condition of materials during extraction, transportation, and storage. These devices provide continuous streams of data that are recorded on the blockchain, ensuring that stakeholders have real-time visibility into the status of resources.

For example, IoT-enabled sensors mounted on subsea collection vehicles can record the exact coordinates and depth of extraction for each batch of polymetallic nodules. This data is uploaded to the blockchain, providing an immutable record of the nodule's origin. During transportation, IoT devices on transport vessels can monitor the environmental conditions of the cargo hold, ensuring that materials are stored under appropriate conditions and flagging any deviations for immediate attention. Once the materials arrive at processing facilities, IoT systems can track their movement through refining and manufacturing processes, verifying that they are handled in accordance with sustainability and safety standards.

The integration of blockchain and IoT technologies creates a powerful synergy, enabling a level of traceability and provenance that is both detailed and secure. Blockchain provides the immutable record-keeping necessary for trust and accountability, while IoT enhances the granularity and immediacy of the data. Together, these technologies address key challenges in supply chain management, such as preventing fraud, ensuring compliance, and reducing inefficiencies.

One of the most significant benefits of blockchain-driven traceability is its ability to demonstrate compliance with environmental and ethical standards. Deep sea mining operations are subject to stringent regulations designed to minimize ecological impacts and protect marine ecosystems. By recording compliance data on the blockchain, operators can provide verifiable evidence of their adherence to these standards. For instance, a mining company might document the implementation of low-impact collection techniques, the avoidance of sensitive habitats, or the mitigation of sediment plumes. These records can be made accessible to regulators, certification bodies, and other stakeholders, fostering transparency and trust.

Similarly, blockchain technology supports ethical sourcing by verifying that materials are extracted and processed responsibly. Concerns about human rights violations, environmental degradation, and the exploitation of vulnerable communities have heightened the demand for ethically sourced materials. Blockchain enables operators to provide detailed provenance data, such as the certification of sustainable practices, the fair treatment of workers, and the equitable distribution of benefits to affected communities. This information is particularly valuable for downstream industries, such as electric vehicle manufacturers and technology companies, which are increasingly required to demonstrate the ethical credentials of their supply chains.

Another advantage of blockchain and IoT-driven traceability systems is their ability to enhance efficiency and reduce costs in supply chain operations. Traditional supply chains often suffer from inefficiencies due to fragmented data, manual record-keeping, and limited visibility across different stages. Blockchain provides a centralized and standardized platform for recording and sharing data, eliminating redundancies and streamlining processes. IoT devices further enhance efficiency by automating data collection and monitoring, reducing the need for manual intervention and ensuring that information is accurate and up to date.

For example, a blockchain-based platform for deep sea mining could automate the documentation of material transfers, reducing administrative overhead and minimizing the risk of errors. IoT sensors on transport vessels could monitor fuel consumption and route efficiency, providing insights that optimize logistics and reduce costs. These efficiencies not only improve the economic viability of mining operations but also contribute to sustainability goals by reducing waste and resource use.

Despite their transformative potential, the implementation of blockchain and IoT-driven traceability systems in deep sea mining faces several challenges. The integration of these technologies requires significant investment in infrastructure, including IoT devices, blockchain platforms, and data processing systems. Ensuring the interoperability of these systems with existing supply chain processes is also critical, requiring careful planning and collaboration among stakeholders.

Another challenge is the scalability and accessibility of these systems, particularly for smaller operators or those in developing regions. Blockchain and IoT technologies must be designed to accommodate diverse participants with varying levels of technical capacity and resources. Collaborative efforts, such as industry consortia or public-private partnerships, can help address these challenges by providing shared platforms and resources that lower barriers to entry.

The future of blockchain and IoT-driven traceability in deep sea mining is closely tied to ongoing advancements in technology and governance. Emerging innovations, such as decentralized finance (DeFi) and smart contracts, promise to further enhance the functionality and accessibility of blockchain systems. Smart contracts, for instance, can automate the execution of agreements between parties, such as payments for material transfers or penalties for non-compliance, based on predefined conditions recorded on the blockchain. These capabilities streamline operations and ensure that transactions are conducted transparently and efficiently.

Blockchain and IoT-driven traceability and provenance represent a paradigm shift in supply chain management for deep sea mining. By providing transparent, secure, and verifiable records of resource extraction, transportation, and processing, these technologies address key challenges in accountability, compliance, and efficiency. They enable stakeholders to demonstrate their commitment to sustainability and ethical sourcing, fostering trust and confidence in the industry. As deep sea mining continues to evolve, the adoption of blockchain and IoT technologies will be essential for building a responsible and resilient supply chain, ensuring that the ocean's mineral wealth is harnessed for the benefit of all while safeguarding the integrity of marine ecosystems. Through innovation, collaboration, and a commitment to transparency, blockchain and IoT offer a pathway to a more sustainable and equitable future for deep sea resource management.

The intricacies of supply chain and logistics optimization in deep sea mining present significant challenges, necessitating sophisticated analytical tools to balance economic performance, operational efficiency, and environmental sustainability. Cost-benefit analysis (CBA) has long been a cornerstone of decision-making in industrial operations, but its application in the deep sea mining sector requires innovative approaches capable of addressing the complex dynamics of this nascent industry. Agent-based modeling (ABM) offers a transformative method for conducting CBAs, enabling the simulation of individual entities, agents, within the supply chain and their interactions, thereby providing a granular and dynamic perspective on costs and benefits.

Agent-based modeling is a computational approach that represents individual components of a system, such as vessels, vehicles, infrastructure, and regulatory bodies, as autonomous agents with distinct attributes, behaviors, and decision-making rules. These agents interact within a simulated environment, producing emergent outcomes that reflect the complexities of real-world systems. In the context of deep sea mining, ABM enables stakeholders to simulate the behavior of supply chain actors under varying conditions, evaluate the impacts of alternative strategies, and optimize operations in light of both economic and environmental considerations.

The foundation of a cost-benefit analysis using ABM lies in the identification and characterization of agents within the supply chain. Each agent represents a specific entity or process, such as a mining vessel, a subsea crawler, a processing facility, or a regulatory authority. Agents are assigned attributes that define their capabilities, constraints, and objectives. For example, a subsea crawler might be characterized by its extraction capacity, energy consumption, and maintenance requirements, while a processing facility might be defined by its throughput, operating costs, and emissions profile. These attributes are informed by empirical data, technical specifications, and industry benchmarks, ensuring that the model reflects the realities of deep sea mining operations.

Interactions between agents are governed by rules and algorithms that capture the dynamics of the supply chain. For instance, a mining vessel may communicate with subsea crawlers to coordinate the collection and transfer of materials, while regulatory agents may impose constraints on operational intensity to mitigate environmental impacts. These interactions generate complex feedback loops that influence the overall performance of the system, making ABM particularly well-suited to modeling the interconnected and dynamic nature of deep sea mining supply chains.

One of the most valuable applications of ABM in cost-benefit analysis is the evaluation of alternative operational strategies. By simulating different scenarios, ABM enables stakeholders to compare the costs and benefits of various approaches, such as centralized versus decentralized processing, single-site versus multi-site operations, or low-impact versus high-intensity extraction techniques. For example, a model might simulate the logistics of transporting polymetallic nodules from multiple extraction sites to a centralized processing facility, comparing the economic efficiency of this approach with a decentralized system where processing occurs closer to the extraction sites. By quantifying key metrics such as transportation costs, energy consumption, and material throughput, the model provides insights into the trade-offs associated with each strategy.

Environmental considerations are a critical component of ABM-driven CBAs, reflecting the growing emphasis on sustainability in deep sea mining. Environmental impacts, such as sediment plumes, habitat disturbance, and greenhouse gas emissions, are incorporated into the model as costs that must be weighed against the economic benefits of resource extraction. For instance, a model might simulate the dispersal of sediment plumes generated by different nodule collection techniques, evaluating their effects on benthic ecosystems and water quality. By assigning monetary values to these impacts—based on mitigation costs, regulatory penalties, or societal willingness to pay for ecosystem services, the model quantifies the environmental trade-offs of alternative approaches.

ABM also excels at capturing the uncertainties and variabilities inherent in deep sea mining operations. Supply chains in this industry are influenced by a wide range of factors, including fluctuating commodity prices, unpredictable seafloor conditions, and evolving regulatory requirements. ABM addresses these uncertainties by incorporating probabilistic elements into the behavior of agents and the parameters of the model. For example, the availability and performance of mining equipment may vary based on stochastic failure rates, while the demand for critical minerals may fluctuate in response to technological advancements or policy shifts. By simulating these uncertainties, ABM provides a robust framework for evaluating the resilience of supply chain strategies and identifying options that perform well under a range of conditions.

The use of ABM in CBAs extends to the optimization of resource allocation and scheduling, ensuring that supply chain operations are both efficient and adaptable. For example, an ABM might optimize the deployment of a fleet of transport vessels, balancing the trade-offs between fuel consumption, travel time, and cargo volume. Similarly, the model could determine the optimal sequence of mining activities across multiple sites, minimizing operational costs while maximizing resource recovery. These optimizations are achieved through iterative simulations and heuristic algorithms, which explore the solution space and identify strategies that achieve the desired balance of costs and benefits.

Collaboration and stakeholder engagement are key advantages of ABM-driven CBAs, as the models provide a transparent and interactive platform for exploring supply chain dynamics. Stakeholders, including mining operators, regulators, and environmental organizations, can use the model to visualize the impacts of different scenarios, evaluate trade-offs, and build consensus around preferred strategies. For instance, a collaborative modeling exercise might involve simulating the cumulative impacts of multiple mining operations within a region, identifying strategies to coordinate logistics and mitigate environmental effects. By fostering dialogue and shared understanding, ABM supports the development of supply chain strategies that align with diverse stakeholder priorities.

Despite its many benefits, the implementation of ABM in cost-benefit analysis for deep sea mining requires careful attention to model design and validation. The complexity of ABM necessitates a thorough understanding of the system being modeled, including the interactions between agents and the external factors influencing their behavior. Ensuring the accuracy and reliability of the model requires high-quality data, robust calibration methods, and sensitivity analyses to evaluate the influence of key assumptions. Additionally, the computational demands of ABM can be significant, particularly for large-scale or high-resolution simulations. Advances in computing power, cloud-based platforms, and parallel processing are addressing these challenges, enabling the efficient execution of complex models.

The future of ABM in deep sea mining holds great promise, driven by ongoing advancements in modeling techniques, data collection technologies, and stakeholder collaboration. Emerging innovations, such as digital twins and AI-driven decision support systems, are enhancing the capabilities of ABM, enabling more precise and dynamic simulations. Digital twins, which create virtual replicas of physical systems, provide a powerful tool for integrating real-time data into ABM, ensuring that the models remain relevant and responsive to changing conditions. Similarly, AI algorithms are improving the efficiency of model calibration and optimization, accelerating the identification of cost-effective and sustainable supply chain strategies.

Agent-based modeling represents a powerful and versatile tool for conducting cost-benefit analyses in the deep sea mining industry. By simulating the behavior and interactions of supply chain agents, ABM provides a dynamic and granular perspective on the trade-offs associated with different operational strategies. It enables stakeholders to evaluate the economic, environmental, and logistical impacts of their decisions, ensuring that supply chain operations are both efficient and responsible. As the industry continues to evolve, the adoption of ABM-driven CBAs will be essential for optimizing supply chain performance, balancing economic and environmental objectives, and fostering collaboration among stakeholders.

6.3. Investment, Financing, and Risk Management

Investment decision-making in the deep sea mining industry is inherently fraught with uncertainty, stemming from volatile commodity markets, evolving regulatory frameworks, and the technological challenges of operating in one of the Earth's most extreme environments. The scale of capital required for exploration, infrastructure, and operations, coupled with the long timelines for project execution and return on investment, amplifies the stakes for stakeholders. To navigate these complexities, decision-makers increasingly rely on advanced investment frameworks that integrate artificial intelligence (AI) with probabilistic tools such as Monte Carlo simulations and scenario analysis. These methods enable a deeper understanding of risks and opportunities, equipping stakeholders with the insights needed to optimize capital allocation and mitigate uncertainties.

AI has transformed investment decision frameworks by enhancing the capacity to analyze vast datasets, identify patterns, and predict outcomes. At its core, AI facilitates the modeling of complex, high-dimensional systems that characterize deep sea mining projects, where variables such as resource availability, operational costs, market dynamics, and environmental impacts interact in intricate and often unpredictable ways. By integrating AI into Monte Carlo simulations and scenario analysis, decision-makers gain access to dynamic, data-driven tools that provide a comprehensive view of potential investment outcomes under varying conditions.

Monte Carlo simulations, a staple of probabilistic modeling, are particularly well-suited to the deep sea mining sector, where uncertainty is both pervasive and multifaceted. These simulations rely on repeated random sampling to generate a distribution of possible outcomes, capturing the variability and interdependencies of key factors that influence investment performance. AI enhances the utility of Monte Carlo methods by improving the accuracy and efficiency of simulations. Machine learning algorithms analyze historical and real-time data to refine probability distributions, ensuring that simulations reflect the most up-to-date information about market trends, technological developments, and project-specific parameters.

For example, a Monte Carlo simulation for a polymetallic nodule mining project might model the net present value (NPV) of the investment under a range of scenarios for cobalt and nickel prices, resource recovery rates, and operational costs. AI algorithms process data from commodity markets, exploration surveys, and equipment performance to estimate the likelihood of achieving specific financial outcomes. By generating thousands of possible scenarios, the simulation provides decision-makers with a probabilistic view of risks and rewards, highlighting the range of potential returns and the likelihood of surpassing key financial thresholds.

AI further enhances Monte Carlo simulations by enabling real-time updates and adaptive modeling. Traditional Monte Carlo methods often rely on static inputs, limiting their relevance as conditions evolve. AI systems, by contrast, continuously incorporate new data, such as changes in market prices, regulatory developments, or operational performance, ensuring that simulations remain dynamic and responsive. For instance, if a surge in global demand for electric vehicles drives up cobalt prices, an AI-enhanced Monte Carlo simulation can immediately adjust its probability distributions, providing stakeholders with an updated assessment of the project's financial prospects.

Scenario analysis complements Monte Carlo simulations by exploring the implications of specific, well-defined conditions or events. While Monte Carlo simulations focus on probabilistic distributions, scenario analysis examines discrete scenarios that represent alternative futures, such as regulatory changes, technological breakthroughs, or geopolitical shifts. AI enhances scenario analysis by generating and evaluating a broader range of scenarios, uncovering insights that might be overlooked through manual analysis. Natural language processing (NLP) algorithms, for example, can analyze news articles, policy documents, and market reports to identify emerging trends and potential disruptions. These insights inform the development of scenarios that capture the full spectrum of uncertainties facing deep sea mining investments.

For instance, an AI-driven scenario analysis for a seafloor massive sulfide (SMS) project might evaluate the financial and operational implications of different regulatory regimes. One scenario could assume stringent environmental regulations that increase compliance costs but enhance market access for responsibly sourced materials, while another scenario might explore the impacts of delayed regulatory approvals or shifting geopolitical alliances. By quantifying the financial outcomes and trade-offs associated with each scenario, decision-makers gain a clearer understanding of the risks and opportunities inherent in their investments.

The integration of AI into scenario analysis also supports the identification of early warning signals for emerging risks and opportunities. Machine learning models analyze historical patterns and leading indicators to predict the likelihood of specific scenarios materializing. For example, an AI system might identify a correlation between increasing investments in renewable energy infrastructure and rising demand for rare earth elements, flagging this trend as a key driver of future market dynamics. By providing these predictive insights, AI-enabled scenario analysis empowers stakeholders to proactively adjust their investment strategies, positioning themselves to capitalize on favorable developments or mitigate potential threats.

One of the most significant advantages of AI-enhanced investment frameworks is their ability to evaluate the interdependencies and cumulative effects of multiple variables. Deep sea mining projects operate within a complex web of economic, environmental, and social factors, where changes in one area can cascade across others. Traditional investment analyses often struggle to capture these dynamics, relying on simplified assumptions that overlook critical interactions. AI addresses this limitation by modeling the system as a whole, capturing the feedback loops and nonlinear relationships that influence investment outcomes.

For example, an AI-enabled framework might analyze the interplay between operational costs, environmental mitigation measures, and market demand. The model could simulate how investments in low-impact extraction technologies affect both operating expenses and the market premium for sustainably sourced materials, providing a holistic view of the trade-offs involved. This integrated perspective ensures that investment decisions are informed by a comprehensive understanding of the project's economic and environmental dimensions.

AI-driven investment frameworks also enhance collaboration and transparency among stakeholders, supporting more informed and inclusive decision-making processes. Interactive visualization tools, powered by AI, enable stakeholders to explore the results of Monte Carlo simulations and scenario analyses in an intuitive and accessible format. For example, dynamic dashboards might display probability distributions of financial outcomes, heatmaps of risk factors, or comparative analyses of alternative investment strategies. These visualizations facilitate dialogue among investors, operators, regulators, and conservation groups, fostering consensus and alignment around shared objectives.

Despite their transformative potential, the implementation of AI-enhanced investment frameworks in deep sea mining faces challenges that must be addressed to fully realize their benefits. The accuracy and reliability of these frameworks depend on the quality and completeness of the input data, which can be limited by gaps in exploration surveys, market intelligence, or environmental monitoring. Addressing these challenges requires investment in data collection technologies, such as high-resolution sonar, satellite imagery, and long-term monitoring platforms, as well as collaboration among industry stakeholders to build robust datasets.

Another challenge lies in the interpretability and transparency of AI models, particularly those based on complex machine learning algorithms. While these models excel at processing large datasets and generating predictions, understanding the rationale behind their outputs can be difficult. Developing explainable AI techniques is essential for building trust and ensuring that stakeholders can evaluate the validity of model results. For instance, tools that highlight the key drivers of model predictions or provide visual explanations of decision paths can enhance transparency and accountability.

AI-enhanced investment decision frameworks, leveraging tools such as Monte Carlo simulations and scenario analysis, represent a shift in the way deep sea mining projects are evaluated and managed. By integrating probabilistic modeling, real-time data analysis, and predictive insights, these frameworks provide stakeholders with a comprehensive understanding of risks and opportunities, enabling more informed and adaptive investment strategies. As the deep sea mining industry continues, the adoption of AI-driven investment frameworks will be essential for navigating the complexities of global markets, ensuring that resource extraction contributes to sustainable economic growth while safeguarding the health and resilience of marine ecosystems.

Dynamic hedging strategies and risk mitigation mechanisms are indispensable in managing the uncertainties inherent in deep sea mining investments. The industry's exposure to fluctuating commodity prices, regulatory shifts, environmental constraints, and technological challenges requires an agile and sophisticated approach to risk management. Dynamic hedging, a strategy that adjusts financial positions in response to changing market conditions, plays a central role in protecting against adverse movements while enabling stakeholders to capitalize on favorable trends. Combined with advanced risk mitigation techniques, these strategies ensure that deep sea mining projects remain financially viable and resilient in the face of complex and evolving challenges.

At the heart of dynamic hedging is the principle of adaptability. Unlike static hedging strategies, which involve fixed financial positions that remain unchanged over the life of an investment, dynamic hedging continuously adjusts these positions based on real-time data and market conditions. This flexibility is critical in the volatile markets that characterize deep sea mining commodities such as cobalt, nickel, and rare earth elements. Prices for these materials are influenced by a wide range of factors, including technological advancements, geopolitical tensions, supply chain disruptions, and shifts in demand from key industries such as renewable energy and electric vehicles. Dynamic hedging strategies leverage advanced analytics and predictive models to anticipate these changes and adjust positions accordingly, minimizing losses and maximizing gains.

One of the most common instruments used in dynamic hedging is derivatives, such as options, futures, and swaps. These financial contracts enable stakeholders to lock in prices, limit exposure to adverse movements, or gain exposure to favorable trends without directly owning the underlying asset. For example, a deep sea mining operator anticipating a potential decline in nickel prices might purchase put options, which grant the right to sell nickel at a predetermined price. If prices fall below this level, the options provide a financial buffer that offsets the impact of the decline. Conversely, if prices rise, the operator can choose not to exercise the options and instead benefit from the higher market price.

Dynamic hedging strategies are enhanced by the integration of artificial intelligence (AI) and machine learning, which provide the analytical tools needed to model market behavior, assess risks, and optimize financial positions. AI algorithms analyze historical price data, market trends, and macroeconomic indicators to identify patterns and correlations that inform hedging decisions. For instance, a machine learning model might detect that cobalt prices tend to rise following announcements of increased electric vehicle production targets, prompting the operator to increase long positions in cobalt futures. These models also incorporate real-time data, enabling continuous updates to hedging strategies as market conditions evolve.

In addition to managing price risks, dynamic hedging strategies address a broader spectrum of financial uncertainties, including currency fluctuations, interest rate changes, and credit risks. Deep sea mining projects often involve multinational operations and cross-border transactions, exposing stakeholders to exchange rate volatility. For example, a mining company extracting resources in the Pacific Ocean might incur costs in US dollars but sell its products in euros, creating exposure to currency risk. Dynamic hedging strategies mitigate this risk through instruments such as currency swaps or forward contracts, which lock in exchange rates for future transactions.

Interest rate risk is another critical consideration, particularly for projects financed through debt. Changes in interest rates can affect the cost of borrowing and the financial performance of mining projects. Dynamic hedging strategies use interest rate swaps and caps to manage this exposure, ensuring that financing costs remain predictable and aligned with project budgets. For example, an operator might use an interest rate swap to convert a variable-rate loan into a fixed-rate loan, providing stability in a rising interest rate environment.

Credit risk, which arises from the potential default of counterparties or customers, is also addressed through dynamic hedging and risk mitigation mechanisms. For instance, mining companies can use credit default swaps to protect against the default of key buyers or partners, ensuring that revenue streams remain secure even in adverse scenarios. AI-enhanced credit risk models further improve the accuracy of these strategies by analyzing financial statements, payment histories, and macroeconomic factors to assess the creditworthiness of counterparties.

While financial instruments play a central role in dynamic hedging, risk mitigation in deep sea mining extends beyond the financial domain to encompass operational, environmental, and reputational risks. The integration of real-time monitoring and data analytics is critical in managing these risks effectively. Advanced sensor networks, autonomous monitoring platforms, and predictive maintenance systems provide continuous insights into equipment performance, environmental conditions, and operational efficiency. These technologies enable operators to identify potential issues before they escalate, reducing downtime, minimizing environmental impacts, and ensuring compliance with regulatory requirements.

For example, a predictive maintenance system might analyze data from sensors on a subsea crawler to detect early signs of wear or malfunction, prompting preemptive repairs that avoid costly breakdowns. Similarly, environmental monitoring systems equipped with AI algorithms can track sediment plumes, water quality, and marine biodiversity, ensuring that mining activities remain within permitted thresholds and mitigating the risk of regulatory penalties or reputational damage.

Dynamic hedging strategies and risk mitigation also benefit from scenario analysis, a powerful tool for evaluating the implications of different market and operational conditions. Scenario analysis models simulate a range of possible futures, allowing stakeholders to assess the resilience of their strategies under varying circumstances. For instance, a scenario analysis might evaluate the financial and operational impacts of a sudden increase in demand for rare earth elements, driven by geopolitical tensions or advances in green technologies. By quantifying the potential risks and rewards associated with these scenarios, stakeholders can refine their hedging strategies and allocate resources more effectively.

The application of blockchain technology further enhances risk mitigation by providing transparency and security in supply chain transactions. Blockchain creates an immutable ledger of transactions, enabling stakeholders to verify the authenticity and provenance of materials, track compliance with regulatory standards, and ensure accountability across the supply chain. For example, a blockchain-based system could document the environmental credentials of extracted resources, ensuring that they meet sustainability criteria and reducing the risk of disputes or penalties.

Collaboration and stakeholder engagement are essential components of dynamic hedging and risk mitigation in deep sea mining. The complex and interconnected nature of the industry requires coordination among operators, regulators, investors, and environmental organizations to address shared risks and challenges. Collaborative initiatives, such as industry consortia or public-private partnerships, provide a platform for sharing data, best practices, and innovative solutions. For example, an industry-wide effort to develop standardized hedging protocols or environmental monitoring frameworks could enhance the resilience and credibility of the sector as a whole.

Despite their many advantages, the implementation of dynamic hedging strategies and risk mitigation mechanisms in deep sea mining faces challenges that must be addressed to fully realize their potential. The effectiveness of these strategies depends on the availability and quality of data, which can be limited by gaps in monitoring coverage, inconsistencies in reporting, or the inherent difficulties of operating in remote and extreme environments. Addressing these challenges requires investment in data collection technologies, such as high-resolution sensors, satellite systems, and autonomous monitoring platforms, as well as collaboration among industry stakeholders to establish standardized data-sharing protocols.

Another challenge lies in the complexity and scalability of dynamic hedging strategies, particularly for smaller operators or new entrants to the industry. Ensuring that these strategies are accessible and practical for a diverse range of stakeholders requires the development of user-friendly tools, automated decision-support systems, and training programs that build capacity and expertise.

Dynamic hedging strategies and risk mitigation mechanisms represent a cornerstone of financial and operational resilience in the deep sea mining industry. By integrating advanced analytics, real-time data, and innovative technologies, these strategies enable stakeholders to navigate the uncertainties of volatile markets, complex supply chains, and stringent regulatory environments.

The adoption of dynamic hedging and risk mitigation will be essential for ensuring that deep sea mining projects are not only economically viable but also environmentally responsible and socially equitable.

The viability of deep sea mining projects is deeply intertwined with the complex and evolving landscape of global regulatory frameworks. Operating in the international waters of the deep ocean, where jurisdictional boundaries blur and environmental stakes are high, demands adherence to an intricate web of rules and standards established by international bodies, national governments, and regional organizations. These regulations, aimed at balancing economic development with environmental preservation and equitable benefit-sharing, play a decisive role in shaping the feasibility, profitability, and sustainability of mining operations. For investors and operators, understanding and navigating these global regulatory influences is not merely a compliance issue but a strategic necessity that directly impacts project viability.

At the heart of the regulatory landscape governing deep sea mining is the United Nations Convention on the Law of the Sea (UNCLOS), a comprehensive treaty that defines the rights and responsibilities of nations in their use of the world's oceans. UNCLOS designates the seabed beyond national jurisdictions as the "common heritage of mankind," mandating that its resources be managed for the benefit of all humanity. This principle is operationalized through the International Seabed Authority (ISA), a UN-established body tasked with regulating mineral exploration and extraction in the international seabed area. The ISA's role includes issuing licenses, setting environmental standards, and ensuring equitable sharing of financial and other benefits derived from mining activities.

The regulatory regime established by UNCLOS and the ISA profoundly influences the viability of deep sea mining projects. For operators, securing an ISA exploration or exploitation license is a prerequisite for legal operation in international waters. This process entails meeting stringent requirements, including submitting detailed environmental impact assessments (EIAs), demonstrating technical and financial capability, and adhering to a complex set of rules governing resource extraction and environmental protection. The costs and timelines associated with these requirements can significantly impact project planning and financing, necessitating careful consideration in feasibility studies and investment strategies.

One of the most critical aspects of the ISA's regulatory framework is its focus on environmental stewardship. Deep sea ecosystems are among the least understood and most vulnerable on the planet, with unique species and habitats adapted to extreme conditions. The potential impacts of mining activities, including habitat destruction, sediment plumes, and noise pollution, have raised significant concerns among scientists, conservationists, and policymakers. To address these concerns, the ISA has established environmental management regulations that mandate the use of precautionary approaches, adaptive management, and robust monitoring systems. These requirements influence the design and operation of mining projects, often necessitating investment in low-impact technologies, comprehensive baseline studies, and ongoing environmental monitoring.

In addition to environmental regulations, the ISA's benefit-sharing mechanism represents a key regulatory influence on project economics. Under UNCLOS, revenues generated from mining activities in international waters must be shared equitably among member states, particularly developing nations. This principle is operationalized through royalty payments, financial contributions to the ISA, and capacity-building initiatives. While these obligations promote fairness and inclusivity, they also add to the cost structure of mining projects, impacting profitability and financial modeling. Investors must account for these costs in their assessments, evaluating how benefit-sharing requirements align with project returns and long-term sustainability.

Beyond the ISA, national governments play a critical role in shaping the regulatory environment for deep sea mining, particularly in areas within their exclusive economic zones (EEZs). Countries such as Japan, South Korea, and Papua New Guinea have developed their own legal frameworks for seabed mining, establishing rules for exploration, licensing, and environmental management. These frameworks vary widely in their stringency, scope, and enforcement, creating a patchwork of regulatory environments that operators must navigate. For instance, Japan has prioritized the development of technologies for extracting cobalt-rich ferromanganese crusts within its EEZ, while Papua New Guinea has faced criticism over the environmental and social impacts of its Solwara 1 project. Understanding the nuances of national regulations is essential for assessing project viability and managing risks associated with regulatory compliance.

Regional organizations and multilateral agreements further complicate the regulatory landscape, particularly in areas where transboundary impacts or shared resources are at stake. For example, the Pacific Islands Forum, a regional organization of Pacific island nations, has called for a moratorium on deep sea mining until its environmental and economic impacts are better understood. Similarly, agreements such as the Convention on Biological Diversity (CBD) influence mining activities by establishing global targets for biodiversity conservation and sustainable use. These regional and international initiatives reflect growing concerns about the cumulative impacts of deep sea mining and the need for coordinated governance to address shared challenges.

The evolving nature of global regulations presents both challenges and opportunities for deep sea mining projects. On one hand, regulatory uncertainty can hinder investment and project development, as operators grapple with changing rules, unclear timelines, and potential delays. For instance, the ISA has yet to finalize its exploitation regulations, creating uncertainty about the specific requirements and standards that will apply to commercial mining operations. This regulatory ambiguity complicates feasibility studies, financial modeling, and risk assessments, requiring operators and investors to adopt flexible strategies that account for a range of potential outcomes.

On the other hand, regulatory frameworks that prioritize sustainability and transparency can enhance the credibility and long-term viability of deep sea mining. By demonstrating compliance with rigorous environmental standards and equitable benefit-sharing principles, operators can build trust with stakeholders, secure social license to operate, and access premium markets for responsibly sourced materials. For example, downstream industries such as electric vehicle manufacturers and technology companies increasingly demand traceable and ethically sourced minerals to meet consumer expectations and regulatory requirements. Aligning with global sustainability standards can provide a competitive advantage, opening new opportunities for collaboration and market access.

Artificial intelligence (AI) and advanced analytics play a transformative role in navigating global regulatory influences on deep sea mining. AI-driven tools enable operators and investors to model regulatory scenarios, evaluate compliance costs, and assess the financial implications of different regulatory outcomes. For instance, a machine learning model could analyze historical regulatory trends, stakeholder positions, and market dynamics to predict the likelihood of specific policy changes, such as the adoption of stricter environmental standards or the introduction of new benefit-sharing mechanisms.

These insights inform strategic planning, ensuring that projects are resilient to regulatory shifts and aligned with long-term objectives.

AI also enhances the efficiency and accuracy of compliance monitoring and reporting, a critical requirement under global regulatory frameworks. Advanced sensor networks, autonomous monitoring platforms, and blockchain-based traceability systems enable real-time data collection and analysis, ensuring that operators meet environmental and operational standards. For example, AI algorithms can analyze data from environmental monitoring systems to detect anomalies, assess compliance with sediment plume thresholds, and generate reports for regulatory authorities. By automating these processes, AI reduces administrative burdens, improves transparency, and strengthens stakeholder confidence in the industry's commitment to responsible practices.

Collaboration and stakeholder engagement are essential for navigating the complexities of global regulations and securing social license to operate. Deep sea mining projects involve a diverse array of stakeholders, including governments, international organizations, local communities, environmental groups, and downstream industries. Building trust and alignment among these stakeholders requires transparent communication, inclusive decision-making, and a demonstrated commitment to sustainability. For instance, collaborative efforts to develop regional environmental management plans or shared benefit-sharing mechanisms can enhance governance, reduce conflicts, and promote equitable outcomes.

Global regulatory influences are a defining factor in the viability of deep sea mining projects, shaping their economic, environmental, and social dimensions. The complex and evolving nature of these regulations requires a proactive and adaptive approach to compliance, risk management, and strategic planning. By leveraging AI, fostering collaboration, and aligning with sustainability principles, operators and investors can navigate the challenges of the regulatory landscape while maximizing opportunities for long-term success. Its ability to operate within a robust and transparent regulatory framework will be essential for ensuring that the benefits of resource extraction are realized responsibly and equitably, safeguarding the health and resilience of marine ecosystems for future generations.

7. Expanded Oil and Gas Production in the Deep Sea

7.1. The Evolution of Deep Sea Oil and Gas Exploration

The transition from conventional onshore and shallow-water oil and gas production to the expansive and challenging frontier of deepwater resources represents one of the most transformative shifts in the energy sector's history. This evolution has been driven by a combination of technological advancements, economic imperatives, and the relentless global demand for hydrocarbons. While onshore and shallow-water fields were once sufficient to meet the world's energy needs, declining reserves, geopolitical pressures, and the need for energy diversification have pushed exploration and production further offshore and into deeper waters.

The early days of oil and gas production were dominated by onshore fields, which offered accessible and relatively straightforward opportunities for resource extraction. Regions such as the United States, Russia, and the Middle East became synonymous with oil booms, their abundant reserves fueling industrial growth and geopolitical power. By the mid-20th century, shallow-water offshore production emerged as a natural progression, with fields located on continental shelves providing significant new reserves. Offshore platforms in places like the Gulf of Mexico, the North Sea, and Southeast Asia marked the beginning of marine-based oil and gas exploration, benefiting from proximity to established markets and comparatively shallow depths that allowed for relatively straightforward engineering solutions.

However, by the latter half of the 20th century, the limitations of conventional resources became increasingly apparent. Many of the world's largest onshore and shallow-water fields were reaching maturity, with production rates declining and reserves becoming more difficult and expensive to exploit. At the same time, global energy demand continued to grow, driven by rapid industrialization in developing nations and the expansion of energy-intensive industries. These dynamics created a pressing need to explore new frontiers, leading the energy sector to look beyond the continental shelves and into the vast, untapped potential of deepwater basins.

Deepwater resources, located at depths exceeding 500 meters and often stretching to ultra-deepwaters beyond 2,000 meters, represent an immense and largely untapped reservoir of hydrocarbons. Regions such as Brazil's pre-salt basins, the Gulf of Mexico, and West Africa's offshore fields emerged as key targets for exploration, offering the promise of vast reserves capable of sustaining global energy needs for decades. However, accessing these resources required a fundamental rethinking of exploration and production techniques. The harsh conditions of the deep ocean, intense pressures, frigid temperatures, and the absence of natural light, posed challenges that far exceeded those encountered in conventional operations.

The transition to deepwater exploration was made possible by a series of technological breakthroughs that redefined the limits of what was achievable in offshore engineering. One of the most significant advancements was the development of floating production systems, including floating production storage and offloading units (FPSOs) and semi-submersible platforms. Unlike fixed platforms used in shallow waters, these floating systems could operate in deep and ultra-deepwaters, where the seafloor was too distant and unstable for traditional foundations. Equipped with dynamic positioning systems and advanced mooring technologies, floating production systems provided the flexibility and stability needed to extract resources from some of the world's most challenging environments.

Subsea technologies also played a pivotal role in enabling deepwater exploration. The deployment of remotely operated vehicles (ROVs) and autonomous underwater vehicles (AUVs) allowed operators to survey and map the seafloor, inspect infrastructure, and perform maintenance tasks at depths that would be inaccessible to human divers. Advanced drilling technologies, such as riserless drilling and extended-reach wells, enabled operators to tap into reservoirs located far below the seabed with precision and efficiency. These innovations were complemented by improvements in seismic imaging and data analysis, which provided detailed insights into the geology of deepwater basins, reducing the risks and uncertainties associated with exploration.

The transition to deepwater resources also required significant advancements in safety and environmental management, as the stakes of operating in such extreme environments were higher than ever before. High-profile incidents, such as the Deepwater Horizon disaster in 2010, underscored the catastrophic consequences of equipment failures or mismanagement in deepwater operations. In response, the industry implemented stricter safety protocols, developed advanced blowout preventers, and invested in real-time monitoring systems that could detect and address anomalies before they escalated into full-scale crises. These measures not only enhanced operational safety but also reinforced public confidence in the industry's ability to manage the risks associated with deepwater exploration.

The economic implications of the transition to deepwater resources have been profound, reshaping the global energy landscape and the strategic priorities of oil and gas companies. Deepwater projects, while capital-intensive and technologically demanding, offer the potential for substantial returns due to the sheer scale of the reserves they target. For example, Brazil's pre-salt fields, discovered beneath thick layers of salt and rock, are estimated to hold billions of barrels of recoverable oil, making them one of the most significant discoveries of the 21st century. Similarly, the Gulf of Mexico and West Africa have emerged as key hubs of deepwater production, attracting investment from major oil companies and driving economic growth in surrounding regions.

Geopolitically, the shift to deepwater resources has also had far-reaching consequences. Countries with access to deepwater basins have gained new opportunities to strengthen their energy independence, reduce reliance on imports, and assert greater influence in global energy markets. At the same time, the high costs and technical complexities of deepwater projects have encouraged international collaboration, with operators, technology providers, and governments working together to share knowledge, resources, and risks. These partnerships have facilitated the development of global standards and best practices, ensuring that deepwater exploration is conducted responsibly and sustainably.

Looking ahead, the transition to deepwater resources is expected to continue evolving, driven by advancements in technology, shifts in energy demand, and the growing imperative to balance resource extraction with environmental stewardship. Artificial intelligence (AI) and machine learning are poised to play an increasingly central role in deepwater operations, enabling operators to optimize drilling techniques, predict equipment failures, and monitor environmental impacts with unprecedented accuracy. Meanwhile, innovations in carbon capture and storage (CCS) and renewable energy integration are providing new pathways for deepwater projects to align with global efforts to mitigate climate change.

The transition from conventional to deepwater resources represents a defining chapter in the evolution of the oil and gas industry. It has been marked by extraordinary technological achievements, significant economic impacts, and the ability to overcome seemingly insurmountable challenges in pursuit of the world's energy needs. As the industry continues to push the boundaries of what is possible in deepwater exploration, it must also address the ethical, environmental, and geopolitical dimensions of its activities, ensuring that this transition contributes to a more sustainable and equitable energy future.

The discovery and extraction of oil and gas reserves beneath the ocean floor have always been at the forefront of technological innovation. Over the past century, as conventional onshore and shallow-water resources have declined, the focus has shifted to deeper and more challenging offshore reserves. The role of advanced technologies in unlocking these resources cannot be overstated. From the development of sophisticated drilling techniques to the deployment of autonomous systems and data-driven analytics, technological advancements have redefined the possibilities of deep sea oil and gas exploration, enabling operators to overcome physical, logistical, and environmental challenges while maximizing efficiency and safety.

One of the most transformative technological breakthroughs in deep sea exploration has been the development of floating production systems. These systems, which include floating production storage and offloading units (FPSOs), semi-submersible platforms, and tension-leg platforms, have enabled the industry to access hydrocarbon reserves in waters too deep or unstable for fixed structures. Unlike traditional platforms anchored to the seabed, floating systems are equipped with dynamic positioning systems and advanced mooring technologies that allow them to remain stable in the face of strong currents, high winds, and rough seas. This flexibility is essential for operating in ultra-deepwater environments, where water depths often exceed 2,000 meters and seafloor conditions are highly variable.

Floating production systems are complemented by advancements in subsea technologies, which have transformed the way operators interact with the seafloor. Remotely operated vehicles (ROVs) and autonomous underwater vehicles (AUVs) are at the heart of this revolution, providing the tools needed to map, inspect, and maintain subsea infrastructure. ROVs, controlled from surface vessels, are equipped with high-resolution cameras, sonar systems, and robotic arms that allow them to perform tasks such as pipeline inspection, valve adjustment, and debris removal. AUVs, operating independently, use advanced sensors and AI algorithms to conduct detailed surveys of the seafloor, collecting data on geological formations, sediment composition, and environmental conditions.

The ability to accurately map and model the subsurface is critical for identifying and evaluating offshore reserves. Seismic imaging has long been a cornerstone of oil and gas exploration, and recent advancements in this field have significantly enhanced its capabilities. Three-dimensional (3D) and four-dimensional (4D) seismic imaging techniques provide detailed, dynamic representations of subsurface structures, allowing geoscientists to identify hydrocarbon reservoirs with greater precision. These techniques rely on the use of sophisticated seismic sources and receivers, which generate and capture acoustic waves that travel through the Earth's layers. The data collected from these surveys is processed using high-performance computing systems, which apply complex algorithms to create visualizations of the subsurface.

The integration of artificial intelligence (AI) and machine learning has further revolutionized seismic imaging and data interpretation. AI algorithms analyze seismic datasets to identify patterns and anomalies that may indicate the presence of hydrocarbons. These systems can process vast amounts of data quickly and accurately, reducing the time and cost associated with exploration while minimizing the risk of drilling dry wells. For example, machine learning models can analyze seismic waveforms to distinguish between oil-bearing and water-bearing formations, providing geoscientists with actionable insights that guide exploration decisions.

Drilling technology has also advanced significantly, enabling operators to reach and extract resources from increasingly challenging environments. Extended-reach drilling (ERD) and horizontal drilling are two techniques that have been instrumental in unlocking offshore reserves. ERD allows operators to drill wells that extend horizontally or at an angle from the vertical, reaching reservoirs located far from the surface entry point. This capability is particularly valuable in deepwater settings, where reservoirs may be located several kilometers away from the drilling rig. Horizontal drilling, which involves drilling wells parallel to the reservoir layer, maximizes contact with the hydrocarbon-bearing formation, enhancing recovery rates and reducing the need for multiple wells.

The development of riserless drilling systems has further expanded the industry's ability to operate in ultra-deepwater environments. These systems eliminate the need for traditional marine risers, which connect the drilling rig to the wellhead on the seafloor. Instead, riserless drilling uses specialized equipment to circulate drilling fluids and manage pressure, reducing the complexity and cost of deepwater operations. This technology is particularly advantageous for exploratory drilling, where the goal is to collect data on the subsurface without committing to full-scale production infrastructure.

Subsea production systems represent another critical advancement in deep sea oil and gas exploration. These systems, which include subsea wellheads, manifolds, and pipelines, allow operators to extract and transport hydrocarbons from the seafloor to surface facilities or directly to onshore processing plants. By eliminating the need for large surface platforms, subsea systems reduce the environmental footprint of offshore operations and improve the safety and reliability of production. Innovations such as subsea compression and boosting technologies have further enhanced the efficiency of these systems, enabling operators to maintain flow rates and recover resources from reservoirs with declining pressure.

Advanced monitoring and control systems are integral to the success of offshore operations, providing real-time data on equipment performance, reservoir conditions, and environmental impacts. These systems leverage a combination of sensors, communication networks, and data analytics to ensure that operations are conducted safely and efficiently. For example, downhole sensors installed in wells monitor parameters such as pressure, temperature, and flow rates, providing operators with the information needed to optimize production and detect potential issues before they escalate. Similarly, environmental monitoring systems track factors such as water quality, sediment disturbance, and marine life activity, ensuring compliance with regulatory standards and minimizing ecological impacts.

The role of AI in operational optimization extends beyond exploration and production to include predictive maintenance and risk management. Predictive maintenance systems use AI algorithms to analyze data from equipment sensors, identifying patterns that indicate wear, corrosion, or other forms of degradation. By predicting when and where equipment is likely to fail, these systems enable operators to schedule repairs proactively, reducing downtime and avoiding costly disruptions. For example, an AI-driven maintenance platform might analyze data from subsea pumps and valves, identifying early signs of fatigue or malfunction and generating alerts for maintenance teams.

Risk management is another area where advanced technologies have had a transformative impact. The integration of AI and data analytics into risk assessment frameworks allows operators to model and simulate a wide range of scenarios, from equipment failures to environmental incidents. These models provide valuable insights into the likelihood and consequences of different risks, enabling operators to develop targeted mitigation strategies and enhance their overall resilience. For instance, a risk assessment platform might simulate the impact of a subsea pipeline leak, providing operators with the information needed to implement containment measures and minimize environmental damage.

Despite these advancements, the deployment of advanced technologies in deep sea oil and gas exploration is not without challenges. The harsh conditions of the deep ocean, including high pressures, low temperatures, and corrosive seawater, place significant demands on equipment and systems, requiring materials and designs that can withstand extreme environments. Additionally, the integration of new technologies into existing workflows and infrastructures often requires significant investment and organizational change. Addressing these challenges requires a commitment to research and development, as well as collaboration among industry stakeholders, technology providers, and regulatory bodies.

Advanced technologies have played a pivotal role in unlocking offshore oil and gas reserves, transforming the possibilities of deep sea exploration and production. From floating production systems and subsea technologies to AI-driven analytics and predictive maintenance, these innovations have enabled operators to overcome the physical, logistical, and environmental challenges of operating in one of the world's most demanding environments. The integration of cutting-edge technologies will remain essential for maximizing efficiency, enhancing safety, and minimizing ecological impacts.

The deep sea represents one of the final frontiers of resource extraction, holding vast reserves of hydrocarbons that are critical to meeting the world's energy demands. As the energy industry transitions toward greater efficiency, sustainability, and safety, artificial intelligence (AI) is emerging as a transformative force in deep sea oil and gas production. AI's ability to process massive datasets, predict outcomes, and optimize complex operations makes it uniquely suited to address the challenges of exploring and extracting resources from the deep ocean. By enhancing precision, reducing risks, and driving innovation, AI is redefining what is possible in the sector and positioning itself as the next frontier in deep sea oil and gas production.

One of the most significant contributions of AI to deep sea oil and gas production is its ability to revolutionize exploration. Identifying hydrocarbon reserves beneath the ocean floor requires the analysis of complex geological and geophysical data, a task that has historically relied on manual interpretation and computationally intensive methods. AI algorithms, particularly those based on machine learning, have the capacity to process vast datasets from seismic surveys, sonar imaging, and core sampling, identifying patterns and anomalies with speed and accuracy that surpasses human capabilities. For example, deep learning models can analyze seismic waveforms to distinguish between oil-bearing and non-productive formations, providing operators with actionable insights that minimize the risk of drilling dry wells.

AI-driven predictive analytics further enhance exploration by reducing uncertainties and improving decision-making. Advanced models integrate data from multiple sources, such as satellite imagery, historical production records, and environmental monitoring systems, to forecast the likelihood of resource presence in specific areas. These predictions enable operators to prioritize exploration targets, allocate resources efficiently, and reduce the time and cost associated with finding new reserves. For instance, an AI system might analyze the tectonic history and sedimentary characteristics of a deepwater basin to predict the location and size of hydrocarbon traps, guiding exploratory drilling efforts with unprecedented precision.

The role of AI extends beyond exploration to encompass the entire lifecycle of deep sea oil and gas production. In drilling operations, AI systems optimize well placement, trajectory, and performance, ensuring that resources are extracted safely and efficiently. Directional drilling, a technique that involves steering the drill bit to follow a predefined path, benefits significantly from AI algorithms that process real-time data on pressure, temperature, and rock properties. These algorithms adjust drilling parameters dynamically, mitigating risks such as blowouts, equipment damage, and environmental contamination.

AI also enhances production efficiency by optimizing the performance of subsea infrastructure. Subsea production systems, including wellheads, manifolds, and pipelines, are equipped with sensors that monitor key parameters such as flow rates, pressure, and temperature. AI systems analyze this data to identify inefficiencies, predict equipment failures, and recommend adjustments that maximize productivity. For example, a machine learning model might detect subtle changes in flow patterns that indicate the onset of hydrate formation or wax buildup, prompting operators to implement corrective measures before these issues escalate.

The integration of AI into maintenance and risk management processes represents another critical advancement in deep sea oil and gas production. Predictive maintenance systems use AI to analyze data from equipment sensors, identifying patterns that indicate wear, corrosion, or other forms of degradation. By predicting when and where failures are likely to occur, these systems enable operators to schedule repairs proactively, reducing downtime and avoiding costly disruptions. For instance, an AI-driven maintenance platform might analyze vibration and acoustic data from subsea pumps to identify early signs of mechanical stress, generating alerts for maintenance teams to address the issue before it compromises production.

Risk management is a key area where AI's capabilities shine, particularly in the high-stakes environment of deep sea operations. AI systems model and simulate a wide range of scenarios, from equipment failures to environmental incidents, providing operators with a probabilistic understanding of risks and their potential consequences. These insights inform the development of targeted mitigation strategies and enhance the overall resilience of operations. For example, an AI-powered risk assessment platform might simulate the impact of a subsea pipeline rupture, helping operators identify the most effective containment measures and response protocols.

AI's ability to monitor and mitigate environmental impacts is particularly valuable in an era of heightened regulatory scrutiny and public awareness. Environmental monitoring systems equipped with AI analyze data from sensors, cameras, and sampling devices to track factors such as water quality, sediment disturbance, and marine life activity. These systems provide real-time insights into the environmental footprint of operations, ensuring compliance with regulatory standards and enabling operators to address potential impacts proactively. For instance, an AI system might detect changes in turbidity levels near a drilling site, indicating the spread of a sediment plume, and recommend adjustments to minimize disturbance to nearby ecosystems.

The role of AI in deep sea oil and gas production extends to enhancing safety and operational resilience. Autonomous systems powered by AI, such as remotely operated vehicles (ROVs) and autonomous underwater vehicles (AUVs), perform critical tasks in hazardous environments, reducing the need for human intervention and minimizing risks to personnel. These systems are equipped with advanced sensors, cameras, and AI-driven navigation algorithms that enable them to inspect subsea infrastructure, identify anomalies, and conduct repairs with precision. For example, an ROV equipped with AI might use computer vision to detect corrosion or damage on a pipeline, generating a detailed report for operators to evaluate and address.

AI's capacity for real-time decision-making is particularly valuable in managing complex and dynamic systems. In deep sea operations, where conditions can change rapidly and unpredictably, AI systems provide operators with the information and recommendations needed to respond effectively. For example, an AI-driven control system might monitor data from multiple sensors to detect early signs of a pressure imbalance in a well, initiating corrective actions such as adjusting flow rates or activating blowout preventers. These systems not only enhance safety but also improve operational efficiency by reducing downtime and optimizing resource recovery.

The economic implications of AI's integration into deep sea oil and gas production are profound, offering significant cost savings and competitive advantages. By optimizing exploration, reducing downtime, and enhancing operational efficiency, AI enables operators to maximize the value of their investments and improve profitability. Additionally, AI-driven insights support the development of more accurate financial models and forecasts, providing stakeholders with the information needed to make informed investment decisions and manage risks effectively.

Despite its transformative potential, the adoption of AI in deep sea oil and gas production is not without challenges. The success of AI systems depends on the quality and comprehensiveness of the data they process, yet deep sea environments often present gaps and inconsistencies in data availability. Addressing this challenge requires investment in data collection technologies, such as high-resolution sensors, satellite systems, and autonomous monitoring platforms. Furthermore, integrating AI into existing workflows and infrastructures demands organizational change, training, and collaboration across departments.

Ethical considerations are also central to the deployment of AI in deep sea oil and gas production. As these systems play an increasingly central role in decision-making, ensuring their transparency, fairness, and accountability is critical for maintaining trust among stakeholders. For example, AI algorithms used for site selection or risk assessment must be free from biases that could lead to inequitable or unsustainable outcomes. Governance frameworks that promote transparency and stakeholder engagement are essential for addressing these ethical concerns and ensuring that AI systems operate responsibly.

Looking ahead, the role of AI in deep sea oil and gas production is expected to expand as technology continues to advance and the industry adapts to evolving demands. The integration of AI with emerging technologies, such as quantum computing and digital twins, will unlock new possibilities for modeling complex systems, optimizing operations, and managing risks. Additionally, the adoption of AI-driven sustainability strategies will enable operators to align their activities with global efforts to mitigate climate change and protect marine ecosystems.

AI represents the next frontier in deep sea oil and gas production, offering transformative capabilities for exploration, production, maintenance, and environmental management. By enabling operators to overcome the challenges of operating in extreme environments, AI is reshaping the industry and paving the way for a more efficient, resilient, and sustainable future. As the integration of AI accelerates, its role in deep sea operations will continue to grow, driving innovation, enhancing safety, and contributing to the responsible utilization of the ocean's resources.

8. Lowering the World's Sea Level Through AI Deep-Sea Operations

8.1. The Concept of Artificial Sea Level Modulation

Throughout history, humanity has sought to reshape the natural world to meet its needs, from irrigating arid lands to building dams that generate power and control water flow. Sea level management, while less explored, represents an extension of this ambition, an effort to address one of the most pressing challenges posed by global climate change: rising sea levels. The concept of artificially modulating sea levels is both historical and speculative, rooted in a combination of ancient practices, modern engineering achievements, and visionary ideas that seek to redefine humanity's relationship with the ocean. This chapter explores these approaches, examining their feasibility, ethical implications, and potential integration with advanced technologies such as artificial intelligence (AI).

The earliest human interactions with sea level management can be traced to ancient civilizations that sought to protect coastal settlements and farmland from tidal surges and flooding. For example, the Mesopotamians, Egyptians, and Chinese constructed dikes, levees, and canals to manage water levels and mitigate the impact of seasonal flooding. These rudimentary systems, while localized in scope, reflected an understanding of hydrodynamics and a desire to exert control over natural forces. In the Netherlands, where large portions of the land lie below sea level, the Dutch have a centuries-long history of building polders, tracts of reclaimed land protected by dikes and drainage systems. This practice exemplifies humanity's early efforts to adapt to the challenges posed by rising and fluctuating sea levels.

In the modern era, the scope of sea level management has expanded dramatically, driven by advances in engineering, hydrology, and environmental science. Large-scale infrastructure projects such as the Thames Barrier in London and the Delta Works in the Netherlands are iconic examples of modern sea level management. These projects are designed to protect cities and regions from storm surges and rising seas, incorporating sophisticated mechanisms to control water flow and mitigate flood risks. The Thames Barrier, for instance, consists of movable gates that can be raised during high tides or storm surges to prevent the River Thames from flooding central London. Similarly, the Delta Works employs a series of dams, sluices, locks, and storm surge barriers to protect the Dutch coastline from the North Sea.

While these projects are monumental in scale and effectiveness, they are fundamentally reactive in nature, addressing the symptoms of rising sea levels rather than their underlying causes. This limitation has prompted researchers, engineers, and visionaries to explore speculative approaches to sea level management that are more proactive and global in scope. These approaches, often referred to as "geoengineering" or "ocean engineering," seek to modulate sea levels by directly manipulating the volume or distribution of ocean water.

One speculative approach involves the large-scale redistribution of ocean water to achieve localized or global reductions in sea level. Proponents of this idea envision the use of massive pumping systems to transfer water from coastal regions to inland basins or underground reservoirs, effectively reducing the volume of water in the oceans. For example, excess water could be stored in depleted aquifers, salt flats, or man-made depressions, creating a buffer against rising seas. While the technical feasibility of such projects remains uncertain, advances in AI and robotics could play a critical role in designing and implementing the necessary infrastructure. AI algorithms could optimize the placement and operation of pumping systems, ensuring that water is transported efficiently and with minimal environmental impact.

Another speculative approach involves the excavation of deep ocean basins or trenches to increase the capacity of the oceans and reduce sea levels. This concept, often referred to as "ocean deepening," would require the removal of vast quantities of sediment and rock from the seafloor, creating additional space for water storage. While the scale of such a project is daunting, the integration of AI-controlled excavation technologies and autonomous underwater vehicles (AUVs) could make it more feasible. AI systems could analyze geological and geophysical data to identify suitable locations for excavation, simulate the impacts of seafloor modification, and coordinate the activities of robotic excavation units.

A more ambitious speculative approach involves the manipulation of the Earth's cryosphere, its frozen water systems, to influence sea levels. Melting glaciers and ice sheets are major contributors to rising seas, and efforts to stabilize or reverse this process could have significant benefits. For instance, researchers have proposed building barriers or artificial islands to prevent warm ocean currents from reaching glaciers, reducing their rate of melting. AI could be instrumental in designing these structures, modeling their interactions with ocean currents, and monitoring their effectiveness over time.

While speculative approaches to sea level management offer intriguing possibilities, they also raise significant ethical and environmental questions. The large-scale manipulation of ocean systems carries inherent risks, including unintended ecological consequences, geopolitical tensions, and disruptions to natural cycles. For example, transferring ocean water to inland basins could alter regional climates, disrupt ecosystems, and affect groundwater supplies. Similarly, excavating ocean basins could release trapped methane, destabilize marine sediments, or disrupt deep-sea habitats. These risks underscore the need for rigorous impact assessments, transparent governance, and broad stakeholder engagement in the development of any sea level modulation project.

The speculative nature of these approaches does not diminish their importance; rather, it highlights the need for continued exploration and innovation. As climate change accelerates and its impacts become more pronounced, the search for bold and transformative solutions will become increasingly urgent. Artificial intelligence, with its unparalleled capacity for data analysis, modeling, and optimization, offers a powerful tool for advancing these efforts. By integrating AI into sea level management projects, researchers and engineers can test hypotheses, refine designs, and predict outcomes with greater precision and confidence.

For example, AI-driven simulations could model the global impacts of transferring ocean water to inland basins, accounting for variables such as evaporation rates, sediment deposition, and regional climate changes. Similarly, machine learning algorithms could analyze historical data on glacier dynamics to identify the most effective strategies for stabilizing ice sheets and reducing their contribution to sea level rise. These insights would not only inform the design of speculative projects but also guide their implementation and monitoring, ensuring that they are conducted responsibly and with a commitment to sustainability.

The historical and speculative approaches to sea level management explored in this chapter demonstrate the enduring human ambition to shape and control the natural world. From ancient dikes and canals to modern barriers and geoengineering concepts, these efforts reflect a recognition of the ocean's power and the necessity of adapting to its challenges. As we look to the future, the integration of advanced technologies such as AI and robotics offers new possibilities for addressing the complex and multifaceted problem of rising sea levels. By building on the lessons of the past and embracing the potential of innovation, humanity can envision a future where sea level management is not only a response to crisis but a proactive and sustainable strategy for global resilience. Through collaboration, foresight, and ethical stewardship, the dream of artificial sea level modulation can become a reality that safeguards both people and the planet.

The concept of redistributing oceanic water masses to modulate global or regional sea levels is both an ambitious and scientifically complex endeavor. It seeks to address one of the most visible and pressing consequences of climate change: rising sea levels that threaten coastal cities, island nations, and low-lying regions worldwide. The idea involves transferring vast quantities of seawater to specific locations, whether inland basins, artificial reservoirs, or deeper parts of the ocean, thereby reducing the water volume in critical coastal areas. While this approach remains speculative, the underlying science offers intriguing possibilities for mitigating the impacts of sea level rise, provided it is guided by rigorous research, advanced technologies, and robust governance frameworks.

At its core, the science of redistributing oceanic water masses draws upon principles of hydrodynamics, geophysics, and climatology. Water is a fluid that responds dynamically to gravitational forces, thermal expansion, and the rotation of the Earth, making its redistribution a highly sensitive process. Small changes in the distribution of water masses can have significant effects on global sea levels, ocean currents, and regional climates. Understanding these dynamics is critical for designing interventions that achieve desired outcomes without causing unintended consequences.

One proposed method for redistributing oceanic water masses involves pumping seawater from coastal areas to inland basins or aquifers. This approach would require the construction of large-scale pumping systems capable of moving water over vast distances and elevations. For example, seawater could be transferred to arid regions with natural basins or man-made reservoirs, such as salt flats or former lakebeds. By reducing the volume of water in specific coastal areas, this strategy could lower regional sea levels and alleviate the risk of flooding in vulnerable zones.

The feasibility of this method depends on several scientific and engineering factors. The energy requirements for pumping seawater across significant distances and elevations are substantial, necessitating the use of renewable energy sources, such as solar or wind power, to minimize greenhouse gas emissions. The design of the pumping and storage infrastructure must also account for the chemical and physical properties of seawater, including its salinity, temperature, and potential for corrosion. Advanced materials and coatings may be needed to prevent damage to pipelines, reservoirs, and associated equipment.

Artificial intelligence (AI) offers transformative potential for optimizing the redistribution of oceanic water masses. AI algorithms can analyze vast datasets on topography, hydrology, and climate to identify optimal locations for water storage, taking into account factors such as capacity, evaporation rates, and ecological impacts. For example, machine learning models could simulate how transferring seawater to a desert basin might affect local temperatures, humidity, and vegetation, providing valuable insights for assessing the feasibility and sustainability of the project. AI-driven systems can also monitor and manage the operation of pumping infrastructure in real time, ensuring efficiency and responsiveness to changing conditions.

Another approach to redistributing oceanic water masses involves the excavation and deepening of ocean basins or trenches. This concept, often referred to as "ocean deepening," seeks to create additional capacity for seawater storage by removing sediment and rock from the seafloor. By increasing the depth and volume of specific ocean regions, this strategy could lower global sea levels while maintaining the overall distribution of water masses. Potential sites for ocean deepening include tectonically stable regions with minimal ecological sensitivity, such as certain abyssal plains or inactive submarine plateaus.

The science of ocean deepening presents unique challenges and opportunities. Excavating the seafloor on a large scale requires advanced robotics, autonomous systems, and material-handling technologies capable of operating in extreme pressures and low temperatures. Soft robotics and bio-inspired designs could play a critical role in minimizing ecological disturbance during excavation, while AI-driven mapping and modeling systems could optimize the selection and design of deepening sites. For instance, an AI system might analyze geophysical and ecological data to identify areas with minimal biodiversity and stable geological conditions, ensuring that excavation activities avoid sensitive habitats and geohazards.

Redistributing oceanic water masses through deepening or pumping raises important questions about the long-term stability and resilience of these interventions. Water is a dynamic and interconnected resource, and changes in its distribution can influence global ocean currents, thermal gradients, and atmospheric patterns. For example, transferring large volumes of seawater to inland basins could alter regional precipitation patterns or disrupt groundwater systems, with implications for agriculture, ecosystems, and human settlements. Similarly, modifying the seafloor could affect the flow of deep ocean currents, which play a critical role in regulating global climate by transporting heat and nutrients between regions.

To address these complexities, the science of redistributing oceanic water masses must be grounded in comprehensive modeling and simulation. Advanced computational tools, including digital twins and high-performance computing platforms, enable researchers to simulate the interactions between water masses, ocean currents, and climate systems under different scenarios. These models can incorporate variables such as evaporation rates, sediment transport, and feedback loops, providing a holistic understanding of the potential impacts of water redistribution. For example, a digital twin of a proposed water storage basin could simulate how its presence would affect local and regional hydrology, informing decisions about its design and management.

Environmental and ecological considerations are central to the science of redistributing oceanic water masses. Large-scale interventions in the ocean system carry inherent risks of unintended consequences, including habitat loss, biodiversity shifts, and changes in ecosystem dynamics. For instance, transferring seawater to inland basins could increase salinity levels in soils and aquifers, rendering them unsuitable for agriculture or human consumption. Similarly, excavating the seafloor could release trapped methane, destabilize sediments, or disrupt deep-sea habitats, with potential cascading effects on marine ecosystems and the global carbon cycle.

To mitigate these risks, any effort to redistribute oceanic water masses must be guided by rigorous environmental impact assessments (EIAs) and adaptive management frameworks. These assessments should evaluate the direct and indirect effects of proposed interventions, considering both immediate and long-term consequences for ecosystems, biodiversity, and human communities. Adaptive management frameworks, supported by AI-driven monitoring systems, enable real-time tracking of environmental indicators and allow for dynamic adjustments to project operations based on observed impacts.

The governance of ocean water redistribution presents additional challenges, requiring collaboration among governments, international organizations, and local stakeholders. Water redistribution projects often cross national and regional boundaries, necessitating agreements on resource sharing, impact mitigation, and cost distribution. For example, a project that transfers seawater to a transboundary basin would need to address the rights and interests of all affected nations, as well as the communities and ecosystems within the basin's watershed. Transparent governance structures, underpinned by international treaties and guidelines, are essential for ensuring that these projects are conducted equitably and with accountability.

The potential of redistributing oceanic water masses to address sea level rise is both ambitious and uncertain, reflecting the interplay between scientific innovation, technological capability, and societal values. While the feasibility of large-scale water redistribution remains unproven, the underlying science offers valuable insights for exploring its possibilities and limitations. Advances in AI, robotics, and environmental modeling provide powerful tools for assessing the risks and benefits of these interventions, ensuring that they are grounded in evidence and aligned with principles of sustainability.

The vision of artificially modulating sea levels to combat the consequences of rising seas has long been considered an ambitious and speculative concept. However, recent advancements in artificial intelligence (AI) have transformed this vision from a theoretical exercise into a plausible and potentially actionable strategy. AI's capacity to analyze vast datasets, optimize complex systems, and adapt dynamically to changing conditions positions it as a critical enabler for large-scale, precise interventions aimed at modulating sea levels. By integrating AI into the design, implementation, and monitoring of these interventions, researchers and engineers can overcome technical, logistical, and environmental challenges, ensuring that such efforts are both effective and sustainable.

One of the most significant contributions of AI to sea level modulation lies in its ability to process and interpret massive datasets. The oceans are highly dynamic systems influenced by a wide range of variables, including temperature, salinity, currents, and atmospheric conditions. These variables interact in complex and nonlinear ways, making it difficult to predict the outcomes of large-scale interventions using traditional analytical methods. AI algorithms, particularly those based on machine learning, excel at identifying patterns and relationships within these datasets, enabling researchers to develop accurate models of ocean behavior.

For example, an AI system might analyze historical data on ocean circulation, sediment transport, and sea surface temperatures to predict how redistributing water masses would affect global and regional sea levels. By incorporating real-time data from satellite sensors, autonomous underwater vehicles (AUVs), and ocean buoys, the system can continuously update its predictions, ensuring that interventions are informed by the latest information. These capabilities allow for the design of targeted and adaptive strategies that minimize risks and maximize benefits.

AI also plays a critical role in optimizing the logistics and operations of sea level modulation projects. Large-scale interventions, such as transferring seawater to inland basins or excavating ocean basins, involve complex systems of pumps, pipelines, and robotic excavators that must operate in coordination to achieve desired outcomes. AI-powered optimization algorithms can model and simulate these systems, identifying the most efficient configurations and schedules for their operation. For instance, an AI system might determine the optimal placement of pumping stations and the most energy-efficient routes for transporting seawater, taking into account factors such as topography, energy availability, and environmental constraints.

The dynamic nature of the ocean environment requires that interventions be highly adaptable to changing conditions. AI's capacity for real-time decision-making and autonomous control ensures that large-scale projects can respond dynamically to unexpected developments. For example, if an AI system detects a sudden change in ocean currents or weather patterns, it can adjust the operation of pumps and pipelines to prevent disruptions or inefficiencies. Similarly, AI-driven excavation systems equipped with advanced sensors and machine vision can identify and avoid ecologically sensitive areas, ensuring that seafloor modifications are conducted responsibly.

One of the most transformative applications of AI in sea level modulation is its ability to integrate and coordinate multiple interventions across different regions and scales. Sea level rise is a global challenge that requires coordinated efforts to address its diverse impacts on coastal cities, island nations, and vulnerable ecosystems. AI systems can act as central decision-making hubs, analyzing data from interconnected projects and optimizing their interactions to achieve overarching objectives. For instance, an AI platform might manage a network of water redistribution projects in different parts of the world, ensuring that their combined effects contribute to stabilizing global sea levels while minimizing regional disparities.

Environmental monitoring is another critical area where AI enables precise interventions. Large-scale sea level modulation projects have the potential to impact marine ecosystems, regional climates, and hydrological cycles, necessitating continuous monitoring and adaptive management. AI-powered monitoring systems leverage data from a wide array of sources, including satellite imagery, underwater sensors, and eDNA analysis, to track the environmental impacts of interventions in real time. For example, an AI system might analyze changes in sediment composition, water quality, and marine biodiversity to assess the effects of transferring seawater to inland basins or deepening ocean trenches. These insights inform dynamic adjustments to project operations, ensuring that environmental risks are mitigated and ecosystem health is preserved.

The integration of AI into sea level modulation also enhances its alignment with sustainability and climate adaptation goals. By optimizing the use of resources and minimizing energy consumption, AI systems reduce the environmental footprint of interventions. For example, an AI-driven pumping system might prioritize the use of renewable energy sources, such as solar or wind power, and adjust its operation to align with periods of peak energy availability. Similarly, AI algorithms can identify opportunities for synergy between sea level modulation projects and other climate adaptation initiatives, such as enhancing coastal defenses or restoring blue carbon ecosystems.

AI's ability to simulate and predict long-term outcomes is particularly valuable for assessing the feasibility and sustainability of sea level modulation strategies. Advanced modeling platforms powered by AI enable researchers to explore a wide range of scenarios, accounting for variables such as future climate trends, population growth, and economic development. These models provide a probabilistic understanding of the risks and benefits associated with different interventions, allowing stakeholders to make informed decisions about their design and implementation. For instance, an AI system might simulate the long-term impacts of transferring seawater to inland basins on regional hydrology, agriculture, and ecosystems, providing valuable insights for evaluating the trade-offs of such an approach.

Ethical and social considerations are central to the deployment of AI in sea level modulation. Large-scale interventions in the ocean system carry inherent risks of unintended consequences, including disruptions to ecosystems, regional inequalities, and geopolitical tensions. Ensuring that AI systems operate transparently and accountably is critical for addressing these challenges and maintaining public trust. Governance frameworks must include mechanisms for oversight, stakeholder engagement, and ethical review, ensuring that AI-driven interventions align with principles of equity, sustainability, and intergenerational responsibility.

For example, an AI system used to manage a global network of water redistribution projects should provide stakeholders with clear explanations of its decision-making processes, including the criteria and data used to prioritize actions. Public-facing dashboards and interactive tools can enhance transparency, allowing stakeholders to visualize the impacts of interventions and provide feedback. Additionally, the development and deployment of AI systems must be guided by international agreements and standards, ensuring that their use respects the rights and interests of affected communities and ecosystems.

The potential of AI to enable large-scale, precise interventions in sea level modulation is matched by the challenges associated with its implementation. The development of AI systems requires access to high-quality data, robust computational infrastructure, and interdisciplinary expertise. Collaborations between governments, industry, and academia are essential for building the capabilities needed to design, deploy, and monitor AI-driven interventions. Investment in research and development, as well as in training programs for AI specialists and environmental scientists, is critical for advancing the field and ensuring its alignment with global sustainability goals.

Looking to the future, the role of AI in sea level modulation will continue to evolve as technologies and methodologies advance. Emerging innovations, such as quantum computing and edge AI, offer new possibilities for enhancing the scale, speed, and accuracy of AI-driven interventions. For example, quantum algorithms could enable more complex simulations of ocean systems, while edge AI systems deployed on autonomous platforms could process data locally, reducing latency and improving responsiveness.

8.2. Technologies for Large-Scale Deep-Sea Operations

The concept of redistributing deep-sea water to manage rising sea levels and adapt to climate change represents a bold and complex challenge that demands advanced technological solutions. Among the critical components of such an undertaking are AI-controlled pumping systems, which leverage the power of artificial intelligence to optimize the transfer of vast quantities of water across oceans, basins, or inland reservoirs. These systems hold transformative potential, enabling precise, efficient, and sustainable water redistribution while minimizing environmental impacts and operational costs. By integrating cutting-edge AI technologies, these pumping systems are poised to become the linchpins of large-scale interventions aimed at mitigating the effects of rising seas.

At the heart of AI-controlled pumping systems is the ability to process and respond to vast datasets in real time. The redistribution of deep-sea water involves a multitude of dynamic variables, including ocean currents, water pressure, sediment loads, and regional climatic conditions. These factors interact in complex and nonlinear ways, making it impossible to design effective pumping strategies using static or pre-programmed methods. AI systems, particularly those based on machine learning and deep learning, excel at analyzing such data, identifying patterns, and optimizing operations in dynamic and uncertain environments.

For instance, an AI-controlled pumping system might analyze data from satellite imagery, underwater sensors, and real-time weather forecasts to determine the optimal timing, location, and intensity of water transfer operations. By continuously updating its models based on incoming data, the system can adapt to changing conditions, such as shifts in ocean currents or unexpected weather events, ensuring that pumping operations remain efficient and effective. This adaptability is particularly valuable in the context of large-scale interventions, where even small inefficiencies or disruptions can have significant financial and environmental implications.

The design of AI-controlled pumping systems also incorporates advanced robotics and automation technologies that enable precise and autonomous operation. These systems are equipped with sensors, actuators, and control algorithms that allow them to monitor and adjust their performance in real time.

For example, a network of AI-controlled pumps deployed across a coastal region might use machine vision and sonar to detect sediment accumulation in pipelines, triggering automated cleaning processes to maintain optimal flow rates. Similarly, the pumps could adjust their energy consumption and flow rates based on the availability of renewable energy sources, such as solar or wind power, minimizing their environmental footprint and operational costs.

Energy efficiency is a critical consideration for AI-controlled pumping systems, as the energy demands of moving large volumes of water over long distances and elevations can be substantial. AI algorithms play a central role in optimizing energy use, leveraging predictive analytics and real-time monitoring to identify opportunities for reducing energy consumption. For example, an AI system might schedule pumping operations to coincide with periods of low electricity demand, taking advantage of lower energy costs and reducing strain on the grid. Additionally, the system could use AI-driven simulations to evaluate the trade-offs between different pumping configurations, identifying the most energy-efficient options for specific scenarios.

AI-controlled pumping systems also enable precise management of water quality, ensuring that the redistribution process does not introduce harmful substances or disrupt ecosystems. By analyzing data from chemical sensors and environmental monitoring systems, AI algorithms can detect changes in water composition, such as variations in salinity, temperature, or nutrient levels. These insights allow operators to implement measures that prevent contamination or mitigate impacts on marine and terrestrial ecosystems. For instance, the system might adjust the flow rate or direction of water transfer to avoid introducing high-salinity water into freshwater reservoirs or ecologically sensitive areas.

One of the most innovative aspects of AI-controlled pumping systems is their potential for integration with broader climate adaptation and resource management strategies. These systems can be connected to global monitoring networks, providing real-time data on sea level changes, water availability, and environmental impacts. By collaborating with other technologies, such as AI-driven sediment plume monitoring or renewable energy grids, pumping systems can become part of an interconnected ecosystem that addresses multiple challenges simultaneously. For example, an AI-controlled pumping system might work in tandem with a floating solar farm, using excess energy generated by the farm to power water transfer operations, while simultaneously contributing to regional energy supply and emissions reduction goals.

The scalability of AI-controlled pumping systems is another key advantage, enabling their deployment in a wide range of contexts, from localized flood mitigation projects to global sea level modulation initiatives. Modular designs and autonomous control capabilities allow these systems to be customized and adapted to meet specific needs, whether transferring water to inland reservoirs, deepening ocean basins, or balancing regional sea levels. For instance, a system designed for a densely populated coastal city might prioritize rapid response and high-capacity operations, while a system deployed in a remote or ecologically sensitive area might focus on minimizing environmental impacts and energy use.

Despite their potential, the development and deployment of AI-controlled pumping systems face significant challenges that must be addressed to ensure their success. One of the primary challenges is the need for robust and reliable infrastructure that can withstand the harsh conditions of the deep ocean, including high pressures, corrosive seawater, and extreme weather events. Advances in material science, robotics, and engineering are essential for creating pumps, pipelines, and control systems that can operate effectively in these environments over extended periods.

Another challenge is the integration of AI systems with existing infrastructure and governance frameworks. Effective coordination among stakeholders, including governments, industry leaders, and environmental organizations, is critical for ensuring that pumping systems align with broader sustainability and climate adaptation goals. For example, a project that transfers seawater to an inland reservoir must consider the rights and interests of local communities, as well as the potential impacts on regional ecosystems and water resources. Transparent governance structures, supported by international agreements and ethical guidelines, are essential for addressing these challenges and maintaining public trust.

The role of AI in enabling large-scale, precise interventions through pumping systems also raises important ethical considerations. As these systems become more autonomous and influential, ensuring their transparency, accountability, and alignment with societal values is critical. For example, the algorithms used to optimize pumping operations must be free from biases that could lead to inequitable outcomes, such as disproportionately affecting certain regions or communities. Governance frameworks should include mechanisms for oversight, stakeholder engagement, and ethical review, ensuring that AI-controlled pumping systems operate responsibly and with a commitment to sustainability and equity.

Looking to the future, the potential of AI-controlled pumping systems to address the challenges of sea level rise and water redistribution is immense. Emerging technologies, such as quantum computing and edge AI, offer new possibilities for enhancing the scale, speed, and accuracy of these systems. For instance, quantum algorithms could enable more complex simulations of water flow and distribution, while edge AI systems deployed on autonomous platforms could process data locally, reducing latency and improving responsiveness. These advancements will further strengthen the capabilities of AI-controlled pumping systems, enabling them to meet the demands of increasingly ambitious and interconnected projects.

The concept of deepening ocean basins and expanding trenches to manage sea levels represents a transformative and highly speculative approach to addressing one of the most pressing consequences of climate change: rising seas. This ambitious vision entails the design and deployment of advanced infrastructure capable of modifying the seafloor to increase its capacity for storing water, thereby reducing global or regional sea levels. While such interventions remain in the realm of theoretical and experimental exploration, the rapid advancement of technology, particularly in robotics, artificial intelligence (AI), and material science, has brought this concept closer to feasibility. The design and implementation of infrastructure for ocean basin deepening and trench expansion demand unprecedented engineering ingenuity, environmental foresight, and ethical governance.

At the heart of this approach is the need to excavate vast quantities of sediment and rock from the ocean floor, creating additional storage capacity for seawater. This process requires the development of specialized excavation technologies that can operate autonomously and efficiently in the extreme conditions of the deep ocean, including high pressures, low temperatures, and corrosive environments. Autonomous underwater vehicles (AUVs) and remotely operated vehicles (ROVs) equipped with advanced excavation tools are central to this effort. These systems, controlled by AI algorithms, can navigate complex seafloor terrains, identify optimal excavation sites, and perform precision digging with minimal disruption to surrounding ecosystems.

One of the primary challenges in designing infrastructure for ocean basin deepening is the need to handle and transport the vast amounts of sediment and rock generated during excavation. The logistics of removing and relocating this material are formidable, requiring the development of modular and scalable systems that can operate continuously and autonomously. For example, pipelines and conveyor systems might be deployed to transport excavated material to designated deposition sites, such as inactive basins or geological formations capable of supporting the additional load. AI-driven optimization algorithms play a critical role in managing these systems, ensuring that material is transported efficiently and that deposition sites are selected based on geophysical stability, environmental sensitivity, and proximity to excavation areas.

The materials used in excavation tools and infrastructure must be carefully selected to withstand the harsh conditions of the deep ocean. Corrosion-resistant alloys, advanced ceramics, and composite materials are essential for ensuring the durability and longevity of these systems. Additionally, the tools themselves must be designed to operate with high precision and adaptability, minimizing ecological disturbance while maximizing efficiency. For instance, soft robotic excavators inspired by marine organisms might be used to handle delicate sediments or navigate narrow trenches, reducing the risk of habitat destruction and sediment plumes.

The integration of AI into the design and operation of excavation infrastructure is a transformative innovation, enabling real-time decision-making, adaptive control, and predictive modeling. AI algorithms analyze data from a wide array of sources, including sonar imaging, geophysical surveys, and environmental monitoring systems, to identify the most suitable locations for excavation and predict the impacts of seafloor modification. For example, an AI system might simulate the effects of deepening a specific basin on regional currents, sediment transport, and marine biodiversity, providing valuable insights for planning and mitigation. These simulations ensure that interventions are informed by robust scientific evidence and aligned with sustainability goals.

Environmental monitoring and impact assessment are integral components of infrastructure design for ocean basin deepening. Large-scale modifications to the seafloor carry significant risks, including habitat loss, changes in sediment dynamics, and potential disruptions to deep-sea ecosystems. To address these risks, infrastructure must incorporate advanced environmental monitoring systems that track key indicators such as water quality, sediment composition, and species diversity. These systems, powered by AI and autonomous sensors, provide real-time data that informs adaptive management strategies, allowing operators to adjust excavation activities in response to observed impacts.

For example, an environmental monitoring system might detect an increase in turbidity or a decline in benthic biodiversity near an excavation site, triggering adjustments to the operation of excavation tools or the selection of deposition sites. By enabling dynamic and responsive management, these systems ensure that infrastructure for ocean basin deepening operates within acceptable environmental thresholds and minimizes harm to marine ecosystems.

The scale and complexity of ocean basin deepening require a high degree of coordination and integration across multiple components of infrastructure. Modular design principles are essential for achieving this coordination, allowing systems to be customized and scaled to meet specific project requirements. For instance, a modular excavation platform might include interchangeable tools for different sediment types, as well as adaptable transport systems that can be reconfigured based on the location and scale of excavation activities. This flexibility ensures that infrastructure can respond to the diverse and dynamic conditions of the deep ocean, enhancing both efficiency and resilience.

Energy efficiency is another critical consideration in the design of infrastructure for ocean basin deepening. The energy demands of excavation, transport, and monitoring systems are substantial, particularly in remote and deepwater locations where access to traditional power sources is limited. Renewable energy solutions, such as underwater turbines, floating solar arrays, and wave energy converters, offer promising alternatives for powering these systems sustainably. AI algorithms can optimize the integration and operation of these energy sources, ensuring that infrastructure operates efficiently and with minimal carbon footprint.

The governance and regulation of ocean basin deepening are as important as the technical and environmental considerations. Large-scale interventions in the ocean system carry inherent risks of unintended consequences, including geopolitical tensions, economic disparities, and ethical dilemmas. Ensuring that infrastructure for ocean basin deepening is designed and operated responsibly requires transparent governance frameworks, international cooperation, and robust ethical oversight. For example, agreements under the United Nations Convention on the Law of the Sea (UNCLOS) could provide a basis for regulating the activities of excavation infrastructure, ensuring that they align with global sustainability goals and respect the rights and interests of affected communities.

Stakeholder engagement is also critical for the success of ocean basin deepening projects, providing a platform for dialogue, collaboration, and consensus-building. Coastal communities, environmental organizations, and scientific researchers have valuable perspectives and expertise that can inform the design and operation of excavation infrastructure. For instance, local knowledge of marine ecosystems and sediment dynamics can enhance the accuracy and relevance of environmental impact assessments, while stakeholder input can guide the selection of deposition sites and mitigation measures.

Despite the challenges, the potential benefits of infrastructure for ocean basin deepening are immense, offering a transformative solution to the global challenge of rising sea levels. By increasing the capacity of the oceans to store water, these interventions could alleviate flooding risks for coastal cities, protect vulnerable ecosystems, and provide a buffer against the impacts of climate change. Additionally, the technologies and methodologies developed for ocean basin deepening could have broader applications, from enhancing deep-sea resource management to advancing marine research and exploration.

The excavation and deposition of sediments in the deep ocean are critical components of large-scale interventions such as basin deepening, trench expansion, and water redistribution. These activities, while essential for addressing challenges like rising sea levels and resource extraction, pose significant environmental, logistical, and operational challenges. Managing sediment excavation and deposition at such a scale demands precision, adaptability, and sustainability, qualities that are increasingly enabled by artificial intelligence (AI). AI's ability to process vast datasets, optimize operations, and make real-time decisions positions it as a transformative tool for managing these complex processes with minimal ecological and logistical impact.

At the heart of sediment excavation is the need to understand and work within the physical and ecological constraints of the seafloor. Sediments in the deep ocean vary widely in composition, density, and stability, ranging from fine silts and clays to compacted sands and rocky substrates. The excavation of these materials requires tools and techniques that are tailored to their specific properties, as well as an understanding of how their removal will affect the surrounding environment. AI systems, particularly those powered by machine learning and deep learning, excel at analyzing geophysical data to characterize sediment properties and predict their behavior during excavation.

For example, an AI-driven system might process data from sonar imaging, core samples, and in situ sensors to create detailed maps of sediment composition and stability across a target area. These maps enable operators to identify optimal excavation sites, minimizing the risk of slope failure, sediment plumes, or other disturbances. By continuously updating these models with real-time data from autonomous underwater vehicles (AUVs) and remotely operated vehicles (ROVs), AI systems ensure that excavation activities remain adaptive and responsive to changing conditions.

AI also plays a central role in optimizing the operation of excavation equipment. Large-scale sediment excavation involves the use of advanced robotic systems, such as dredges, suction excavators, and cutter heads, which must be precisely controlled to achieve efficient and environmentally responsible results. AI algorithms analyze data from equipment sensors, such as pressure gauges, flow meters, and vibration monitors, to optimize parameters like cutting speed, suction power, and tool angle. These adjustments ensure that excavation tools operate at peak efficiency while minimizing energy consumption and wear.

One of the most significant challenges in sediment excavation is managing the generation and dispersion of sediment plumes. These plumes, created when fine particles are suspended in the water column during excavation, can spread over large areas, affecting water quality, light penetration, and marine life. AI systems provide powerful tools for predicting, monitoring, and mitigating the impacts of sediment plumes. For instance, machine learning models can analyze historical and real-time data on ocean currents, sediment properties, and excavation activities to predict the trajectory and dispersion of plumes. These predictions inform the design of mitigation measures, such as adjusting excavation techniques, deploying sediment curtains, or scheduling activities to coincide with favorable current conditions.

The deposition of excavated sediments is an equally complex and critical process, requiring careful planning and management to avoid ecological harm and ensure long-term stability. AI systems enable precise control over deposition activities by analyzing data on seafloor topography, sediment composition, and environmental sensitivity. For example, an AI-driven deposition system might use bathymetric maps and ecological data to identify suitable sites for sediment placement, ensuring that deposition does not disrupt sensitive habitats or alter natural hydrodynamic patterns.

The integration of AI into sediment deposition systems also enhances their efficiency and adaptability. Autonomous platforms equipped with AI algorithms can transport and distribute sediments with precision, using sensors and machine vision to monitor deposition in real time. For instance, an autonomous barge or subsea vehicle might adjust its trajectory and release mechanisms based on feedback from sonar and environmental sensors, ensuring that sediments are placed evenly and within designated areas. These systems can also detect and respond to anomalies, such as unexpected shifts in seafloor conditions or deviations from deposition plans, reducing the risk of unintended impacts.

AI's role in managing sediment excavation and deposition extends beyond operational optimization to include environmental monitoring and impact assessment. Large-scale sediment management activities have the potential to affect marine ecosystems in complex and interconnected ways, from altering benthic habitats to influencing nutrient cycles and carbon storage. AI-powered monitoring systems provide real-time data on key environmental indicators, such as turbidity, oxygen levels, and species diversity, enabling operators to track the impacts of their activities and implement adaptive management strategies.

For example, an AI system might analyze data from underwater cameras, hydrophones, and eDNA samplers to assess the presence and behavior of marine life near excavation and deposition sites. If the system detects a decline in biodiversity or an increase in stress indicators, it can recommend adjustments to operational parameters or trigger mitigation measures, such as relocating deposition activities or reducing excavation intensity. These insights ensure that sediment management practices are guided by a commitment to minimizing ecological harm and maintaining the health of marine ecosystems.

The scalability and interconnectedness of AI systems offer additional benefits for managing sediment excavation and deposition across multiple sites or regions. By integrating data from a network of monitoring stations, autonomous platforms, and remote sensing technologies, AI systems can provide a comprehensive and real-time view of sediment dynamics on a large scale. This capability is particularly valuable for projects that involve multiple excavation and deposition sites, where the interactions between activities can create cumulative impacts that are difficult to predict or manage using traditional methods.

For instance, an AI platform might coordinate the activities of multiple dredging units and deposition barges operating in different parts of a basin, ensuring that their combined effects do not exceed environmental thresholds or disrupt hydrodynamic patterns. By optimizing the timing, location, and intensity of these activities, the system can achieve project objectives while maintaining ecological balance and operational efficiency.

Despite its transformative potential, the integration of AI into sediment excavation and deposition systems faces several challenges. The success of these systems depends on the availability and quality of data, which can be limited by gaps in monitoring coverage, inconsistencies in data collection methods, or the inherent difficulties of operating in remote and extreme environments. Addressing these challenges requires investment in advanced sensors, autonomous monitoring platforms, and data-sharing frameworks that enable the collection and integration of high-resolution, high-frequency data from diverse sources.

Ethical and governance considerations are also central to the deployment of AI in sediment management. The large-scale modification of seafloor environments carries inherent risks of unintended consequences, including habitat loss, biodiversity shifts, and changes in ecosystem dynamics. Ensuring that AI systems operate transparently and accountably is critical for addressing these risks and maintaining public trust. Governance frameworks must include mechanisms for oversight, stakeholder engagement, and ethical review, ensuring that AI-driven sediment management practices align with principles of sustainability, equity, and intergenerational responsibility.

Looking to the future, the role of AI in managing sediment excavation and deposition is expected to expand as technologies and methodologies continue to advance. Emerging innovations, such as quantum computing and edge AI, offer new possibilities for enhancing the scale, speed, and accuracy of sediment management systems. For example, quantum algorithms could enable more complex simulations of sediment dynamics and environmental impacts, while edge AI systems deployed on autonomous platforms could process data locally, reducing latency and improving responsiveness.

8.3. Environmental and Geophysical Implications

The vast and largely uncharted expanse of the deep ocean harbors some of the most fragile and biodiverse ecosystems on Earth. These ecosystems, ranging from hydrothermal vent communities to abyssal plains and seamounts, are not only critical to marine biodiversity but also play essential roles in global processes such as carbon sequestration, nutrient cycling, and climate regulation. Large-scale deep-sea operations, including excavation, water redistribution, and infrastructure development, pose significant risks to these ecosystems and the species that inhabit them. Understanding and mitigating these potential impacts is a central challenge for ensuring that human interventions in the deep ocean are conducted responsibly and sustainably.

Marine ecosystems in the deep ocean are characterized by unique adaptations to extreme conditions, including high pressures, low temperatures, and complete darkness. Organisms in these environments have evolved over millions of years, often in isolation, resulting in high levels of endemism. Hydrothermal vent communities, for example, support complex food webs based on chemosynthesis, where microorganisms convert chemical compounds into energy in the absence of sunlight. These communities host a variety of specialized species, such as giant tube worms, vent crabs, and heat-tolerant shrimp, many of which are found nowhere else on the planet. Similarly, abyssal plains, despite their seemingly barren appearance, are home to a diverse array of life forms, including echinoderms, bivalves, and meiofauna, that thrive in the nutrient-rich sediments.

The potential impacts of deep-sea operations on these ecosystems are multifaceted and interrelated. One of the most direct threats is habitat destruction, which occurs when seafloor habitats are physically altered or removed during excavation, dredging, or construction activities. For example, the excavation of sediment for water redistribution or basin deepening can result in the loss of habitat for benthic organisms, many of which are sessile or slow-moving and unable to escape disturbance. The destruction of seafloor habitats disrupts the structure of marine communities, leading to declines in population density, species richness, and ecosystem functionality.

Another significant impact of deep-sea operations is the generation of sediment plumes, which occur when fine particles are suspended in the water column during excavation or dredging. These plumes can spread over large areas, depending on ocean currents and sediment properties, affecting water quality and light penetration. Sediment plumes have the potential to smother benthic organisms, clogging their feeding and respiratory structures and altering their behavior. Filter-feeding species, such as sponges and bivalves, are particularly vulnerable to sedimentation, as their feeding efficiency is directly affected by suspended particles. Additionally, sediment plumes can reduce light availability in the water column, affecting photosynthetic organisms such as phytoplankton and impacting the broader food web.

The release of contaminants trapped in seafloor sediments is another concern associated with deep-sea operations. Sediments in certain regions may contain heavy metals, hydrocarbons, or other pollutants that have accumulated over time. Disturbance of these sediments can release these substances into the water column, where they may be taken up by marine organisms and bioaccumulate through the food web. This poses risks not only to the health of individual species but also to the integrity of entire ecosystems, as toxins can affect reproduction, growth, and survival rates.

Noise pollution is an often-overlooked but significant impact of deep-sea operations. The use of heavy machinery, dredges, and excavation tools generates underwater noise that can disrupt the behavior and communication of marine species. Many deep-sea organisms rely on sound for navigation, predator avoidance, and mating, and noise pollution can interfere with these vital functions. For example, the low-frequency noise generated by excavation equipment may mask the communication signals of certain fish and marine mammals, leading to disorientation or behavioral changes.

Chemical changes in the surrounding environment are another potential impact of deep-sea operations. Activities such as sediment excavation and water redistribution can alter the chemical composition of seawater, including changes in oxygen levels, pH, and nutrient concentrations. For example, the redistribution of deep-sea water, which is typically low in oxygen and high in dissolved carbon dioxide, can lead to localized deoxygenation or acidification in affected areas. These changes can have cascading effects on marine ecosystems, particularly for species that are already living near their physiological limits.

The cumulative impacts of these stressors raise concerns about the long-term resilience of deep-sea ecosystems and their ability to recover from disturbances. Many deep-sea organisms have slow growth rates, late maturity, and low reproductive output, making them particularly vulnerable to habitat loss and population declines. For instance, deep-sea corals, which provide important habitat for a variety of species, can take decades or even centuries to recover from physical damage. Similarly, the microbial communities that drive nutrient cycling and carbon sequestration in sediments may be disrupted by changes in sediment composition or hydrodynamic conditions, affecting ecosystem services on a broader scale.

Addressing the potential impacts of deep-sea operations on marine ecosystems and biodiversity requires a comprehensive and multidisciplinary approach. Environmental impact assessments (EIAs) are a critical tool for evaluating the risks and trade-offs associated with specific activities, providing a foundation for decision-making and mitigation. EIAs must be rigorous, science-based, and inclusive, incorporating data on species distributions, habitat characteristics, and ecological interactions. For example, baseline studies conducted prior to excavation or water redistribution can provide valuable information on the composition and function of local ecosystems, enabling operators to identify sensitive areas and design interventions that minimize harm.

Adaptive management is another key strategy for mitigating impacts and ensuring that deep-sea operations align with sustainability goals. Adaptive management involves continuous monitoring and adjustment of activities based on observed outcomes, allowing operators to respond dynamically to environmental changes and unforeseen impacts. AI-powered monitoring systems play a central role in this approach, providing real-time data on environmental indicators such as turbidity, oxygen levels, and species behavior. For instance, an AI system might detect a decline in biodiversity near an excavation site and recommend adjustments to operational parameters, such as reducing the intensity of sediment removal or relocating activities to less sensitive areas.

The integration of mitigation measures into the design and operation of deep-sea technologies is essential for minimizing ecological impacts. Technologies such as soft robotics, bio-inspired excavation tools, and sediment containment systems can reduce the physical and chemical disturbances associated with deep-sea operations. For example, robotic excavators equipped with adaptive sensors and machine learning algorithms can target specific sediments with precision, avoiding unnecessary disruption to surrounding habitats. Similarly, sediment curtains or deflectors can be deployed to contain plumes and prevent their spread to ecologically sensitive areas.

Collaboration among scientists, policymakers, industry stakeholders, and conservation organizations is critical for advancing sustainable practices and ensuring that deep-sea operations are conducted responsibly. International agreements and guidelines, such as those under the United Nations Convention on the Law of the Sea (UNCLOS) and the International Seabed Authority (ISA), provide a framework for regulating activities in the deep ocean and protecting marine biodiversity. However, these frameworks must be continually updated and strengthened to address emerging challenges and incorporate the latest scientific knowledge.

Large-scale deep-sea operations, such as excavation, trench expansion, and ocean basin deepening, carry significant geophysical risks that must be carefully considered and managed. Among the most critical concerns are the potential for inducing or exacerbating seismic activity and the broader tectonic consequences of modifying the seafloor. These risks arise from the inherent interconnectedness of Earth's geophysical systems, where even localized changes to the seafloor can propagate through tectonic plates, sediment layers, and fault zones. Understanding these dynamics is essential for ensuring that deep-sea interventions are conducted responsibly, minimizing the risk of unintended consequences while aligning with global sustainability goals.

Seismic activity in the deep ocean is often driven by the movement of tectonic plates, which are in constant motion due to forces generated within Earth's mantle. This activity is most pronounced along plate boundaries, where plates converge, diverge, or slide past one another. These boundaries are home to many of the world's most geologically active regions, including subduction zones, mid-ocean ridges, and transform faults. Large-scale excavation or sediment removal in such areas has the potential to destabilize tectonic systems, either by altering stress distributions or by directly impacting fault zones and lithospheric stability.

One of the primary geophysical risks associated with deep-sea operations is the potential to induce seismic events by redistributing stresses within the Earth's crust. Excavation activities, particularly those that involve the removal of significant volumes of sediment or rock, can alter the weight and pressure exerted on underlying tectonic plates. This redistribution of loads can influence the balance of forces acting on fault lines, potentially triggering slip events that manifest as earthquakes. For example, the removal of dense sediments from a trench or basin could reduce the compressive forces along a fault zone, increasing the likelihood of a sudden rupture.

These induced seismic events are not limited to the immediate vicinity of the excavation site. Stress changes in the Earth's crust can propagate over large distances, affecting tectonic systems in adjacent regions. For instance, changes in stress along a subduction zone might influence the behavior of nearby faults, potentially increasing seismic activity in surrounding areas. This interconnectedness underscores the need for a holistic approach to assessing geophysical risks, one that considers not only localized impacts but also the broader implications for regional and global tectonic systems.

Another significant concern is the potential for deep-sea operations to influence submarine landslides, which are a common trigger for tsunamis. Submarine landslides occur when the stability of seafloor sediments is compromised, often due to natural processes such as earthquakes, sediment deposition, or gas hydrate dissociation. Large-scale excavation or sediment redistribution can exacerbate these conditions by destabilizing slopes or altering sediment composition. For example, the removal of sediments from a slope may reduce its shear strength, increasing the likelihood of a landslide. Similarly, deposition activities that add weight to an unstable slope could act as a trigger for collapse.

The consequences of submarine landslides can be catastrophic, particularly when they generate tsunamis that affect coastal regions. The 1998 Papua New Guinea tsunami, which resulted in significant loss of life and property, was caused by a submarine landslide triggered by an earthquake. Large-scale deep-sea operations, if not managed carefully, could contribute to similar events, highlighting the importance of rigorous geotechnical assessments and monitoring.

Tectonic consequences of deep-sea operations extend beyond seismic activity and landslides to include potential impacts on the long-term behavior of plate boundaries and mantle dynamics. For example, the removal or redistribution of large volumes of sediment could alter the heat flow and pressure conditions within the lithosphere, potentially affecting the processes that drive plate tectonics. These changes, while difficult to predict, could influence the rates of plate movement, the development of new fault systems, or the behavior of existing tectonic features.

To address these geophysical risks, the integration of advanced technologies and methodologies is critical. Artificial intelligence (AI) plays a central role in modeling and monitoring the impacts of deep-sea operations on seismic and tectonic systems. AI algorithms can analyze vast datasets from seismographs, satellite observations, and geophysical surveys to identify patterns and predict the potential consequences of specific activities. For example, machine learning models might simulate the stress changes associated with sediment removal in a trench, providing insights into the likelihood of induced seismic events and their potential magnitude.

Real-time monitoring systems equipped with AI further enhance the ability to detect and respond to geophysical risks. Networks of sensors placed on the seafloor and within excavation equipment can collect data on parameters such as ground motion, pressure changes, and fault activity. AI-driven analytics can process this data in real time, identifying anomalies that may indicate the onset of seismic activity or slope instability. For instance, a sudden increase in microseismic events near an excavation site might prompt operators to halt activities, allowing for further assessment and mitigation measures.

The role of geophysical modeling is also critical in understanding the broader tectonic implications of deep-sea operations. High-performance computing and AI-powered simulations enable researchers to explore the interactions between excavation activities, sediment dynamics, and tectonic systems. These models can account for a wide range of variables, including sediment composition, fault geometry, and heat flow, providing a comprehensive view of the potential impacts. For example, a model might simulate how the removal of sediments from a subduction zone affects the coupling between tectonic plates, offering valuable insights for risk assessment and project planning.

Mitigation measures are essential for minimizing the geophysical risks of deep-sea operations. One approach involves designing excavation and deposition activities to distribute loads evenly and avoid excessive stress concentrations. For instance, sediments removed from a trench could be deposited in a manner that offsets the weight changes caused by excavation, reducing the likelihood of triggering seismic events. Advanced robotics and AI systems enable the precise control needed to implement such strategies, ensuring that sediments are handled in a way that balances operational goals with geophysical stability.

Collaboration among scientists, engineers, and policymakers is critical for addressing the geophysical risks of deep-sea operations. Interdisciplinary research efforts can advance the understanding of tectonic systems and the potential impacts of human activities, while international agreements and guidelines provide a framework for regulating these activities responsibly. For example, the United Nations Convention on the Law of the Sea (UNCLOS) and other multilateral treaties could establish protocols for assessing and mitigating geophysical risks, ensuring that deep-sea operations align with global sustainability goals.

Despite the challenges, the exploration of geophysical risks in deep-sea operations offers valuable opportunities for advancing science and technology. The data collected from monitoring and modeling efforts contributes to a deeper understanding of tectonic processes, informing not only resource management but also broader fields such as earthquake science and climate resilience. Moreover, the development of innovative tools and methodologies for mitigating geophysical risks can enhance the safety and sustainability of a wide range of marine activities, from renewable energy projects to subsea infrastructure development.

The concept of artificially reducing sea levels through large-scale oceanic interventions, such as water redistribution or basin deepening, is both ambitious and fraught with complexity. At its core, this idea aims to alleviate the immediate threats posed by rising seas, such as flooding, erosion, and habitat loss, while mitigating their broader socioeconomic impacts. However, the interconnectedness of Earth's climate systems demands a careful consideration of how such interventions might influence global climate patterns. Balancing sea level reduction with the stability of climate systems is not only a scientific and technical challenge but also an ethical and geopolitical one, requiring rigorous analysis, advanced technology, and international cooperation.

Sea levels are not merely a function of the volume of water in the ocean; they are intricately linked to global climate patterns through mechanisms such as thermal expansion, salinity gradients, and ocean circulation. Changes in these patterns can have far-reaching consequences, influencing weather systems, heat distribution, and nutrient cycles. Interventions aimed at reducing sea levels, such as transferring water to inland basins or creating deeper ocean basins, could alter these mechanisms in ways that are difficult to predict and manage.

One of the primary considerations in balancing sea level reduction with global climate patterns is the potential impact on ocean currents. Ocean currents, driven by temperature and salinity gradients as well as the Earth's rotation, play a critical role in regulating the planet's climate by redistributing heat and energy. For example, the Atlantic Meridional Overturning Circulation (AMOC), which includes the Gulf Stream, transports warm water from the tropics to the North Atlantic, influencing temperatures and precipitation patterns across Europe and North America. Similarly, the thermohaline circulation, often referred to as the "global conveyor belt," connects surface and deep waters across all major ocean basins, facilitating the exchange of heat, carbon, and nutrients.

Artificially redistributing ocean water through large-scale pumping or excavation could disrupt these currents by altering the distribution of temperature and salinity. For instance, transferring large volumes of warm, saline water from coastal regions to inland basins might reduce the density contrasts that drive thermohaline circulation, weakening its flow and potentially destabilizing regional climates. Such disruptions could lead to cooler temperatures in northern Europe, more intense hurricanes in the Atlantic, or shifts in monsoon patterns across Asia and Africa. These cascading effects underscore the need for detailed modeling and scenario analysis to anticipate and mitigate potential impacts on ocean circulation.

Thermal expansion, a key driver of sea level rise, presents another dimension of complexity in balancing sea level reduction with climate patterns. As seawater absorbs heat from the atmosphere, it expands, contributing to rising sea levels. Artificial interventions that alter the depth or volume of ocean basins could influence the rate and distribution of thermal expansion. For example, deepening ocean basins might increase their capacity to store heat, potentially moderating thermal expansion in the short term. However, this increased heat storage could also exacerbate ocean warming over time, affecting marine ecosystems and accelerating the melting of ice sheets and glaciers.

The role of salinity in ocean dynamics further complicates efforts to balance sea level reduction with climate stability. Salinity influences the density of seawater, which in turn affects buoyancy and circulation. Large-scale water redistribution projects, such as transferring seawater to inland basins, could alter regional salinity levels, with implications for both local and global systems. For instance, introducing high-salinity water into a freshwater-dominated estuary could disrupt local ecosystems, while changes in salinity gradients in the open ocean could affect the strength and direction of currents. AI-driven monitoring and predictive models are essential for understanding these interactions and designing interventions that minimize unintended consequences.

Another critical factor to consider is the potential impact of sea level reduction on the carbon cycle. The ocean is the largest active reservoir of carbon on Earth, storing approximately 50 times more carbon than the atmosphere. This storage is facilitated by biological and physical processes, including the uptake of carbon dioxide (CO_2) from the atmosphere, the sequestration of organic carbon in sediments, and the circulation of carbon-rich deep waters. Interventions that modify ocean basins or redistribute water masses could influence these processes, affecting the ocean's capacity to act as a carbon sink.

For example, deepening ocean basins might increase the burial of organic carbon in sediments, enhancing long-term carbon sequestration. However, it could also disrupt the upwelling of nutrient-rich waters that support marine productivity and the biological carbon pump. Similarly, changes in ocean circulation could alter the distribution of CO_2 between the surface and deep ocean, influencing atmospheric carbon levels and feedback loops in the climate system. Understanding these dynamics is critical for ensuring that sea level reduction strategies align with global efforts to mitigate climate change.

The integration of advanced technologies, particularly artificial intelligence (AI), is vital for balancing sea level reduction with global climate patterns. AI-driven models and simulations provide powerful tools for exploring the interactions between sea level interventions and climate systems, enabling researchers to test hypotheses, evaluate risks, and optimize strategies. For example, machine learning algorithms can analyze data from satellite observations, ocean buoys, and climate models to identify patterns and predict outcomes. These predictions inform the design of interventions, ensuring that they are both effective and compatible with climate stability.

AI's ability to process vast amounts of data in real time also supports the monitoring and management of sea level reduction projects. Autonomous sensor networks equipped with AI algorithms can track key environmental indicators, such as ocean temperature, salinity, and current velocities, providing continuous feedback on the impacts of interventions. For instance, an AI system might detect deviations in thermohaline circulation following the implementation of a water redistribution project, triggering adjustments to operational parameters or mitigation measures. These insights ensure that interventions are guided by evidence and adapt dynamically to changing conditions.

Collaboration and governance are essential for addressing the challenges of balancing sea level reduction with global climate patterns. The interconnected nature of the ocean and atmosphere means that interventions in one region can have far-reaching effects, necessitating international cooperation and coordination. Governance frameworks must include mechanisms for assessing the global implications of sea level reduction projects, ensuring that they are aligned with the broader goals of climate adaptation and mitigation. For example, agreements under the United Nations Framework Convention on Climate Change (UNFCCC) or the United Nations Convention on the Law of the Sea (UNCLOS) could provide a basis for regulating and monitoring these activities.

Ethical considerations are central to the governance of sea level reduction, particularly in terms of equity and accountability. Vulnerable communities, such as those in low-lying island nations and coastal regions, are among the most affected by rising seas and have the most to gain from effective interventions. However, they also face the greatest risks if these interventions disrupt climate systems or exacerbate existing inequalities. Ensuring that the benefits and burdens of sea level reduction are distributed fairly requires inclusive decision-making processes that prioritize the voices and interests of these communities.

Despite the challenges, the exploration of sea level reduction as a strategy for climate adaptation offers valuable opportunities for advancing scientific understanding and technological innovation. The data and insights generated by modeling and monitoring efforts contribute to a deeper understanding of ocean and climate dynamics, informing not only sea level management but also broader efforts to address climate change. Moreover, the development of innovative tools and methodologies for balancing sea level reduction with climate stability can enhance the resilience and sustainability of a wide range of environmental and industrial activities.

8.4. Ethical and Policy Dimensions

The management of sea levels through large-scale interventions is not merely a scientific or technical endeavor; it is a profoundly geopolitical issue. The interconnected nature of the world's oceans means that any action taken to manipulate sea levels, whether through water redistribution, ocean basin deepening, or other strategies, has the potential to affect regions far beyond the site of the intervention. When these efforts are guided and executed by artificial intelligence (AI), the implications become even more complex, raising questions about sovereignty, equity, accountability, and global cooperation. The geopolitical dimensions of AI-managed sea level projects must be carefully considered, as they will play a defining role in shaping the future of climate adaptation and ocean governance.

One of the most immediate geopolitical concerns associated with AI-managed sea level projects is the potential for unequal impacts and benefits among nations and regions. Sea level rise is not a uniform phenomenon; it varies based on local factors such as land subsidence, ocean currents, and gravitational effects. Similarly, interventions to manage sea levels are likely to produce localized outcomes that may benefit some areas while adversely affecting others. For example, a project designed to lower sea levels in a densely populated coastal region could inadvertently raise water levels elsewhere, exacerbating flooding risks for vulnerable island nations or low-lying countries. These disparities have the potential to create tensions between nations, particularly if the affected parties perceive the project as serving the interests of wealthier or more powerful states at their expense.

The use of AI to manage sea level projects adds another layer of complexity to these dynamics. AI systems are capable of processing vast amounts of data and optimizing operations with unparalleled precision, but their decisions are only as equitable as the algorithms and data that underpin them. If the criteria used to prioritize interventions are biased, whether intentionally or unintentionally, AI-managed projects could reinforce existing inequalities and geopolitical divides. For example, an AI system that prioritizes interventions based on economic value might favor protecting major industrial hubs and trade routes over less economically significant but equally vulnerable regions. Such outcomes could deepen distrust among nations and undermine efforts to achieve global cooperation on climate adaptation.

Transparency and accountability are critical for addressing these concerns and ensuring that AI-managed sea level projects are perceived as legitimate and fair. The algorithms and decision-making processes used by AI systems must be transparent and accessible to all stakeholders, allowing them to evaluate how priorities are set and impacts are distributed. For instance, an AI system managing a water redistribution network should clearly communicate the criteria it uses to determine the allocation of resources, such as population density, economic activity, or ecological sensitivity. Public-facing dashboards and interactive tools can enhance transparency, enabling governments, scientists, and affected communities to monitor the system's decisions and provide feedback.

The geopolitical implications of AI-managed sea level projects also extend to issues of sovereignty and control. The oceans are governed by a complex patchwork of national jurisdictions, international agreements, and customary laws. Projects that involve large-scale manipulation of oceanic systems are likely to straddle these boundaries, raising questions about who has the authority to initiate, manage, and regulate such interventions. For example, a project that redistributes water from one region of the ocean to another might require cooperation from multiple coastal states, each with its own interests and priorities. Ensuring that these projects respect national sovereignty while addressing global challenges requires robust governance frameworks that balance local autonomy with collective responsibility.

The role of international organizations and agreements is particularly important in navigating these challenges. Institutions such as the United Nations, the International Seabed Authority (ISA), and regional ocean governance bodies have a critical role to play in fostering dialogue, establishing norms, and resolving disputes related to AI-managed sea level projects. For instance, the United Nations Convention on the Law of the Sea (UNCLOS) provides a framework for regulating activities in international waters, but its provisions may need to be updated or expanded to address the unique challenges posed by AI-driven interventions. Similarly, new agreements or protocols may be required to ensure that the benefits and risks of sea level projects are shared equitably among nations.

Geopolitical competition is another factor that must be considered in the context of AI-managed sea level projects. As climate adaptation becomes an increasingly urgent priority, the ability to implement large-scale interventions may be seen as a source of strategic advantage. Nations with advanced AI capabilities, technical expertise, and financial resources are likely to play a leading role in these projects, potentially enhancing their influence in global governance and climate negotiations. This dynamic could create tensions between technologically advanced countries and those with fewer resources, particularly if the latter feel excluded from decision-making processes or disadvantaged by the outcomes of interventions.

To address these concerns, it is essential to promote international collaboration and capacity-building. Developing countries and small island states, which are often the most vulnerable to sea level rise, should have a meaningful voice in the planning and implementation of AI-managed projects. This requires not only including them in decision-making processes but also providing access to the tools, data, and expertise needed to participate effectively. For example, partnerships between developed and developing nations could support the transfer of AI technologies, training programs for scientists and policymakers, and the creation of regional monitoring networks that enhance local capacity for climate adaptation.

The ethical dimensions of AI-managed sea level projects are also deeply intertwined with their geopolitical implications. Decisions about how, where, and when to implement these projects involve trade-offs between competing values and priorities, such as economic development, environmental protection, and social equity. These trade-offs must be navigated in a way that respects the rights and interests of all stakeholders, particularly those who are most affected by sea level rise. For instance, indigenous communities and small-scale fishers who depend on marine ecosystems for their livelihoods should have a voice in decisions that may alter these ecosystems, whether through changes in salinity, currents, or habitat structures.

Public engagement and transparency are critical for addressing these ethical and geopolitical challenges. Open dialogue with affected communities, civil society organizations, and international stakeholders can help build trust and legitimacy, ensuring that AI-managed sea level projects are guided by shared values and informed by diverse perspectives. For example, public consultations and participatory workshops could provide a platform for stakeholders to express their concerns, propose alternatives, and contribute to the development of ethical guidelines and governance frameworks. These processes not only enhance the inclusivity and accountability of sea level projects but also foster a sense of shared ownership and responsibility.

Despite the challenges, the geopolitical implications of AI-managed sea level projects also present opportunities for fostering global cooperation and innovation. The shared threat of rising seas provides a powerful incentive for nations to work together, transcending political and economic divides to address a common challenge. Collaborative efforts to develop and deploy AI technologies for sea level management could strengthen international partnerships, advance scientific knowledge, and build the foundations for a more resilient and equitable future. For instance, a global consortium of nations, research institutions, and private sector actors could pool resources and expertise to create an open-access platform for monitoring and managing sea level projects, supported by AI-driven analytics and real-time data sharing.

The manipulation of the ocean on a large scale, whether for mitigating rising sea levels, enhancing carbon sequestration, or redistributing water masses, raises profound ethical questions that extend far beyond technical feasibility or operational success. These considerations are deeply rooted in humanity's relationship with the natural world, encompassing issues of environmental justice, intergenerational responsibility, and the balance between human needs and ecological integrity. As technologies such as artificial intelligence (AI), robotics, and advanced engineering make large-scale ocean interventions increasingly plausible, it is essential to navigate the ethical dimensions of these projects with care, transparency, and a commitment to sustainability.

One of the central ethical considerations in large-scale ocean manipulation is the principle of environmental stewardship. The ocean is a shared resource that supports a vast array of ecosystems and provides essential services to humanity, including climate regulation, food supply, and cultural significance. Intervening in this complex and interconnected system carries inherent risks of unintended consequences, such as habitat destruction, biodiversity loss, and disruptions to natural processes. Ethical stewardship demands that these risks be minimized through rigorous environmental impact assessments, transparent governance, and the adoption of precautionary approaches that prioritize the protection of marine ecosystems.

For example, projects that involve the redistribution of ocean water or the deepening of basins must consider the potential impacts on marine life, including changes in salinity, temperature, and nutrient availability. These factors influence the health and behavior of marine organisms, from microscopic plankton to apex predators, and any disruption to these dynamics could have cascading effects on ecosystem stability. Ethical decision-making requires that such interventions be guided by comprehensive scientific understanding, ensuring that the benefits of the project outweigh its ecological costs.

Another key ethical consideration is the principle of equity, which addresses the distribution of benefits and burdens associated with large-scale ocean manipulation. Sea level rise and other climate-related challenges disproportionately affect vulnerable populations, including low-lying island nations, coastal communities, and marginalized groups that depend on marine resources for their livelihoods. Ethical interventions must ensure that these populations are not further disadvantaged by the actions taken to address global challenges. For instance, a project designed to lower sea levels in one region must not exacerbate flooding or ecological degradation in another, particularly if the affected areas lack the resources or political influence to advocate for their interests.

The integration of artificial intelligence (AI) into ocean manipulation projects introduces additional ethical considerations, particularly regarding transparency, accountability, and inclusivity. AI systems are often used to optimize decision-making, model outcomes, and monitor impacts, but their outputs are only as reliable as the data and algorithms that inform them. If these systems are biased, opaque, or poorly calibrated, they could produce outcomes that are inequitable or misaligned with project objectives. For example, an AI system tasked with identifying optimal sites for water redistribution might prioritize economically valuable regions over ecologically sensitive ones, reinforcing existing inequalities or undermining conservation efforts.

Ensuring the ethical use of AI in ocean manipulation requires a commitment to transparency and accountability. The criteria and algorithms used to guide AI-driven decisions must be openly shared and subject to scrutiny, enabling stakeholders to evaluate their fairness, accuracy, and alignment with ethical principles. Public-facing platforms that visualize the inputs, processes, and outputs of AI systems can enhance transparency, fostering trust among stakeholders and empowering them to participate meaningfully in decision-making processes. Additionally, mechanisms for oversight and ethical review, such as advisory boards or independent audits, are essential for holding AI systems and their operators accountable for their impacts.

The precautionary principle, which advocates for caution in the face of scientific uncertainty, is a cornerstone of ethical decision-making in large-scale ocean manipulation. The ocean is a vast and poorly understood system, and the potential for unintended consequences is significant. Ethical stewardship demands that interventions proceed only when the risks are well-characterized and mitigation measures are in place. For instance, a project that involves modifying ocean basins to store additional water must include detailed modeling of its impacts on ocean currents, sediment dynamics, and marine ecosystems, as well as contingency plans for addressing adverse outcomes.

Intergenerational responsibility is another critical dimension of ethical decision-making, reflecting the obligation to consider the long-term consequences of ocean manipulation for future generations. Many of the impacts of large-scale interventions, such as changes to ocean circulation or the release of sequestered carbon, may not manifest fully for decades or centuries. Ethical stewardship requires that these projects be designed with a long-term perspective, ensuring that they contribute to the resilience and sustainability of the planet rather than creating new challenges for future generations. For example, a project that deepens ocean basins to mitigate sea level rise must account for its potential effects on global heat distribution, carbon cycling, and marine biodiversity over the coming centuries.

Public engagement and inclusivity are essential for addressing the ethical dimensions of large-scale ocean manipulation. Decisions about how, where, and when to intervene in the ocean system affect a wide range of stakeholders, including coastal communities, indigenous peoples, conservation organizations, and industries such as fishing and shipping. Ensuring that these stakeholders have a voice in the decision-making process is critical for building trust, legitimacy, and shared ownership of the outcomes. For example, participatory workshops, public consultations, and advisory committees can provide platforms for stakeholders to express their concerns, propose alternatives, and contribute to the development of governance frameworks.

The ethical dimensions of ocean manipulation also intersect with questions of governance and sovereignty. The ocean is a global commons that transcends national boundaries, and interventions in one region can have far-reaching effects on others. Ethical decision-making requires that these projects be governed by transparent and inclusive frameworks that respect the rights and interests of all affected parties. For example, international agreements such as the United Nations Convention on the Law of the Sea (UNCLOS) and emerging frameworks for biodiversity beyond national jurisdiction (BBNJ) provide a foundation for regulating activities in the high seas, ensuring that they are conducted responsibly and equitably.

Despite the challenges, the exploration of large-scale ocean manipulation offers opportunities for advancing ethical and sustainable practices in environmental management. The data and insights generated by these projects can contribute to a deeper understanding of ocean systems, informing not only specific interventions but also broader efforts to address climate change, biodiversity loss, and resource management. Moreover, the development of innovative tools and methodologies for ocean manipulation can set new standards for ethical governance, demonstrating how technological progress can be aligned with the principles of sustainability, equity, and intergenerational responsibility.

8.5. A Vision for Future Applications

As the impacts of climate change intensify, humanity is confronted with the urgent need to transition from reactive responses to proactive adaptation strategies. Artificial intelligence (AI) offers a transformative toolset for addressing this challenge, enabling societies to anticipate, mitigate, and adapt to climate-related risks with unprecedented precision and efficiency. In the context of ocean systems and sea level management, AI's capacity for data analysis, prediction, and optimization positions it at the forefront of proactive climate adaptation efforts. By integrating AI into decision-making processes, infrastructure design, and environmental monitoring, humanity can move toward a future where climate resilience is not merely an aspiration but a reality.

One of AI's most significant contributions to proactive climate adaptation lies in its ability to predict and model complex environmental systems. The ocean, as the Earth's largest and most dynamic system, plays a central role in regulating the climate, influencing weather patterns, and mitigating the effects of global warming. However, the interactions between ocean currents, atmospheric conditions, and ecological processes are highly intricate and nonlinear, making them challenging to understand and predict. AI-powered models, particularly those based on machine learning and deep learning, excel at processing vast datasets to identify patterns, correlations, and trends that may not be apparent through traditional methods.

For example, AI systems can analyze satellite data, ocean buoy measurements, and historical climate records to predict the trajectory and intensity of tropical cyclones, providing early warnings that enable coastal communities to prepare and respond effectively. These models can also simulate the potential impacts of rising sea levels on specific regions, incorporating variables such as land subsidence, storm surges, and urban infrastructure vulnerabilities. By providing granular and actionable insights, AI enables policymakers, planners, and communities to implement targeted adaptation measures, such as strengthening coastal defenses, relocating critical infrastructure, or enhancing natural barriers like mangroves and wetlands.

AI's role in proactive climate adaptation extends beyond prediction to include real-time monitoring and management of environmental systems. Autonomous sensor networks equipped with AI algorithms offer continuous and high-resolution data on key climate indicators, such as sea surface temperatures, salinity levels, and carbon concentrations. These networks provide a comprehensive picture of ocean health, enabling stakeholders to detect emerging threats and implement timely interventions. For instance, an AI-driven monitoring system might identify areas of coral bleaching caused by elevated temperatures, triggering actions to mitigate stressors such as pollution or overfishing and support reef recovery.

In the realm of sea level management, AI plays a critical role in optimizing large-scale interventions aimed at mitigating the impacts of rising seas. Projects such as water redistribution, ocean basin deepening, and sediment management require precise coordination and adaptability to achieve desired outcomes without causing unintended consequences. AI algorithms excel at managing these complexities, using predictive analytics and real-time feedback to optimize operations and ensure that interventions align with broader climate adaptation goals. For example, an AI-controlled pumping system might dynamically adjust water transfer rates based on weather forecasts, energy availability, and environmental constraints, maximizing efficiency while minimizing ecological impacts.

The integration of AI into urban planning and infrastructure design further enhances its potential as a tool for proactive climate adaptation. Coastal cities, which are particularly vulnerable to sea level rise and extreme weather events, can benefit from AI-driven models that assess the resilience of existing infrastructure and identify opportunities for adaptation. These models consider a wide range of factors, including population density, economic activity, and environmental risks, providing a comprehensive framework for designing sustainable and resilient cities. For instance, an AI system might recommend the construction of elevated roads, permeable pavements, or flood-resistant buildings in areas prone to inundation, ensuring that urban development is both adaptive and sustainable.

AI's capacity for optimization and resource management is also critical for addressing the economic dimensions of climate adaptation. The financial costs of climate-related damages and adaptation measures are substantial, requiring careful allocation of resources to maximize their impact. AI-powered decision-support systems enable governments and organizations to prioritize investments based on cost-benefit analyses, risk assessments, and social equity considerations. For example, an AI platform might analyze the relative costs and benefits of building a seawall versus restoring a wetland in a specific region, providing stakeholders with the information needed to make informed and equitable decisions.

The potential of AI as a tool for proactive climate adaptation is not limited to mitigating risks; it also extends to leveraging opportunities for innovation and collaboration. The challenges posed by climate change are global in scope, necessitating coordinated efforts across nations, industries, and disciplines. AI facilitates these efforts by providing a common platform for data sharing, modeling, and decision-making, fostering collaboration and enabling stakeholders to align their strategies. For instance, an international consortium could use AI to develop a shared database of climate adaptation projects, documenting best practices, lessons learned, and performance metrics that inform future initiatives.

Despite its promise, the integration of AI into climate adaptation efforts is not without challenges. One of the primary obstacles is the need for high-quality and comprehensive data, which serves as the foundation for AI models. In many regions, particularly in developing countries and remote areas, data on environmental conditions, infrastructure vulnerabilities, and social dynamics is limited or inconsistent. Addressing this challenge requires investment in data collection technologies, such as satellite systems, autonomous sensors, and community-based monitoring programs, as well as initiatives to standardize and share data across platforms.

Ethical considerations are also central to the deployment of AI in climate adaptation. Decisions about how to allocate resources, prioritize interventions, and balance trade-offs between competing objectives have profound implications for social equity and environmental justice. AI systems must be designed and governed in ways that reflect these values, ensuring that their outputs are transparent, accountable, and inclusive. For instance, algorithms used to optimize adaptation measures should be free from biases that disproportionately favor wealthier regions or marginalized groups. Mechanisms for stakeholder engagement and public oversight are essential for addressing these concerns and building trust in AI-driven processes.

The future of AI as a tool for proactive climate adaptation will be shaped by ongoing advancements in technology, governance, and interdisciplinary collaboration. Emerging innovations, such as quantum computing and federated learning, offer new possibilities for enhancing the scale, speed, and accuracy of AI systems. For example, quantum algorithms could enable more complex simulations of ocean and climate dynamics, while federated learning models could facilitate data sharing and collaboration across jurisdictions without compromising privacy or security. These advancements will further strengthen the capabilities of AI to address the multifaceted challenges of climate adaptation.

The integration of advanced technologies, particularly artificial intelligence (AI), into global coastal management strategies represents a transformative approach to addressing the multifaceted challenges posed by rising sea levels, coastal erosion, and climate change. Coastal regions are among the most densely populated and economically significant areas of the world, yet they are increasingly vulnerable to environmental pressures. The convergence of AI with coastal management not only enhances the precision and effectiveness of interventions but also provides the tools needed to harmonize local and global efforts, creating resilient and sustainable coastlines for future generations.

Coastal management encompasses a wide range of activities aimed at balancing the needs of human communities with the preservation of natural ecosystems. These activities include flood prevention, habitat restoration, erosion control, and the sustainable use of marine resources. Traditional approaches to coastal management often rely on static models and manual processes that struggle to account for the dynamic and interconnected nature of coastal systems. AI, with its capacity for real-time data analysis, predictive modeling, and optimization, offers a paradigm shift in how these challenges are addressed.

One of the most significant contributions of AI to coastal management is its ability to process and analyze vast amounts of data from diverse sources. Coastal environments are shaped by a complex interplay of factors, including tides, currents, sediment transport, and human activities. These factors are influenced by both local conditions and broader climatic trends, making it essential to integrate data from satellite observations, ocean buoys, weather stations, and community-based monitoring programs. AI algorithms excel at synthesizing this information, identifying patterns and correlations that inform decision-making.

For example, an AI-driven system might analyze data on tidal patterns, wave heights, and sediment movement to predict areas at high risk of erosion or flooding. By providing early warnings and actionable insights, these systems enable coastal managers to implement targeted interventions, such as reinforcing seawalls, restoring mangroves, or relocating vulnerable infrastructure.

The ability to predict and mitigate risks proactively not only reduces the social and economic costs of coastal hazards but also enhances the resilience of coastal communities.

AI also plays a critical role in optimizing the design and implementation of coastal infrastructure. Traditional engineering solutions, such as breakwaters, groynes, and levees, often involve trade-offs between effectiveness, cost, and environmental impact. AI-powered design tools enable engineers to evaluate multiple scenarios, optimizing infrastructure to achieve desired outcomes while minimizing unintended consequences. For instance, an AI system might simulate the impact of different seawall configurations on wave energy, sediment deposition, and habitat connectivity, providing stakeholders with a comprehensive understanding of the trade-offs involved.

The integration of AI into nature-based solutions is another promising avenue for advancing coastal management. Nature-based solutions, such as restoring wetlands, coral reefs, and dune systems, leverage the natural resilience of ecosystems to provide protection against coastal hazards. AI enhances these efforts by analyzing ecological and environmental data to identify areas where nature-based interventions are most likely to succeed. For example, an AI system might evaluate the suitability of a site for mangrove restoration based on factors such as soil composition, tidal range, and historical vegetation patterns. By ensuring that interventions are informed by robust scientific evidence, AI helps maximize their ecological and protective benefits.

Global coastal management strategies must also address the challenges of coordinating efforts across jurisdictions and scales. Coastal systems are interconnected, with changes in one region often affecting others through processes such as sediment transport, water flow, and species migration. Effective management requires collaboration among local, regional, and international stakeholders, supported by tools that enable data sharing, communication, and joint decision-making. AI facilitates this coordination by providing a common platform for monitoring and managing coastal systems.

For instance, an AI-driven coastal management network might integrate data from multiple countries bordering a shared sea or ocean, such as the Mediterranean or the Caribbean. This network could analyze cross-border impacts of coastal activities, such as the construction of dams or the dredging of navigation channels, and recommend strategies that align with regional goals. By fostering a holistic and cooperative approach, AI ensures that coastal management efforts are not only effective but also equitable and sustainable.

The role of AI in global coastal management extends beyond technical optimization to include social and economic considerations. Coastal communities often face difficult trade-offs between development, conservation, and adaptation, particularly in regions where resources are limited or competing priorities exist. AI-powered decision-support systems enable stakeholders to evaluate these trade-offs in a transparent and inclusive manner. For example, an AI platform might analyze the costs and benefits of different flood protection measures, incorporating data on economic activity, population density, and ecosystem services. By presenting this information in an accessible format, the platform empowers stakeholders to make informed choices that reflect their values and priorities.

The integration of AI into coastal management also enhances the ability to monitor and enforce compliance with environmental regulations and conservation goals. AI-driven surveillance systems can analyze data from drones, satellites, and underwater sensors to detect illegal activities, such as unregulated fishing, habitat destruction, or pollution. These systems provide real-time alerts to enforcement agencies, enabling swift responses and deterring harmful practices. For instance, an AI-powered monitoring network in a marine protected area might identify unauthorized trawling activity based on vessel movements, generating actionable intelligence for patrols.

Despite its potential, the integration of AI into global coastal management strategies is not without challenges. One of the primary obstacles is the availability and quality of data, particularly in developing regions where monitoring infrastructure may be limited. Addressing this challenge requires investment in data collection technologies, capacity-building programs, and partnerships that support knowledge-sharing and innovation. For example, collaborations between governments, research institutions, and private companies could establish regional data hubs that aggregate and standardize coastal monitoring data, ensuring that it is accessible and actionable for all stakeholders.

Ethical considerations are also central to the integration of AI into coastal management. Decisions about how and where to deploy AI systems have profound implications for equity, transparency, and accountability. For instance, algorithms used to prioritize coastal adaptation projects must be designed to reflect diverse perspectives and avoid biases that disproportionately favor wealthier or more influential communities. Mechanisms for public engagement and oversight are essential for addressing these concerns, ensuring that AI-driven coastal management strategies are guided by shared values and inclusive governance.

The future of AI in global coastal management will be shaped by ongoing advancements in technology, policy, and interdisciplinary collaboration. Emerging innovations, such as quantum computing and federated learning, offer new possibilities for enhancing the scale and accuracy of AI systems. For example, quantum algorithms could enable more complex simulations of coastal dynamics, while federated learning models could facilitate data sharing across jurisdictions without compromising privacy or security. These advancements will further strengthen the capabilities of AI to address the multifaceted challenges of coastal management.

As the challenges of climate change, rising sea levels, and ocean sustainability grow increasingly urgent, the need for collaborative innovation and interdisciplinary research has never been more critical. The complexity and scale of these global challenges require solutions that transcend individual sectors, disciplines, and nations. Collaborative innovation and research pathways serve as the foundation for developing and deploying transformative technologies, fostering knowledge exchange, and aligning efforts to address shared goals. By uniting governments, academia, industry, and civil society, these pathways enable the creation of adaptive and equitable strategies for safeguarding the future of our oceans and coastal communities.

At the heart of collaborative innovation and research is the recognition that no single entity possesses all the expertise, resources, or perspectives needed to tackle the multifaceted problems associated with ocean management and climate adaptation. Coastal and marine systems are interconnected, spanning ecological, social, and economic dimensions, and the interventions required to manage them must reflect this interconnectedness. Collaborative efforts bridge the gaps between fields such as oceanography, artificial intelligence (AI), engineering, environmental science, and policymaking, creating synergies that amplify the impact of individual contributions.

One of the most promising pathways for collaborative innovation lies in the development of global research consortia dedicated to addressing ocean and climate challenges. These consortia bring together experts from diverse disciplines and geographies to conduct large-scale studies, share data, and develop integrated solutions. For example, an international consortium focused on sea level management might combine the expertise of climatologists, oceanographers, and AI specialists to model the impacts of water redistribution projects on regional climates and ecosystems. Such collaborative frameworks enable researchers to leverage the latest advancements in AI and computational modeling, providing deeper insights into complex phenomena and informing evidence-based decision-making.

Data sharing and interoperability are essential components of collaborative research, as they ensure that stakeholders have access to the information needed to develop and implement effective solutions. Advances in technology have made it possible to collect vast amounts of data on ocean systems, from satellite imagery and underwater sensors to climate models and biodiversity surveys. However, this data often resides in siloed repositories, limiting its accessibility and usefulness. Collaborative innovation pathways must prioritize the creation of open-access platforms that aggregate, standardize, and disseminate ocean-related data. These platforms, powered by AI and cloud computing, allow researchers, policymakers, and industry leaders to analyze and visualize data on a global scale, identifying patterns, correlations, and opportunities for intervention.

For example, an open-access platform might integrate data from coastal monitoring stations, marine protected areas, and fishing vessels to provide a real-time snapshot of ocean health. AI algorithms could analyze this data to identify hotspots of ecological stress, such as areas experiencing coral bleaching or overfishing, and recommend targeted actions to mitigate these impacts. By enabling data-driven collaboration, these platforms ensure that efforts to protect and manage the ocean are informed by the best available science and shared across stakeholders.

The integration of AI into collaborative research further enhances its potential to drive innovation and address pressing challenges. AI-powered tools and models enable researchers to simulate complex ocean and climate systems, optimize resource management strategies, and predict the outcomes of interventions. For instance, machine learning algorithms can process historical and real-time data to forecast the impacts of sea level rise on coastal infrastructure, informing decisions about where to invest in flood defenses or relocate communities. Similarly, AI can be used to optimize the design and operation of large-scale interventions, such as sediment redistribution or ocean basin deepening, ensuring that they achieve desired outcomes with minimal environmental impact.

Collaboration between academia and industry is another key pathway for innovation, as it bridges the gap between scientific research and practical applications. Universities and research institutions serve as hubs of knowledge and creativity, conducting foundational studies that expand our understanding of ocean systems and climate dynamics. Industry partners, in turn, bring the resources, expertise, and infrastructure needed to translate this knowledge into deployable technologies and solutions. For example, a partnership between an academic marine research center and a technology company might lead to the development of autonomous underwater vehicles (AUVs) equipped with AI-driven monitoring systems, enabling more efficient and precise data collection in the deep ocean.

Public-private partnerships also play a critical role in advancing collaborative research, particularly in the context of funding and scaling innovative solutions. Climate adaptation and ocean management initiatives often require substantial investment, which can be challenging to secure through public or private channels alone. By pooling resources and sharing risks, public-private partnerships enable the development and deployment of technologies that might otherwise remain out of reach. For instance, a partnership between a national government, an AI startup, and a global environmental organization might support the creation of a predictive model for coastal flooding, combining cutting-edge technology with localized expertise and on-the-ground implementation.

International collaboration is equally vital for addressing the transboundary nature of ocean and climate challenges. The oceans are a shared resource, and the actions taken in one region can have far-reaching effects on others. Collaborative innovation pathways must therefore emphasize the importance of global cooperation, fostering dialogue and alignment among nations, regions, and sectors. Multilateral agreements and frameworks, such as the United Nations Framework Convention on Climate Change (UNFCCC) and the United Nations Convention on the Law of the Sea (UNCLOS), provide a foundation for coordinating efforts and ensuring that they are guided by principles of equity, sustainability, and shared responsibility.

Capacity-building initiatives are a critical aspect of international collaboration, as they ensure that all nations, particularly those most vulnerable to climate impacts, have the tools and expertise needed to participate in and benefit from collaborative research. Training programs, technology transfer, and knowledge-sharing networks empower developing countries and small island states to engage in cutting-edge research and contribute to global solutions. For example, a capacity-building initiative might provide funding and training for scientists in the Pacific Islands to use AI-driven monitoring systems for tracking sea level changes and assessing their impacts on local ecosystems and communities.

The ethical dimensions of collaborative innovation and research cannot be overlooked, as they shape the legitimacy and inclusivity of efforts to address ocean and climate challenges. Collaborative pathways must prioritize transparency, inclusivity, and accountability, ensuring that all stakeholders have a voice in decision-making processes and that their perspectives are respected. For instance, indigenous communities and small-scale fishers, who often have deep knowledge of marine ecosystems and a direct stake in their health, should be actively involved in research initiatives and governance frameworks. Mechanisms for stakeholder engagement, such as advisory boards and participatory workshops, provide valuable opportunities for dialogue and co-creation.

Despite its promise, collaborative innovation and research face challenges that must be addressed to realize their full potential. These include barriers to data sharing, differences in priorities and capacities among stakeholders, and the need for sustained funding and political will. Overcoming these challenges requires a commitment to fostering trust, building partnerships, and aligning efforts with shared goals. For example, establishing clear agreements on data ownership, access, and use can address concerns about intellectual property and ensure that data is shared equitably and responsibly.

9. Sealing CO_2 Emissions in the Deep Sea

9.1. The Role of the Deep Ocean in Carbon Sequestration

The ocean is one of Earth's most vital systems for regulating the global carbon cycle, acting as both a sink and a reservoir for carbon dioxide (CO_2). Through complex physical, chemical, and biological processes, the ocean absorbs, stores, and redistributes carbon, mitigating the impacts of anthropogenic emissions and playing a crucial role in stabilizing the planet's climate. Among its many functions, the deep ocean serves as a critical long-term storage system for carbon, where it is sequestered away from the atmosphere for centuries to millennia. Understanding these natural processes is essential for leveraging the ocean's potential in addressing climate change while safeguarding its ecological integrity.

Carbon enters the ocean primarily through the atmosphere, where CO_2 dissolves into surface waters and reacts with seawater to form carbonic acid. This chemical reaction sets off a series of transformations, leading to the formation of bicarbonate and carbonate ions. These dissolved inorganic carbon compounds make up the majority of the ocean's carbon reservoir, buffering changes in atmospheric CO_2 levels. The efficiency of this process depends on factors such as water temperature, salinity, and ocean circulation. Colder waters, for instance, can hold more dissolved CO_2, which is why polar regions are often significant sites for carbon uptake.

Once absorbed at the surface, carbon can follow several pathways, with some remaining in the upper layers of the ocean while the rest is transported to the deep ocean. The biological pump is one of the key mechanisms facilitating this transfer. Through photosynthesis, phytoplankton, the microscopic plants at the base of the marine food web, convert dissolved CO_2 into organic carbon, producing oxygen as a byproduct. This organic carbon forms the foundation of oceanic life, feeding zooplankton, fish, and other marine organisms. As these organisms die or produce waste, organic matter sinks through the water column, carrying carbon to deeper layers of the ocean.

Not all sinking organic material reaches the deep ocean. Much of it is decomposed or consumed by bacteria and other organisms along the way, releasing CO_2 back into the water. However, a portion of the organic carbon does reach the seafloor, where it is buried in sediments and sequestered for long periods. This sequestration is critical for long-term carbon storage, effectively removing CO_2 from the atmosphere and mitigating its contribution to global warming. The efficiency of the biological pump varies across regions, influenced by factors such as nutrient availability, temperature, and the structure of the marine food web.

The solubility pump is another major pathway for carbon cycling in the ocean, driven by physical processes rather than biological activity. This mechanism involves the transport of dissolved inorganic carbon from surface waters to the deep ocean via thermohaline circulation. In regions where surface waters cool and sink, such as the North Atlantic and Southern Ocean, CO_2-rich water is carried to the deep ocean as part of the global conveyor belt. Once in the deep ocean, this carbon can remain isolated from the atmosphere for centuries, effectively serving as a long-term storage system. The solubility pump works in tandem with the biological pump, reinforcing the ocean's capacity to act as a carbon sink.

A lesser-known but equally important mechanism is the carbonate pump, which involves the formation and deposition of calcium carbonate ($CaCO_3$) by marine organisms such as corals, foraminifera, and coccolithophores. These organisms use carbonate ions and calcium from seawater to build their shells and skeletons. When they die, their remains sink to the seafloor, where they accumulate as sediments. Over time, these carbonate sediments can form limestone and other carbonate rocks, locking away carbon for millions of years. However, the carbonate pump also releases CO_2 during the process of calcification, creating a complex interplay between carbon storage and release that varies across ecosystems.

The storage of carbon in deep ocean sediments is a key component of the ocean's role as a long-term carbon sink. Sediments act as a repository for both organic and inorganic carbon, preserving it in anoxic or low-oxygen conditions that slow down decomposition. The rate of carbon burial depends on factors such as sedimentation rates, the composition of organic material, and the presence of stabilizing compounds like clays and biogenic minerals. In some regions, such as the continental margins and deep-sea basins, the accumulation of organic-rich sediments forms an important part of the global carbon budget, sequestering significant amounts of CO_2 over geological timescales.

The deep ocean's capacity for carbon storage is vast, estimated to contain approximately 37,000 gigatons of carbon, more than 50 times the amount in the atmosphere. This makes it the largest active carbon reservoir on Earth, capable of buffering fluctuations in atmospheric CO_2 and mitigating the impacts of climate change. However, this capacity is not limitless, and its effectiveness depends on the health and stability of the processes that sustain it. Disruptions to ocean circulation, nutrient cycles, or ecosystem dynamics could impair the ocean's ability to absorb and store carbon, with far-reaching consequences for the global climate system.

One of the critical factors influencing the ocean's role in carbon cycling is the availability of nutrients such as nitrogen, phosphorus, and iron, which are essential for phytoplankton growth. In many regions, nutrient limitations constrain the efficiency of the biological pump, reducing the amount of carbon that can be sequestered in the deep ocean. Upwelling zones, where nutrient-rich deep waters are brought to the surface, play a vital role in supporting marine productivity and enhancing carbon uptake. However, changes in upwelling patterns due to climate change could alter these dynamics, affecting the ocean's carbon sequestration potential.

The deep ocean's ability to store carbon is also influenced by physical changes in the climate system, such as warming, acidification, and deoxygenation. Rising sea temperatures reduce the solubility of CO_2 in surface waters, weakening the solubility pump and decreasing the efficiency of carbon uptake. Ocean acidification, caused by the absorption of excess CO_2, affects the carbonate pump by reducing the availability of carbonate ions needed for shell and skeleton formation. Deoxygenation, driven by warming and nutrient runoff, can impact the microbial processes that regulate carbon burial in sediments, potentially altering the balance between carbon storage and release.

Despite these challenges, the natural processes of carbon cycling and storage in the ocean remain a cornerstone of the global climate system. They highlight the importance of protecting and restoring marine ecosystems, which are integral to sustaining these processes. Efforts to enhance the ocean's capacity for carbon sequestration must be guided by a deep understanding of its natural dynamics, ensuring that interventions align with ecological principles and do not compromise the integrity of marine systems.

The deep ocean, with its vast capacity for long-term carbon storage, represents a critical frontier in the fight against climate change. While the natural processes of carbon cycling and sequestration play an essential role in regulating atmospheric CO_2 levels, the accelerating pace of global emissions has prompted scientists and policymakers to explore artificial interventions to enhance the ocean's capacity as a carbon sink. These interventions range from direct CO_2 injection into the deep sea to leveraging biological and chemical processes to capture and store carbon. As these technologies advance, they offer significant potential for mitigating climate change, but they also raise complex scientific, ethical, and environmental considerations.

One of the most direct approaches to artificial CO_2 sequestration in the deep sea is the injection of captured CO_2 into deep ocean waters or sediments. In this method, CO_2 is compressed into a liquid state and transported via pipelines or specialized vessels to depths exceeding 1,000 meters, where high pressures and low temperatures allow it to remain in a dense or supercritical state. At these depths, CO_2 can form stable pools on the seafloor or dissolve into the surrounding water, effectively isolating it from the atmosphere for centuries to millennia. This process takes advantage of the natural density gradients in the ocean, which prevent CO_2 from readily mixing with surface waters.

The feasibility of deep-sea CO_2 injection depends on several factors, including the availability of suitable storage sites, the properties of the injected CO_2, and the potential environmental impacts. Geologically stable regions with thick sediment layers, such as subduction zones or continental margins, are considered ideal locations for CO_2 storage. These sediments can act as a barrier, preventing the upward migration of CO_2 and ensuring its long-term containment. Advanced geophysical surveys and modeling tools, powered by artificial intelligence (AI), are critical for identifying and characterizing these sites, ensuring that they meet the necessary criteria for safety and stability.

AI also plays a key role in optimizing the injection process itself. Machine learning algorithms can analyze data from sensors and simulations to determine the optimal pressure, temperature, and flow rates for CO_2 injection, minimizing the risk of leakage or disturbance to the surrounding environment. For example, an AI-driven control system might adjust the injection parameters in real time based on feedback from seafloor monitors, ensuring that the process remains within safe operational limits. These systems enable precise and adaptive management of CO_2 storage, reducing the likelihood of unintended consequences.

Another promising avenue for artificial CO_2 sequestration in the deep sea is the enhancement of natural biological processes, such as the biological pump.

This mechanism relies on phytoplankton to convert atmospheric CO_2 into organic carbon through photosynthesis, which is then transported to the deep ocean as sinking particulate matter. Artificial interventions, such as ocean fertilization, aim to stimulate phytoplankton growth by adding nutrients like iron, nitrogen, or phosphorus to nutrient-limited regions of the ocean. By increasing phytoplankton productivity, these interventions can enhance the uptake and sequestration of CO_2 in deep ocean sediments.

Ocean fertilization has been the subject of extensive research and debate, with studies demonstrating both its potential benefits and its limitations. While it can increase carbon export to the deep ocean, the efficiency and permanence of this process depend on factors such as the type of phytoplankton stimulated, the depth of carbon export, and the retention of nutrients within the ecosystem. Additionally, ocean fertilization carries risks of unintended ecological consequences, including harmful algal blooms, oxygen depletion, and disruptions to food webs. AI-driven monitoring and modeling systems are essential for managing these risks, providing real-time data on phytoplankton dynamics, nutrient cycling, and ecosystem health.

Artificial upwelling is another innovative approach to enhancing biological carbon sequestration in the deep sea. This method involves the use of pumps or other technologies to bring nutrient-rich deep waters to the surface, stimulating primary production and increasing the export of organic carbon to the deep ocean. Like ocean fertilization, artificial upwelling seeks to leverage the natural processes of the biological pump while addressing the limitations of nutrient availability in surface waters. AI systems play a crucial role in optimizing artificial upwelling, analyzing data on ocean currents, temperature gradients, and nutrient concentrations to determine the most effective locations and strategies for implementation.

Chemical approaches to CO_2 sequestration in the deep sea also hold significant potential. One such method involves the dissolution of alkaline minerals, such as calcium hydroxide or magnesium silicate, in seawater to enhance its capacity to absorb and store CO_2. This process, known as ocean alkalinity enhancement, increases the pH and carbonate ion concentration of seawater, enabling it to take up more atmospheric CO_2 and convert it into stable bicarbonate and carbonate ions. These chemical transformations reduce the partial pressure of CO_2 at the ocean surface, promoting additional uptake of atmospheric carbon.

Ocean alkalinity enhancement has the advantage of addressing both climate change and ocean acidification, as the addition of alkaline substances counteracts the effects of acidification caused by excess CO_2. However, its large-scale implementation requires careful consideration of the environmental and logistical challenges, including the sourcing, transportation, and dissolution of alkaline minerals. AI-powered logistics and optimization tools are critical for managing these challenges, ensuring that the process is conducted efficiently and with minimal ecological disruption.

Hybrid approaches to CO_2 sequestration in the deep sea combine multiple methods to maximize their effectiveness and address their respective limitations. For example, a hybrid strategy might integrate direct CO_2 injection with ocean alkalinity enhancement, leveraging the chemical buffering capacity of alkalinity to stabilize injected CO_2 and reduce the risk of leakage. Similarly, combining artificial upwelling with ocean fertilization could enhance both primary production and nutrient recycling, increasing the efficiency of the biological pump. AI systems are instrumental in designing and managing these hybrid approaches, providing the analytical and predictive capabilities needed to balance trade-offs and optimize outcomes.

Despite their potential, artificial interventions for CO_2 sequestration in the deep sea face significant ethical, environmental, and governance challenges. The introduction of large-scale technologies into marine ecosystems carries inherent risks of unintended consequences, including disruptions to biodiversity, nutrient cycling, and food webs. These risks underscore the need for comprehensive environmental impact assessments (EIAs) and adaptive management frameworks, supported by real-time monitoring and data-driven decision-making.

Ethical considerations also extend to issues of equity and justice, particularly regarding the distribution of benefits and burdens associated with CO_2 sequestration projects. Coastal and island communities, which are often the most vulnerable to climate impacts, must be actively involved in the planning and governance of these projects to ensure that their perspectives and interests are represented. International collaboration and transparency are essential for addressing these concerns, fostering trust and accountability among stakeholders.

The governance of artificial CO_2 sequestration in the deep sea requires robust regulatory frameworks that align with international agreements such as the

United Nations Convention on the Law of the Sea (UNCLOS) and the emerging Biodiversity Beyond National Jurisdiction (BBNJ) treaty. These frameworks must establish clear guidelines for the design, implementation, and monitoring of sequestration projects, ensuring that they are conducted responsibly and in accordance with global sustainability goals. Mechanisms for oversight, stakeholder engagement, and ethical review are critical for maintaining the legitimacy and effectiveness of these efforts.

Artificial intelligence (AI) is emerging as a transformative force in the fight against climate change, with the potential to revolutionize how carbon is managed in the deep ocean. The ocean, as the largest active carbon sink on Earth, plays a crucial role in mitigating global warming by absorbing and storing vast amounts of carbon dioxide (CO_2). However, the complex, dynamic, and poorly understood nature of ocean systems poses significant challenges for managing carbon sequestration effectively and sustainably. AI, with its unparalleled capabilities for data analysis, modeling, and optimization, offers powerful tools for addressing these challenges and unlocking new possibilities for enhancing the ocean's role in global carbon management.

One of AI's most significant contributions to carbon management in the deep ocean is its ability to analyze vast amounts of environmental data and generate actionable insights. The ocean is an incredibly dynamic system influenced by numerous interacting variables, including temperature, salinity, currents, nutrient availability, and biological activity. These variables determine the efficiency of natural carbon sequestration processes, such as the biological pump and the solubility pump, as well as the potential impacts of artificial interventions. AI algorithms, particularly those based on machine learning, excel at processing and integrating these diverse datasets to identify patterns, correlations, and trends that inform decision-making.

For example, AI systems can analyze satellite observations, ocean buoy measurements, and historical climate data to model how different regions of the ocean contribute to carbon uptake and storage. These models provide valuable insights into the factors that enhance or limit carbon sequestration, such as the availability of nutrients for phytoplankton growth or the efficiency of deepwater circulation. By identifying high-priority areas for intervention, AI enables targeted efforts to enhance the ocean's natural carbon storage capacity, whether through conservation measures, restoration projects, or artificial enhancements like ocean fertilization or alkalinity enhancement.

In addition to its role in analysis and modeling, AI offers transformative capabilities for optimizing the design and implementation of carbon management strategies. Artificial interventions for CO_2 sequestration, such as direct CO_2 injection, ocean alkalinity enhancement, and artificial upwelling, require precise planning and coordination to achieve desired outcomes while minimizing environmental risks. AI-powered optimization algorithms can evaluate multiple scenarios and trade-offs, identifying the most effective and sustainable approaches for specific contexts.

For instance, an AI system tasked with managing a direct CO_2 injection project might analyze data on ocean currents, temperature gradients, and sediment composition to determine the optimal depth, location, and injection rate for storing CO_2 securely. By simulating the potential impacts of different configurations, the system can identify strategies that maximize carbon storage while minimizing the risk of leakage or disturbance to marine ecosystems. Similarly, AI-driven optimization can enhance the efficiency of ocean alkalinity enhancement projects by determining the ideal locations and methods for distributing alkaline substances, ensuring that they achieve maximum carbon uptake with minimal disruption to local habitats.

The role of AI in real-time monitoring and adaptive management further underscores its potential to revolutionize carbon management in the deep ocean. Effective carbon management requires continuous monitoring of environmental conditions and the performance of sequestration projects, enabling operators to respond dynamically to changing circumstances and unforeseen impacts. AI systems equipped with advanced sensors and autonomous platforms provide high-resolution, real-time data on key indicators such as CO_2 concentrations, water chemistry, and biological activity. These systems use machine learning algorithms to detect anomalies, assess trends, and generate alerts or recommendations for intervention.

For example, an AI-driven monitoring system might identify an unexpected increase in dissolved CO_2 levels near a sequestration site, indicating a potential leak or inefficiency in the injection process. By analyzing the source and magnitude of the issue, the system can recommend corrective actions, such as adjusting injection parameters or deploying additional containment measures. These insights ensure that carbon management projects remain effective and accountable, minimizing risks to marine ecosystems and maximizing their contribution to climate goals.

AI's ability to model and predict the long-term impacts of carbon management strategies is another critical advantage. The deep ocean operates on timescales that span centuries to millennia, and the effects of interventions may not become fully apparent for decades. AI-powered simulations enable researchers to explore these long-term dynamics, incorporating variables such as ocean circulation, sediment transport, and ecological feedback loops. These models provide a comprehensive understanding of how carbon management strategies interact with natural processes, informing decisions about their design, implementation, and monitoring.

For example, a predictive AI model might simulate the cumulative impacts of a series of ocean alkalinity enhancement projects on global carbon cycles, evaluating their effectiveness in reducing atmospheric CO_2 levels and mitigating ocean acidification. By testing different scenarios and sensitivity analyses, the model can identify potential risks, such as shifts in nutrient availability or unintended effects on marine ecosystems, and recommend strategies to mitigate them. These insights are invaluable for ensuring that carbon management strategies align with both climate goals and ecological principles.

The integration of AI with other emerging technologies further enhances its potential to transform carbon management in the deep ocean. Autonomous underwater vehicles (AUVs), equipped with AI-driven navigation and sensing capabilities, enable more efficient and precise data collection in remote and extreme environments. These platforms can map seafloor carbon storage sites, monitor the behavior of sinking organic matter, and assess the health of ecosystems that support carbon sequestration. Similarly, the use of AI in conjunction with digital twins, virtual replicas of physical systems, allows researchers to simulate and optimize carbon management projects in a controlled, risk-free environment.

The scalability and interconnectedness of AI systems also support global collaboration and coordination in carbon management. The challenges of climate change and carbon sequestration are inherently global, requiring efforts that transcend national and sectoral boundaries. AI-driven platforms provide a common framework for data sharing, decision-making, and monitoring, enabling stakeholders to align their strategies and leverage shared resources. For example, an international consortium could use an AI platform to coordinate a network of carbon sequestration projects, ensuring that their combined impacts contribute to global climate goals while respecting regional priorities and constraints.

Despite its transformative potential, the deployment of AI in carbon management also raises important ethical, technical, and governance challenges. The use of AI to manage the deep ocean, a vast and poorly understood system, carries inherent risks of unintended consequences, including disruptions to ecosystems, resource inequalities, and conflicts over data ownership. Ensuring that AI systems are designed and governed responsibly is critical for addressing these challenges and building trust among stakeholders.

Transparency and accountability are central to the ethical deployment of AI in carbon management. The algorithms and decision-making processes used by AI systems must be transparent and explainable, enabling stakeholders to evaluate their fairness, reliability, and alignment with project goals. Mechanisms for public engagement and oversight, such as advisory boards or independent audits, are essential for ensuring that AI systems operate in accordance with shared values and ethical principles.

The success of AI-driven carbon management also depends on the availability and quality of data, which is often limited in marine environments. Addressing this challenge requires investment in data collection technologies, such as underwater sensors, satellite systems, and community-based monitoring programs, as well as initiatives to standardize and integrate data across platforms. Collaborative research and capacity-building efforts are also critical for ensuring that all stakeholders, including developing countries and small island states, have access to the tools and expertise needed to participate in and benefit from AI-driven carbon management.

9.2. Technologies for Deep Sea Carbon Sequestration

The search for viable CO_2 storage reservoirs in the deep sea is one of the most challenging and consequential aspects of carbon sequestration technology. The ocean's vastness and its poorly understood dynamics make the identification of suitable sites for long-term CO_2 storage a highly complex task. Artificial intelligence (AI) has emerged as a transformative tool in this endeavor, offering unparalleled capabilities for analyzing massive datasets, modeling geological and environmental conditions, and optimizing site selection processes. By integrating AI into site selection, researchers and engineers can enhance the precision, efficiency, and sustainability of deep sea carbon sequestration projects, ensuring that these efforts align with global climate goals and environmental stewardship.

The selection of CO_2 storage reservoirs in the deep sea is governed by a wide range of criteria, including geological stability, sediment composition, depth, and proximity to infrastructure. These criteria must be assessed in the context of dynamic and interconnected ocean systems, where small changes in one variable can have cascading effects on others. For example, a site with favorable geological conditions for CO_2 storage might be located near sensitive ecosystems, requiring a careful balance between technical feasibility and ecological considerations. AI excels at managing such complexity, using advanced algorithms to process and integrate diverse datasets and identify optimal sites for storage.

One of the most critical factors in site selection is the geological stability of the storage reservoir. Deep sea sediments and rock formations must be capable of securely containing CO_2 over long timescales, preventing leakage and ensuring that the injected carbon remains isolated from the atmosphere. AI-driven geophysical models play a central role in evaluating these conditions, analyzing data from seismic surveys, core samples, and remote sensing technologies to characterize subsurface structures. For example, machine learning algorithms can identify patterns in seismic waveforms that indicate the presence of impermeable caprocks or fault zones, providing insights into the suitability of a given site.

The composition and porosity of sediments are equally important considerations in site selection, as they determine the reservoir's capacity to store CO_2 and the ease with which it can be injected. AI systems equipped with data from sediment cores and geochemical analyses can classify sediment types and predict their behavior under high-pressure conditions. For instance, an AI model might analyze data on grain size, mineral content, and compaction to estimate the porosity and permeability of a sediment layer, ensuring that the site can accommodate the injected CO_2 without compromising its stability.

Depth is another key factor in site selection, as it influences the physical state of CO_2 and its interactions with the surrounding environment. At depths greater than 1,000 meters, the high pressure and low temperature of the ocean allow CO_2 to exist in a dense or supercritical state, which is more stable and less likely to migrate upward. AI-driven optimization algorithms can analyze bathymetric data, pressure-temperature profiles, and ocean circulation patterns to identify regions where these conditions are consistently met. By incorporating real-time data from autonomous underwater vehicles (AUVs) and ocean buoys, these systems can adapt dynamically to changing conditions, refining site selection criteria and enhancing precision.

Environmental considerations are integral to AI-driven site selection, reflecting the need to minimize the ecological impacts of CO_2 storage projects. Deep sea ecosystems are characterized by unique and often fragile species that are highly sensitive to disturbances, such as changes in pH, sediment composition, or nutrient availability. AI-powered environmental models provide the tools needed to assess and mitigate these risks, integrating data on biodiversity, habitat connectivity, and ecosystem services. For example, an AI system might identify areas where the introduction of CO_2 could disrupt chemosynthetic communities that rely on stable chemical gradients, prompting the exclusion of those sites from consideration.

Proximity to existing infrastructure is another important criterion for site selection, as it influences the logistical and economic feasibility of CO_2 storage projects. AI-driven spatial analysis tools can evaluate the connectivity of potential sites to pipelines, storage facilities, and other infrastructure, optimizing the balance between technical requirements and cost-effectiveness. For instance, an AI system might recommend a site that minimizes pipeline length while meeting all other geological and environmental criteria, reducing both project costs and the energy required for transport and injection.

The integration of AI into site selection processes also enhances the ability to evaluate long-term risks and uncertainties associated with CO_2 storage. The deep ocean operates on timescales that span centuries to millennia, and the impacts of storage projects may not become fully apparent for decades. AI-powered predictive models enable researchers to simulate the long-term behavior of CO_2 in subsurface reservoirs, accounting for variables such as sediment compaction, chemical reactions, and ocean circulation changes. These simulations provide a comprehensive understanding of how stored CO_2 will interact with its environment over time, informing decisions about site selection and management.

For example, a predictive AI model might simulate the potential for CO_2 to migrate upward through fractures in the caprock, identifying sites where additional measures, such as secondary containment layers or monitoring systems, are needed. Similarly, the model could assess the impacts of CO_2 storage on surrounding sediment chemistry and microbial communities, ensuring that interventions align with ecological and sustainability goals.

AI-driven site selection is further enhanced by its ability to integrate real-time monitoring and adaptive management into the storage process. Continuous monitoring is essential for verifying the safety and effectiveness of CO_2 storage projects, as well as for detecting and responding to potential issues, such as leaks or changes in reservoir conditions. AI systems equipped with advanced sensors and autonomous platforms provide high-resolution, real-time data on key indicators, such as pressure, temperature, and chemical composition. These systems use machine learning algorithms to identify anomalies, assess risks, and recommend corrective actions, ensuring that storage projects remain within safe operational limits.

For instance, an AI-driven monitoring network might detect an unexpected increase in reservoir pressure, indicating the need to adjust injection rates or redistribute CO_2 to alternative sites. By analyzing the source and magnitude of the issue, the system can provide actionable insights for mitigating risks and maintaining the integrity of the storage project. These capabilities not only enhance the safety and reliability of CO_2 storage but also build trust among stakeholders, demonstrating the accountability and transparency of carbon management efforts.

The integration of AI into site selection processes also supports global collaboration and data sharing, fostering innovation and alignment among stakeholders. The challenges of identifying suitable CO_2 storage reservoirs are inherently global, requiring efforts that transcend national and sectoral boundaries. AI-driven platforms provide a common framework for analyzing data, sharing insights, and coordinating strategies, enabling researchers, policymakers, and industry leaders to work together toward shared goals. For example, an international consortium might use an AI platform to create a global database of potential CO_2 storage sites, documenting their geological, environmental, and logistical attributes to support evidence-based decision-making.

Despite its transformative potential, the deployment of AI-driven site selection systems faces challenges that must be addressed to ensure their success. The availability and quality of data are critical, as accurate site selection depends on comprehensive and high-resolution datasets that capture the complexity of deep ocean systems. Addressing this challenge requires investment in data collection technologies, such as seismic surveys, sediment coring, and autonomous monitoring platforms, as well as initiatives to standardize and integrate data across platforms and regions.

Ethical and governance considerations are also central to the deployment of AI in CO_2 storage site selection. The use of AI to guide decisions about the deep ocean, a vast and poorly understood system, carries inherent risks of unintended consequences, including disruptions to ecosystems, conflicts over resource use, and concerns about transparency and accountability. Ensuring that AI-driven systems operate responsibly requires robust governance frameworks, stakeholder engagement, and mechanisms for oversight and review.

The successful implementation of deep sea carbon sequestration depends on the ability to monitor and verify the injection and containment of CO_2 with precision and reliability. These processes are critical for ensuring that injected CO_2 remains securely stored in the intended reservoirs, mitigating climate change without causing unintended environmental consequences. Monitoring and verification not only provide the technical assurance needed for the safety and efficacy of these projects but also build public trust and regulatory compliance. Advanced technologies, particularly those powered by artificial intelligence (AI), are revolutionizing the way CO_2 injection and containment are monitored, offering tools for real-time data collection, analysis, and adaptive management.

The injection of CO_2 into deep sea reservoirs involves the transportation of captured CO_2 to offshore sites, where it is compressed into a dense or supercritical state and injected into geological formations or sediments beneath the ocean floor. These reservoirs, typically located at depths greater than 1,000 meters, must be geologically stable and capable of containing CO_2 for centuries to millennia. However, the dynamic and poorly understood nature of deep ocean systems introduces uncertainties that necessitate robust monitoring and verification strategies.

Monitoring begins with the injection process itself, where the focus is on ensuring that CO_2 is delivered to the reservoir safely and efficiently. Advanced sensors and control systems, often integrated with AI algorithms, play a critical role in managing this process. For example, flow meters and pressure sensors installed along pipelines and at injection sites provide real-time data on the rate and pressure of CO_2 injection, enabling operators to detect anomalies such as blockages, leaks, or over-pressurization. AI systems analyze this data to identify patterns and deviations, generating alerts and recommendations for corrective actions when necessary.

The containment of CO_2 within the reservoir is a central concern, as any leakage into the surrounding environment could undermine the effectiveness of sequestration efforts and pose risks to marine ecosystems. Verifying containment requires continuous monitoring of the reservoir and its surroundings to detect changes in pressure, temperature, and chemical composition that might indicate CO_2 migration. Seismic imaging is one of the most powerful tools for this purpose, providing detailed maps of subsurface structures and enabling the tracking of CO_2 plumes over time. AI enhances the interpretation of seismic data by automating the identification of features such as fractures, faults, and caprock integrity, reducing the time and expertise required for analysis.

Geochemical monitoring is another essential component of verifying CO_2 containment. Changes in the chemical composition of seawater and sediments around the reservoir can provide early warning signs of leakage or unintended interactions between CO_2 and the surrounding environment. For instance, the dissolution of CO_2 in seawater increases its acidity, which can be detected through changes in pH, dissolved inorganic carbon, and carbonate ion concentrations. AI-driven geochemical models integrate data from sensors, laboratory analyses, and field measurements to predict the behavior of CO_2 under various conditions and identify potential risks to containment.

The integration of autonomous systems, such as underwater drones and remotely operated vehicles (ROVs), further enhances monitoring capabilities by providing high-resolution data from challenging and inaccessible locations. These systems, equipped with cameras, sonar, and chemical sensors, can survey the seafloor and water column to detect signs of CO_2 leakage or environmental impacts. For example, an autonomous underwater vehicle (AUV) might conduct regular inspections of the injection site, capturing images and data that are analyzed by AI algorithms to identify anomalies such as gas bubbles, sediment disturbances, or changes in marine life activity.

Real-time monitoring and adaptive management are critical for addressing the dynamic and evolving conditions of deep sea carbon sequestration projects. AI systems excel at processing large volumes of data in real time, enabling operators to respond quickly to emerging issues. For instance, an AI-driven monitoring network might detect a sudden increase in reservoir pressure, indicating a risk of caprock failure or upward migration of CO_2. By analyzing the source and magnitude of the issue, the system can recommend adjustments to injection parameters or the deployment of containment measures, ensuring the integrity of the storage site.

The verification of CO_2 containment also involves the use of tracer techniques, which provide a means of tracking the movement and behavior of injected CO_2. Tracers are chemical or isotopic compounds that are added to the CO_2 stream during injection, allowing researchers to distinguish it from naturally occurring carbon in the environment. AI systems enhance the analysis of tracer data by identifying patterns and correlations that indicate the pathways and rates of CO_2 migration. For example, a machine learning model might analyze tracer concentrations in water samples collected around the reservoir to determine whether CO_2 is escaping through fractures or interacting with surrounding sediments.

Long-term monitoring is essential for verifying the stability and safety of CO_2 storage sites over decades to centuries. The deep ocean operates on timescales that far exceed those of most human activities, and the impacts of storage projects may not become fully apparent for years. AI-powered predictive models provide invaluable tools for simulating the long-term behavior of CO_2 in subsurface reservoirs, accounting for variables such as sediment compaction, chemical reactions, and changes in ocean circulation. These models enable researchers to evaluate the potential risks and benefits of storage projects, informing decisions about site selection, management, and mitigation.

In addition to technical considerations, the monitoring and verification of CO_2 injection and containment carry significant regulatory and ethical implications. Governments and international organizations require rigorous evidence that storage projects meet safety and environmental standards, ensuring that they contribute to climate goals without causing harm. AI-driven monitoring systems provide the transparency and accountability needed to meet these requirements, generating detailed records of project performance that can be audited and shared with stakeholders.

Public trust is another critical factor in the success of deep sea carbon sequestration efforts. The potential risks of CO_2 leakage and environmental impacts have raised concerns among communities, conservation groups, and policymakers, highlighting the need for transparent and inclusive governance. Monitoring and verification systems powered by AI play a key role in addressing these concerns by providing real-time data and clear evidence of project safety. For example, a public-facing dashboard might display information on reservoir pressure, chemical composition, and environmental indicators, enabling stakeholders to track the performance of storage projects and engage in informed dialogue.

Despite its transformative potential, the deployment of AI-driven monitoring and verification systems faces challenges that must be addressed to ensure their success. The availability and quality of data are critical, as accurate monitoring depends on comprehensive and high-resolution datasets that capture the complexity of ocean systems. Addressing this challenge requires investment in advanced sensors, autonomous platforms, and data-sharing frameworks that enable the collection and integration of high-quality data across platforms and regions.

The ethical and governance dimensions of monitoring and verification also require careful consideration. The use of AI to manage and monitor deep ocean systems carries risks of unintended consequences, including biases in decision-making, conflicts over data ownership, and concerns about accountability. Ensuring that AI systems operate transparently and responsibly requires robust governance frameworks, stakeholder engagement, and mechanisms for oversight and review.

The deployment of advanced robotics in constructing and maintaining subsea storage infrastructure is transforming the field of deep sea carbon sequestration. Subsea infrastructure, such as pipelines, injection wells, and monitoring systems, forms the backbone of efforts to store carbon dioxide (CO_2) safely and effectively in the deep ocean. These facilities must operate in some of the most extreme and challenging environments on Earth, where high pressures, low temperatures, and limited accessibility make human intervention impractical. Advanced robotics offers a solution to these challenges, providing the precision, efficiency, and adaptability needed to build and sustain the infrastructure required for long-term carbon storage.

The construction of subsea storage infrastructure begins with the installation of pipelines and injection wells that transport CO_2 from capture facilities to storage reservoirs beneath the ocean floor. These structures must be precisely positioned and securely anchored to ensure their stability and performance over decades or even centuries. Remotely operated vehicles (ROVs) and autonomous underwater vehicles (AUVs) are central to this effort, equipped with advanced navigation, sensing, and manipulation capabilities that enable them to perform complex tasks in deep ocean environments.

For example, an ROV might use robotic arms to position and secure pipeline segments, ensuring that they are aligned and connected correctly. These arms, often equipped with AI-driven control systems, can perform delicate operations such as tightening bolts, attaching sensors, or welding components, all while compensating for ocean currents and other dynamic conditions. Similarly, AUVs equipped with high-resolution sonar and laser scanning systems can map the seafloor with millimeter precision, identifying suitable locations for wells and monitoring potential hazards such as unstable slopes or geological faults.

Once the basic infrastructure is in place, advanced robotics plays a critical role in the installation and operation of injection wells. These wells must be drilled and reinforced to accommodate the high pressures and corrosive conditions associated with CO_2 injection. Robotic drilling systems, often integrated with AI algorithms, enable precise and efficient drilling in deep ocean sediments, reducing the risk of complications such as wellbore collapse or leakage. These systems analyze data from sensors and seismic surveys to optimize drilling parameters in real time, ensuring that wells are positioned and constructed to meet safety and performance standards.

The maintenance and monitoring of subsea storage infrastructure are equally important for ensuring the long-term security and effectiveness of CO_2 storage projects. Over time, pipelines, wells, and other components are exposed to environmental stresses that can lead to wear, corrosion, or mechanical failures. Advanced robotics provides the tools needed to inspect and repair these structures, minimizing downtime and avoiding costly disruptions.

For example, an AUV equipped with visual and ultrasonic sensors might conduct routine inspections of pipelines, capturing high-resolution images and acoustic data that are analyzed by AI systems to detect signs of corrosion, cracks, or other anomalies. If a problem is identified, an ROV can be deployed to perform repairs, such as applying protective coatings, replacing damaged segments, or sealing leaks. These robotic systems are designed to operate autonomously or with minimal human supervision, reducing the risks and costs associated with deep-sea maintenance.

In addition to their technical capabilities, advanced robotics enhances the adaptability and resilience of subsea storage infrastructure by enabling real-time monitoring and dynamic response to changing conditions. Autonomous sensor networks integrated with robotic platforms provide continuous data on key environmental and operational parameters, such as pressure, temperature, and chemical composition. This data is analyzed by AI algorithms to identify trends, predict potential issues, and recommend corrective actions. For instance, if an injection well experiences an unexpected increase in pressure, the system might recommend adjustments to injection rates or the deployment of additional containment measures.

The integration of robotics with monitoring systems also supports environmental stewardship, ensuring that CO_2 storage projects align with sustainability goals and minimize their impact on marine ecosystems. Robotic platforms equipped with environmental sensors can track changes in water quality, sediment composition, and marine life activity around storage sites, providing valuable insights into the interactions between infrastructure and the surrounding environment. For example, an AUV might survey a storage site to detect changes in pH or dissolved oxygen levels that could indicate CO_2 leakage or other ecological impacts. By enabling continuous monitoring and adaptive management, robotics ensures that storage projects are conducted responsibly and transparently.

The use of bio-inspired and soft robotics further enhances the potential of advanced robotics in subsea carbon storage. These technologies draw inspiration from marine organisms, such as jellyfish, octopuses, and fish, to create robots that are flexible, efficient, and capable of interacting safely with delicate environments. For instance, a bio-inspired robotic arm might mimic the movements of an octopus tentacle, enabling it to handle fragile equipment or navigate complex structures without causing damage. Similarly, a soft robotic vehicle might use undulating motions to traverse the seafloor with minimal disturbance to sediments or marine habitats.

Despite their promise, the deployment of advanced robotics in subsea storage infrastructure faces challenges that must be addressed to ensure their success. One of the primary challenges is the need for reliable and robust systems that can withstand the harsh conditions of the deep ocean, including high pressures, corrosive seawater, and low temperatures. Advances in material science, such as the development of corrosion-resistant alloys and bio-compatible polymers, are essential for creating durable robotic components that can operate effectively over extended periods.

Energy supply is another critical consideration for robotic systems operating in remote and inaccessible locations. Autonomous platforms require reliable and efficient power sources to sustain their operations, whether through onboard batteries, tethered connections, or renewable energy systems such as underwater turbines or wave energy converters. AI-driven energy management systems play a key role in optimizing power consumption, ensuring that robots can perform their tasks without interruptions or inefficiencies.

The integration of robotics into existing infrastructure and workflows also requires careful planning and coordination. Subsea storage projects often involve multiple stakeholders, including energy companies, research institutions, and regulatory agencies, each with their own priorities and requirements. Collaborative efforts are needed to standardize robotic technologies, establish interoperability protocols, and align objectives across stakeholders. For example, a consortium of industry and academic partners might develop shared platforms for robotic inspection and maintenance, ensuring that these systems are compatible with a wide range of infrastructure designs and operational scenarios.

Ethical and governance considerations are also central to the deployment of robotics in subsea carbon storage. The use of autonomous systems in the deep ocean raises questions about accountability, transparency, and the potential for unintended consequences. Ensuring that robotics is used responsibly requires robust governance frameworks, stakeholder engagement, and mechanisms for oversight and review. For instance, governance frameworks might establish guidelines for the environmental impacts of robotic operations, ensuring that they do not disrupt marine ecosystems or infringe on international agreements.

9.3. Environmental and Geochemical Considerations

The deep ocean holds immense potential for long-term carbon dioxide (CO_2) sequestration, offering vast reservoirs where captured emissions can be stored away from the atmosphere for centuries or millennia. However, the success of this strategy hinges on its ability to minimize environmental risks, particularly those associated with ocean acidification and leakage. These risks have profound implications for the health of marine ecosystems, the integrity of carbon storage projects, and the overall effectiveness of ocean-based climate mitigation. A careful assessment of these risks is essential for ensuring that deep sea carbon sequestration is both a viable and responsible solution to the climate crisis.

Ocean acidification, a process driven by the absorption of excess atmospheric CO_2, is one of the most pressing environmental challenges of our time. When CO_2 dissolves in seawater, it reacts with water molecules to form carbonic acid, which dissociates into bicarbonate and hydrogen ions. This increase in hydrogen ions lowers the pH of seawater, making it more acidic and altering its chemical balance. Ocean acidification has already caused a measurable decline in pH levels since the Industrial Revolution, affecting marine organisms that rely on calcium carbonate to build their shells and skeletons, such as corals, mollusks, and certain plankton species.

Deep sea carbon sequestration has the potential to exacerbate ocean acidification, particularly if injected CO_2 escapes from storage reservoirs and dissolves into the surrounding water. The localized effects of such leakage can be significant, as CO_2 concentrations near the injection site would far exceed natural levels, creating "hotspots" of acidification. These hotspots could disrupt the chemical equilibrium of deep ocean waters, impacting processes such as calcification, nutrient cycling, and the balance of microbial communities. For example, a sudden increase in CO_2 concentrations could lower the saturation state of aragonite and calcite, minerals essential for the formation of shells and skeletons, reducing the growth and survival of calcifying organisms.

The potential for leakage also raises concerns about the effectiveness of deep sea sequestration as a climate mitigation strategy. If stored CO_2 escapes from its reservoir and reenters the atmosphere, it undermines the primary goal of reducing greenhouse gas concentrations and mitigating global warming. Leakage rates as low as 1% per year could significantly reduce the long-term efficacy of sequestration projects, making it imperative to design systems that minimize the risk of leakage and provide robust mechanisms for detection and response.

Assessing the risks of ocean acidification and leakage requires a multidisciplinary approach that integrates advanced technologies, environmental monitoring, and predictive modeling. Artificial intelligence (AI) plays a critical role in this effort, providing the tools needed to analyze complex datasets, identify patterns, and simulate potential scenarios. For example, AI algorithms can process data from geophysical surveys, chemical sensors, and ecological studies to evaluate the likelihood of leakage and its potential impacts on marine ecosystems. These insights inform the design and management of sequestration projects, ensuring that they align with environmental and sustainability goals.

One of the key factors influencing the risk of leakage is the geological integrity of the storage reservoir. Deep sea reservoirs, such as depleted oil and gas fields or saline aquifers, must possess impermeable caprocks that act as a barrier to prevent CO_2 from migrating upward. Faults, fractures, or weaknesses in these caprocks can create pathways for leakage, necessitating rigorous site characterization and monitoring. AI-driven geophysical models enable researchers to analyze seismic data and other subsurface imaging techniques, identifying features that could compromise reservoir integrity. For instance, a machine learning algorithm might detect subtle anomalies in seismic wave patterns that indicate the presence of fractures or heterogeneities in the caprock.

Chemical interactions between CO_2 and the reservoir environment also play a critical role in determining the risk of leakage and acidification. When CO_2 is injected into a reservoir, it can react with water, minerals, and organic matter, altering the geochemical composition of the surrounding system. These reactions may enhance the stability of stored CO_2 by converting it into solid carbonate minerals, a process known as mineralization. However, they can also destabilize the reservoir by dissolving minerals or creating secondary porosity, increasing the risk of leakage. AI-powered geochemical models simulate these interactions under various conditions, providing insights into the long-term behavior of CO_2 and informing strategies to mitigate risks.

Real-time monitoring is essential for detecting and responding to leakage, as well as for assessing its impacts on ocean chemistry and ecosystems. Autonomous sensor networks, integrated with AI algorithms, provide high-resolution data on key indicators such as pH, dissolved inorganic carbon, and oxygen levels. These networks can detect changes in water chemistry that may indicate CO_2 leakage, triggering alerts and guiding interventions. For example, an AI-driven monitoring system might identify a localized drop in pH near an injection site, prompting further investigation to determine the source and extent of the issue.

The ecological impacts of leakage and acidification are a critical area of concern, as they have the potential to disrupt marine food webs and ecosystem services. Calcifying organisms, such as corals and foraminifera, are particularly vulnerable to acidification, as reduced carbonate ion availability makes it more difficult for them to build and maintain their shells and skeletons.

These organisms play foundational roles in marine ecosystems, providing habitat, food, and structural support for a wide range of species. A decline in their populations could have cascading effects on biodiversity and ecosystem stability.

Leakage and acidification also have implications for microbial communities, which drive key biogeochemical processes such as nutrient cycling, organic matter decomposition, and methane oxidation. Changes in pH and CO_2 concentrations can alter the composition and activity of these communities, potentially affecting the overall functioning of the deep ocean as a carbon sink. For example, certain microbial populations may become less efficient at breaking down organic carbon, reducing the sequestration capacity of deep ocean sediments.

Mitigating the risks of ocean acidification and leakage requires a combination of technological innovation, rigorous governance, and adaptive management. Advanced materials and engineering techniques can enhance the stability and security of storage reservoirs, while AI-driven monitoring systems provide the real-time data needed to detect and respond to potential issues. Governance frameworks, such as those established under the United Nations Convention on the Law of the Sea (UNCLOS) and regional agreements, provide a foundation for regulating and overseeing sequestration projects, ensuring that they are conducted responsibly and transparently.

Collaboration among scientists, policymakers, and industry stakeholders is essential for advancing our understanding of the risks associated with ocean acidification and leakage and for developing strategies to address them. Interdisciplinary research efforts can provide valuable insights into the interactions between CO_2 storage, ocean chemistry, and marine ecosystems, while public engagement ensures that the concerns and values of affected communities are considered.

The success of deep sea carbon sequestration hinges not only on the initial design and implementation of storage projects but also on the ability to monitor and manage them over the long term. Long-term monitoring ensures that injected CO_2 remains securely contained, provides early detection of potential issues, and evaluates the environmental impacts of storage activities on marine ecosystems. Given the challenges of operating in remote and extreme ocean environments, traditional monitoring methods often fall short in terms of precision, scalability, and cost-effectiveness. This is where artificial intelligence (AI) and autonomous sensor networks emerge as transformative solutions, offering the tools needed to conduct continuous, high-resolution monitoring with unprecedented efficiency and accuracy.

Long-term monitoring of deep sea carbon sequestration begins with the deployment of autonomous sensor networks, which consist of distributed sensing devices capable of collecting data on a wide range of environmental and geophysical parameters. These sensors measure key indicators such as pressure, temperature, pH, dissolved oxygen, and chemical composition, providing a comprehensive picture of the conditions within and around CO_2 storage reservoirs. By integrating these networks with AI-driven analytics, researchers and operators can process and interpret this data in real time, identifying trends, anomalies, and potential risks that require attention.

One of the primary goals of long-term monitoring is to verify the containment of injected CO_2 within designated reservoirs. The deep ocean operates on timescales that span centuries to millennia, and the stability of storage sites must be ensured over these extended periods. AI algorithms are uniquely suited to this task, enabling the analysis of vast and complex datasets to detect subtle changes that may indicate leakage or migration of CO_2. For instance, machine learning models can analyze pressure and temperature profiles to identify deviations from expected patterns, suggesting the movement of CO_2 through fractures or faults in the reservoir.

Geochemical monitoring is another critical component of long-term verification, as changes in the chemical composition of seawater and sediments can provide early warning signs of CO_2 leakage. Autonomous sensors equipped with AI algorithms can measure pH, alkalinity, and concentrations of dissolved inorganic carbon and carbonate ions, all of which are influenced by the presence of CO_2. By detecting localized decreases in pH or increases in dissolved carbon, these systems can pinpoint areas where CO_2 may be escaping from the reservoir, enabling targeted investigations and mitigation measures.

The role of AI in long-term monitoring extends beyond detection to include predictive modeling and risk assessment. AI-powered simulations enable researchers to explore how injected CO_2 interacts with its environment over time, accounting for variables such as sediment compaction, mineralization, and chemical reactions. These models provide insights into the long-term behavior of CO_2 in storage reservoirs, helping to identify potential risks and inform strategies for maintaining containment. For example, a predictive model might simulate the impacts of reservoir pressure changes on caprock integrity, guiding decisions about injection rates and pressure management to reduce the likelihood of leakage.

One of the key advantages of using AI and autonomous sensor networks is their ability to operate in real time and adapt to changing conditions. In the dynamic and unpredictable environment of the deep ocean, the ability to respond quickly to emerging issues is critical for ensuring the safety and effectiveness of sequestration projects. AI systems process data from sensor networks continuously, using machine learning algorithms to identify patterns and correlations that may indicate developing problems. For instance, an AI-driven monitoring system might detect an increase in microseismic activity near a storage site, suggesting potential stress on the reservoir's caprock. By analyzing the source and magnitude of the activity, the system can recommend appropriate actions, such as adjusting injection rates or deploying additional monitoring resources.

The integration of autonomous underwater vehicles (AUVs) and remotely operated vehicles (ROVs) further enhances the capabilities of long-term monitoring systems. These robotic platforms are equipped with advanced sensors, cameras, and sampling tools, allowing them to conduct detailed inspections of storage sites and surrounding environments. For example, an AUV might perform a high-resolution survey of the seafloor to detect changes in sediment composition or the presence of gas bubbles, which could indicate CO_2 leakage. AI algorithms enable these platforms to operate autonomously, navigating complex terrains and prioritizing areas of interest based on real-time data.

Environmental monitoring is an integral part of long-term oversight, ensuring that carbon sequestration projects align with sustainability goals and minimize their impact on marine ecosystems. Autonomous sensor networks collect data on biodiversity, habitat conditions, and ecological processes, providing insights into the interactions between storage activities and the surrounding environment. For instance, changes in water chemistry caused by CO_2 leakage could affect the health and behavior of marine organisms, such as corals, mollusks, and fish. AI-driven environmental models analyze these impacts, identifying potential risks and informing mitigation strategies to protect vulnerable ecosystems.

One of the most innovative applications of AI in long-term monitoring is the use of digital twins, virtual replicas of physical systems that simulate their behavior under various conditions. Digital twins of CO_2 storage sites integrate data from sensor networks, predictive models, and historical records to provide a dynamic and interactive representation of the reservoir and its environment. These tools allow researchers and operators to test scenarios, evaluate the effectiveness of mitigation measures, and optimize storage strategies in a risk-free virtual environment. For example, a digital twin might simulate the effects of increasing injection rates on reservoir pressure and caprock stability, providing actionable insights for maintaining containment.

The scalability and interconnectedness of AI-driven monitoring systems offer additional benefits for managing multiple storage sites or coordinating regional efforts. By integrating data from a network of storage projects, AI systems can provide a comprehensive and real-time view of carbon sequestration activities on a global scale. This capability is particularly valuable for identifying cumulative impacts, sharing best practices, and aligning strategies across stakeholders. For instance, an international monitoring platform powered by AI could track the performance of storage sites in different regions, ensuring that they collectively contribute to climate goals while adhering to environmental standards.

Despite their transformative potential, the deployment of AI and autonomous sensor networks for long-term monitoring faces challenges that must be addressed to ensure their success. The availability and quality of data are critical, as accurate monitoring depends on comprehensive and high-resolution datasets that capture the complexity of deep ocean systems. Addressing this challenge requires investment in advanced sensors, data-sharing frameworks, and collaborative research initiatives that enable the collection and integration of high-quality data across platforms and regions.

Ethical and governance considerations are also central to the deployment of AI in long-term monitoring. The use of autonomous systems to manage deep ocean systems raises questions about accountability, transparency, and the potential for unintended consequences. Ensuring that AI-driven monitoring systems operate responsibly requires robust governance frameworks, stakeholder engagement, and mechanisms for oversight and review. For instance, governance frameworks might establish guidelines for the environmental impacts of monitoring activities, ensuring that they do not disrupt marine ecosystems or infringe on international agreements.

The ocean's natural carbon sinks and ecosystems play a fundamental role in regulating the global carbon cycle, absorbing and storing vast quantities of carbon dioxide (CO_2) from the atmosphere. These processes, driven by mechanisms such as the biological pump, the solubility pump, and the carbonate pump, are critical for mitigating climate change and maintaining the stability of Earth's climate system. As the need for large-scale carbon sequestration becomes increasingly urgent, exploring synergies between artificial interventions and natural oceanic processes offers a promising pathway for enhancing carbon storage while protecting and restoring marine ecosystems.

The biological pump is one of the ocean's most effective natural mechanisms for carbon sequestration. This process begins with phytoplankton, microscopic plants that perform photosynthesis in the sunlit upper layers of the ocean, converting CO_2 into organic carbon. This organic carbon forms the foundation of marine food webs, supporting a diverse array of life. When phytoplankton and other organisms die, or when they excrete waste, much of this organic matter sinks into deeper layers of the ocean, carrying carbon away from the atmosphere. Some of this material eventually reaches the seafloor, where it can be buried in sediments and sequestered for millennia.

Artificial interventions can work in concert with the biological pump to enhance its efficiency and expand its capacity for carbon storage. For example, ocean fertilization involves the deliberate addition of nutrients, such as iron, nitrogen, or phosphorus, to stimulate phytoplankton growth in nutrient-limited regions. By increasing primary production, these interventions can enhance the export of organic carbon to the deep ocean, effectively amplifying the natural processes of the biological pump. However, the design and implementation of such interventions must be guided by rigorous scientific research and environmental monitoring to ensure that they do not disrupt ecosystems or cause unintended consequences, such as harmful algal blooms or oxygen depletion.

Artificial upwelling is another strategy that leverages the natural processes of the biological pump. This technique involves using pumps or other technologies to bring nutrient-rich deep waters to the surface, stimulating phytoplankton growth and increasing the uptake of atmospheric CO_2. AI-powered optimization tools play a critical role in identifying optimal locations and conditions for artificial upwelling, ensuring that interventions achieve maximum carbon sequestration with minimal ecological disruption. For instance, an AI system might analyze data on ocean currents, nutrient concentrations, and phytoplankton dynamics to determine where artificial upwelling would have the greatest impact.

The solubility pump, which transports dissolved inorganic carbon from surface waters to the deep ocean via thermohaline circulation, represents another opportunity for synergies between artificial interventions and natural processes. This mechanism relies on the fact that cold, dense water at high latitudes absorbs more CO_2 than warm water, enabling it to carry large quantities of carbon into the deep ocean. Enhancing the solubility pump could involve strategies such as reducing sea surface temperatures in key regions or altering circulation patterns to increase the downward transport of CO_2-rich water. These approaches must be carefully designed to avoid unintended impacts on ocean circulation and regional climates.

Ocean alkalinity enhancement is a particularly promising strategy for creating synergies with natural carbon sinks. This approach involves adding alkaline substances, such as crushed limestone or magnesium silicate, to seawater to increase its capacity to absorb and store CO_2. By buffering the acidity of seawater, alkalinity enhancement not only enhances carbon uptake but also counteracts the effects of ocean acidification, which poses a significant threat to marine ecosystems. For example, coral reefs and shell-forming organisms that rely on carbonate ions for calcification can benefit from increased alkalinity, improving their resilience in the face of environmental stressors.

The carbonate pump, which involves the formation and deposition of calcium carbonate ($CaCO_3$) by marine organisms, is another key mechanism that can be supported by artificial interventions. Shell-building organisms such as corals, foraminifera, and coccolithophores play a central role in this process, converting dissolved carbonate ions into solid forms that sink to the seafloor. Enhancing the carbonate pump could involve restoring or protecting habitats that support calcifying organisms, such as coral reefs and seagrass meadows. AI-driven monitoring systems can provide real-time data on the health and productivity of these habitats, enabling targeted interventions to support their ecological functions.

The integration of artificial interventions with natural carbon sinks also offers opportunities for habitat restoration and biodiversity conservation. Many of the ecosystems that serve as natural carbon sinks, such as mangroves, salt marshes, and seagrass meadows, have been degraded or lost due to human activities. Restoring these habitats not only enhances their capacity to sequester carbon but also provides a range of co-benefits, including flood protection, water filtration, and support for fisheries. For example, replanting mangroves in coastal regions can increase carbon storage in soils while protecting communities from storm surges and erosion.

AI and robotics play a transformative role in supporting these restoration efforts by providing the tools needed for large-scale, precise, and adaptive interventions. Autonomous underwater vehicles (AUVs) equipped with sensors and mapping technologies can assess the health and extent of marine habitats, identifying areas where restoration is most needed. Robotic platforms can also be used to plant seagrass or deploy artificial substrates for coral growth, ensuring that interventions are conducted efficiently and with minimal disruption to existing ecosystems. AI algorithms enable the analysis of environmental data to optimize restoration strategies, ensuring that they align with ecological conditions and long-term sustainability goals.

The co-location of carbon sequestration projects with marine conservation areas represents another opportunity for creating synergies between artificial interventions and natural ecosystems. By integrating carbon storage sites with marine protected areas (MPAs), stakeholders can align efforts to reduce greenhouse gas emissions with goals for biodiversity conservation and ecosystem resilience. For example, a CO_2 injection site located near a coral reef MPA might incorporate measures to monitor and mitigate potential impacts on reef ecosystems, ensuring that the benefits of carbon sequestration are not offset by ecological harm.

The potential for synergies also extends to the socio-economic benefits of combining artificial interventions with natural carbon sinks. Many coastal and marine ecosystems provide livelihoods for millions of people, particularly in developing countries and small island states. Enhancing these ecosystems through restoration or protection can create economic opportunities in sectors such as tourism, fisheries, and sustainable aquaculture. For example, restoring seagrass meadows not only increases carbon storage but also supports fish populations that are vital for food security and local economies.

Despite their promise, the integration of artificial interventions with natural carbon sinks and ecosystems presents challenges that must be addressed to ensure their success. One of the primary challenges is the need for robust environmental monitoring and governance frameworks to manage potential risks and trade-offs. Artificial interventions must be designed and implemented in a way that respects the integrity of natural processes, avoiding disruptions that could undermine ecosystem functions or create unintended consequences.

Ethical considerations are also central to the integration of artificial interventions with natural carbon sinks. Decisions about where and how to implement these interventions must consider the rights and interests of affected communities, particularly those that depend on marine resources for their livelihoods. Inclusive decision-making processes, supported by public engagement and stakeholder consultation, are essential for building trust and ensuring that interventions align with social and environmental values.

9.4. Economic and Policy Implications

Deep sea carbon sequestration represents a compelling strategy for addressing the growing threat of climate change by leveraging the vast storage capacity of the ocean to isolate carbon dioxide (CO_2) from the atmosphere. However, the viability of this approach hinges on a careful assessment of its economic and environmental trade-offs. Cost-benefit analyses provide a critical framework for evaluating deep sea carbon sequestration projects, offering insights into their financial feasibility, climate impact, and alignment with broader policy goals. These analyses incorporate a wide range of factors, from the direct costs of infrastructure and operations to the long-term benefits of reduced greenhouse gas emissions, ecosystem preservation, and enhanced climate resilience.

The costs associated with deep sea carbon sequestration can be broadly categorized into upfront capital expenditures, ongoing operational expenses, and potential environmental liabilities. Capital costs include the design and construction of infrastructure, such as pipelines, injection wells, and monitoring systems, as well as the development of capture and transportation technologies. These investments are substantial, particularly for projects located in remote or extreme environments, where logistical challenges drive up costs. For example, constructing pipelines capable of transporting CO_2 to offshore reservoirs requires advanced engineering to withstand high pressures, corrosive seawater, and seismic activity, all of which add to the overall expense.

Operational costs encompass the day-to-day activities required to maintain and monitor storage sites, including the injection of CO_2, the operation of autonomous sensor networks, and the deployment of robotic platforms for inspections and repairs. These costs are influenced by factors such as the scale of the project, the depth and location of storage reservoirs, and the energy requirements for CO_2 compression and transport. Energy consumption is a particularly significant consideration, as it directly impacts the carbon footprint of sequestration projects. Efforts to integrate renewable energy sources, such as wind or wave power, can help mitigate these emissions while reducing operational expenses over the long term.

Environmental liabilities represent another dimension of cost that must be accounted for in any comprehensive analysis. The risks of leakage, ocean acidification, and ecological disruption pose potential financial and reputational costs to project operators, particularly if mitigation measures or remediation efforts are required. These liabilities underscore the importance of rigorous environmental impact assessments and long-term monitoring to ensure that sequestration projects are conducted responsibly and with minimal impact on marine ecosystems.

The benefits of deep sea carbon sequestration are equally diverse, encompassing direct climate mitigation, economic opportunities, and co-benefits for ecosystems and communities. The primary benefit is the reduction of atmospheric CO_2 levels, which helps mitigate global warming and its associated impacts, such as extreme weather events, sea level rise, and biodiversity loss. The effectiveness of sequestration projects in achieving these outcomes depends on their capacity to store CO_2 securely over the long term, as well as their ability to complement other climate mitigation strategies, such as renewable energy adoption and energy efficiency improvements.

From an economic perspective, deep sea carbon sequestration can create opportunities for job creation, technological innovation, and industry growth. The development and deployment of advanced technologies, such as AI-driven monitoring systems and autonomous robotic platforms, drive demand for skilled labor and research expertise, fostering economic activity in sectors such as engineering, data analytics, and environmental science. Additionally, sequestration projects can generate revenue through carbon markets and offset mechanisms, providing financial incentives for companies and governments to invest in climate solutions.

Cost-benefit analyses must also consider the potential for co-benefits that extend beyond climate mitigation. For example, sequestration projects that integrate habitat restoration or marine conservation measures can enhance ecosystem services, such as fisheries productivity, water quality, and storm protection. These benefits contribute to the resilience and well-being of coastal communities, particularly in regions that are highly vulnerable to climate impacts. By aligning sequestration efforts with broader environmental and social goals, stakeholders can maximize their value and ensure that they address multiple dimensions of sustainability.

The long-term nature of deep sea carbon sequestration adds complexity to cost-benefit analyses, as many of the benefits accrue over decades or even centuries, while costs are often incurred upfront. Discounting, a method used to compare costs and benefits over time, plays a critical role in these analyses, but it also introduces challenges. High discount rates can undervalue future benefits, such as avoided climate damages, while low rates may overstate the attractiveness of sequestration projects relative to other mitigation options. Striking the right balance requires careful consideration of the ethical and intergenerational dimensions of climate action, ensuring that today's decisions reflect the needs and rights of future generations.

The role of policy and regulation is central to shaping the cost-benefit landscape of deep sea carbon sequestration. Supportive policies, such as subsidies, tax incentives, or carbon pricing, can improve the economic viability of sequestration projects by reducing costs or increasing revenue potential. For example, a robust carbon pricing mechanism that assigns a monetary value to avoided emissions can create a powerful financial incentive for investment in sequestration technologies. Similarly, public funding for research and development can accelerate innovation, driving down costs and improving the performance of storage systems.

International cooperation and governance frameworks are also critical for aligning the economic and environmental dimensions of sequestration projects. The transboundary nature of the ocean means that interventions in one region can have far-reaching effects, necessitating collaboration among nations to establish standards, share knowledge, and coordinate efforts. Agreements under the United Nations Framework Convention on Climate Change (UNFCCC) and the United Nations Convention on the Law of the Sea (UNCLOS) provide a foundation for such cooperation, ensuring that sequestration projects contribute to global climate goals while respecting ecological and social considerations.

Public perception and stakeholder engagement are important factors that influence the outcomes of cost-benefit analyses. The success of sequestration projects depends on building trust and legitimacy among communities, policymakers, and investors. Transparent communication about the costs, benefits, and risks of sequestration efforts is essential for addressing concerns and fostering support. For example, public consultations and participatory decision-making processes can provide a platform for stakeholders to express their values and priorities, ensuring that projects align with local needs and aspirations.

Advances in artificial intelligence (AI) and data analytics play a transformative role in enhancing the accuracy and reliability of cost-benefit analyses. AI algorithms can process large datasets to evaluate the economic, environmental, and social impacts of sequestration projects under various scenarios. For instance, machine learning models might simulate the long-term performance of storage reservoirs, incorporating factors such as leakage risks, operational costs, and climate benefits. These simulations provide stakeholders with a comprehensive understanding of trade-offs and uncertainties, enabling informed decisions about project design and implementation.

Despite their value, cost-benefit analyses are not without limitations. The inherent complexity of deep ocean systems and the uncertainties associated with long-term carbon storage make it challenging to predict outcomes with precision. Additionally, the quantification of non-monetary benefits, such as biodiversity conservation or cultural heritage, requires methodologies that capture their intrinsic and societal value. Addressing these challenges requires interdisciplinary collaboration, continuous learning, and the integration of diverse perspectives into analytical frameworks.

As the urgency to address climate change intensifies, integrating carbon dioxide (CO_2) storage into global emissions reduction targets has become an essential strategy for meeting the goals of the Paris Agreement and achieving net-zero emissions. Deep sea CO_2 sequestration, with its potential to store vast quantities of carbon over long timescales, offers a powerful tool for complementing other climate mitigation efforts, such as transitioning to renewable energy and enhancing energy efficiency. However, successfully integrating CO_2 storage into global targets requires robust policy frameworks, technological innovation, and international cooperation. This chapter explores the economic and policy dimensions of this integration, highlighting the opportunities and challenges of aligning CO_2 storage with global climate goals.

The Paris Agreement, adopted in 2015, established a global framework for limiting global temperature rise to well below 2°C above pre-industrial levels, with efforts to keep it below 1.5°C. Achieving these targets requires a rapid and sustained reduction in greenhouse gas (GHG) emissions across all sectors, as well as the active removal of CO_2 from the atmosphere. While renewable energy, energy efficiency, and behavioral changes are critical components of this effort, CO_2 storage provides an essential solution for addressing residual emissions from hard-to-abate sectors such as heavy industry, aviation, and shipping.

Deep sea CO_2 storage plays a unique role in this context by providing a long-term, scalable solution for managing carbon emissions that cannot be eliminated through decarbonization alone. For example, industries such as cement and steel production emit significant quantities of CO_2 as an inherent byproduct of their processes, making them particularly challenging to decarbonize. By capturing CO_2 from these sources and storing it in deep sea reservoirs, stakeholders can offset emissions that are otherwise unavoidable, enabling these sectors to align with net-zero targets.

The integration of CO_2 storage with global emissions reduction targets requires a comprehensive policy framework that supports the deployment of storage technologies while ensuring their alignment with environmental, social, and economic goals. Carbon pricing mechanisms, such as carbon taxes and cap-and-trade systems, are among the most effective tools for incentivizing CO_2 storage by assigning a monetary value to avoided emissions. For instance, a robust carbon pricing system can create a financial incentive for industries to invest in CO_2 capture and storage technologies, as the cost of emitting CO_2 becomes higher than the cost of storing it.

Subsidies and tax incentives also play a critical role in enhancing the economic viability of CO_2 storage projects. Public funding for research and development, as well as direct support for infrastructure and operations, can reduce the financial barriers to entry and accelerate the deployment of storage technologies. For example, government-backed grants or loan guarantees can support the construction of pipelines, injection wells, and monitoring systems, enabling projects to move forward with greater confidence.

International cooperation and coordination are essential for integrating CO_2 storage into global emissions reduction targets. The transboundary nature of the ocean and the global scale of climate challenges necessitate collaboration among nations to establish shared standards, pool resources, and align efforts. Multilateral agreements, such as the United Nations Framework Convention on Climate Change (UNFCCC), provide a platform for fostering dialogue and cooperation on CO_2 storage, ensuring that it contributes to global climate goals while respecting ecological and social considerations.

One of the key challenges in integrating CO_2 storage with global targets is the need for standardized accounting and reporting frameworks. Accurate and transparent accounting is critical for tracking the contribution of CO_2 storage to emissions reductions and ensuring its inclusion in national and international climate inventories. The Intergovernmental Panel on Climate Change (IPCC) provides guidelines for estimating and reporting CO_2 storage, but further refinement and harmonization are needed to account for the unique characteristics of deep sea storage projects. For example, methodologies must address issues such as the potential for leakage, the permanence of storage, and the environmental impacts of storage activities.

The role of artificial intelligence (AI) and digital technologies in supporting integration efforts cannot be overstated. AI-powered monitoring and verification systems enable real-time tracking of CO_2 storage performance, providing the data needed to demonstrate compliance with climate targets and build public trust. For instance, AI algorithms can analyze data from autonomous sensor networks to detect changes in reservoir conditions, assess the effectiveness of storage operations, and identify potential risks such as leakage. These insights are invaluable for ensuring the reliability and transparency of CO_2 storage projects, which are critical for their acceptance and integration into climate policies.

Public perception and stakeholder engagement are also central to the success of integrating CO_2 storage with global emissions reduction targets. While the technical and economic aspects of storage are essential, the social dimensions of these projects often determine their feasibility and sustainability. Building trust among communities, policymakers, and investors requires clear communication about the benefits, risks, and trade-offs of CO_2 storage. For example, public consultations and participatory decision-making processes can provide a platform for stakeholders to express their concerns, share their perspectives, and contribute to the development of governance frameworks.

The co-benefits of CO_2 storage can also play a significant role in enhancing its integration with global targets. In addition to mitigating climate change, storage projects can deliver economic, environmental, and social benefits that align with broader sustainability goals. For example, the development and deployment of storage technologies can create jobs in engineering, data analytics, and environmental science, driving economic growth and innovation. Similarly, integrating CO_2 storage with habitat restoration or marine conservation efforts can enhance ecosystem services, such as biodiversity, water quality, and storm protection.

Despite its potential, the integration of CO_2 storage with global emissions reduction targets is not without challenges. The high upfront costs of infrastructure, the need for long-term monitoring and governance, and the potential risks to marine ecosystems all pose significant hurdles that must be addressed. However, these challenges also present opportunities for innovation and collaboration. For instance, advances in material science, robotics, and AI can reduce costs and improve the performance of storage systems, while interdisciplinary research can provide insights into the environmental and social impacts of storage projects.

The ethical dimensions of CO_2 storage also warrant careful consideration, particularly in terms of equity and justice. Decisions about where and how to implement storage projects must take into account the rights and interests of affected communities, particularly those in developing countries or regions with high dependence on marine resources. Inclusive decision-making processes, supported by public engagement and stakeholder consultation, are essential for ensuring that storage projects align with social and environmental values.

International cooperation and the development of robust carbon markets are essential for scaling deep sea carbon sequestration as a viable strategy for mitigating climate change. The challenges posed by rising greenhouse gas (GHG) emissions transcend national boundaries, necessitating collaborative approaches that align the interests of diverse stakeholders, including governments, industries, and communities. Carbon markets, which assign economic value to carbon reductions and removals, provide a powerful incentive for investment in carbon sequestration technologies, including deep sea storage. Together, international cooperation and carbon markets create a framework for advancing carbon sequestration efforts, ensuring that they contribute meaningfully to global climate goals.

The transboundary nature of the ocean highlights the need for international cooperation in managing deep sea carbon sequestration projects. The ocean is a shared resource, and activities in one region can have far-reaching effects on marine ecosystems, atmospheric carbon levels, and neighboring states. For instance, a deep sea CO_2 injection site located near a continental margin may affect water chemistry and circulation patterns beyond national jurisdictions. International cooperation is essential for addressing these interconnected impacts, establishing shared standards, and fostering trust among stakeholders.

Multilateral agreements such as the United Nations Framework Convention on Climate Change (UNFCCC) and the Paris Agreement provide a foundation for coordinating global climate efforts, including carbon sequestration. The Paris Agreement, in particular, emphasizes the importance of reaching net-zero emissions by mid-century, creating a strong mandate for the development and deployment of carbon storage technologies. These agreements also facilitate knowledge exchange, capacity building, and financial support for countries that wish to explore carbon sequestration as part of their climate strategies.

Regional frameworks and initiatives play a complementary role in advancing international cooperation on carbon sequestration. For example, regional ocean governance bodies, such as the North-East Atlantic's OSPAR Commission or the Pacific Islands Forum, can address region-specific challenges and opportunities related to deep sea storage. These organizations can develop tailored guidelines, support regional research initiatives, and promote collaboration among member states, ensuring that sequestration projects align with local priorities and ecological conditions.

International cooperation is also critical for addressing the ethical and equity dimensions of carbon sequestration. Many of the countries most vulnerable to climate change, such as small island developing states (SIDS) and low-lying coastal nations, lack the resources and technical capacity to implement large-scale sequestration projects on their own. Ensuring that these countries benefit from sequestration efforts requires mechanisms for technology transfer, financial support, and inclusive decision-making. For instance, developed nations could provide funding and technical expertise to help SIDS establish CO_2 storage infrastructure, enhancing their resilience while contributing to global emissions reductions.

Carbon markets play a central role in mobilizing the financial resources needed to scale deep sea carbon sequestration. By assigning a monetary value to carbon reductions and removals, carbon markets create economic incentives for investment in sequestration technologies. These markets operate through mechanisms such as cap-and-trade systems, carbon taxes, and voluntary carbon credit programs, each of which provides pathways for monetizing the climate benefits of CO_2 storage.

In a cap-and-trade system, governments set a limit on the total amount of emissions that can be produced within a given jurisdiction, allocating or auctioning emission allowances to companies. Companies that reduce their emissions below their allocated limit can sell their excess allowances to others, creating a financial incentive for low-carbon innovation. Deep sea carbon sequestration projects, by removing CO_2 from the atmosphere or preventing its release, can generate credits that are traded within these markets, providing a revenue stream for project developers.

Carbon taxes, which impose a fee on each ton of CO_2 emitted, provide another mechanism for incentivizing sequestration. By increasing the cost of emitting CO_2, carbon taxes encourage industries to invest in carbon capture and storage (CCS) technologies, including deep sea sequestration. For instance, a steel manufacturer facing high carbon tax rates might opt to capture and store its emissions in a deep sea reservoir, reducing its tax liability while contributing to climate mitigation.

Voluntary carbon markets, which operate independently of government mandates, offer additional opportunities for financing sequestration projects. These markets allow companies and individuals to purchase carbon credits to offset their emissions, often as part of corporate social responsibility initiatives or net-zero commitments. Deep sea sequestration projects can generate high-quality carbon credits by demonstrating rigorous monitoring, permanence, and co-benefits for ecosystems and communities. For example, a sequestration project that integrates marine habitat restoration could attract buyers seeking to support biodiversity and climate goals simultaneously.

The integration of carbon markets with international cooperation frameworks enhances their effectiveness and scalability. For example, Article 6 of the Paris Agreement establishes mechanisms for international trading of emissions reductions, enabling countries to achieve their climate targets more cost-effectively by investing in mitigation efforts abroad. This framework creates opportunities for cross-border collaboration on deep sea sequestration, allowing countries with high storage potential, such as those with extensive continental margins or saline aquifers, to host projects financed by other nations or entities.

Ensuring the credibility and integrity of carbon markets is critical for their success. The effectiveness of these markets depends on robust methodologies for measuring, reporting, and verifying emissions reductions and removals. In the context of deep sea sequestration, this requires high-resolution monitoring and modeling systems that provide transparent and accurate data on CO_2 storage performance. Artificial intelligence (AI) and autonomous sensor networks play a transformative role in this effort, enabling real-time tracking of storage conditions, leakage risks, and environmental impacts. For example, an AI-driven monitoring system might analyze data from geochemical sensors to verify the containment of injected CO_2, providing the evidence needed to issue carbon credits.

Ethical considerations are also central to the design and operation of carbon markets. Decisions about how and where to implement sequestration projects must ensure that the benefits and burdens of these efforts are distributed equitably. For instance, developing countries and marginalized communities should have access to the economic opportunities created by carbon markets, such as jobs in project development, monitoring, and maintenance. Mechanisms for stakeholder engagement and public participation are essential for addressing these concerns and building trust in the markets.

The integration of carbon markets with other policy instruments further enhances their potential to drive investment in deep sea sequestration. For example, governments can complement carbon pricing mechanisms with subsidies, grants, or tax incentives that lower the upfront costs of sequestration projects. Similarly, public-private partnerships can leverage the resources and expertise of both sectors to accelerate the deployment of storage technologies. For instance, a partnership between a national government, a technology company, and a research institution could support the development of AI-driven monitoring platforms, reducing costs and improving the performance of sequestration projects.

Despite their promise, carbon markets face challenges that must be addressed to ensure their effectiveness and sustainability. One of the primary challenges is the risk of double counting, where emissions reductions or removals are claimed by multiple entities, undermining the integrity of the markets. Addressing this issue requires robust accounting and reporting frameworks that track the ownership and impact of carbon credits throughout their lifecycle. Additionally, carbon markets must address concerns about the permanence of sequestration, ensuring that stored CO_2 remains securely contained over the long term.

9.5. A Blueprint for Scaling CO_2 Sequestration

As the world grapples with the urgent need to mitigate climate change, piloting AI-controlled carbon dioxide (CO_2) sequestration projects offers a promising avenue for testing and refining technologies that could revolutionize global carbon management. These pilot projects serve as critical stepping stones, enabling researchers, policymakers, and industry leaders to assess the feasibility, safety, and effectiveness of using artificial intelligence (AI) to optimize deep sea CO_2 storage. By combining cutting-edge AI algorithms with advanced monitoring systems and autonomous robotics, these initiatives not only demonstrate the potential of AI-driven sequestration but also lay the groundwork for scaling these efforts to address the climate crisis at a global level.

The central premise of AI-controlled sequestration is to harness the power of machine learning and data analytics to manage the complexities of CO_2 storage in the deep ocean. From site selection and injection processes to long-term monitoring and adaptive management, AI provides the tools needed to navigate the dynamic and multifaceted challenges of deep sea sequestration. Piloting these projects allows stakeholders to test these capabilities in real-world conditions, gathering insights that inform best practices and guide the development of policies and standards.

The initial phase of an AI-controlled sequestration pilot involves the selection of a suitable storage site, a process that is both scientifically rigorous and technologically demanding. AI algorithms play a crucial role in this effort, analyzing large datasets from seismic surveys, geophysical models, and environmental assessments to identify locations that meet key criteria for storage. These criteria include geological stability, sediment composition, and proximity to existing infrastructure. For example, an AI system might process data on rock porosity, fault lines, and caprock integrity to pinpoint a reservoir that can securely contain CO_2 over long timescales. By leveraging AI, pilot projects can streamline site selection and ensure that chosen locations maximize safety and efficacy.

Once a site is selected, the pilot project moves to the design and implementation of injection systems. CO_2 injection is a complex process that requires precise control over parameters such as pressure, flow rate, and temperature to avoid compromising reservoir integrity or triggering seismic activity. AI-driven control systems excel at managing these variables in real time, using data from sensors and monitoring equipment to optimize operations. For instance, an AI system might detect changes in reservoir pressure during injection and adjust flow rates accordingly, preventing over-pressurization and ensuring that CO_2 is distributed evenly throughout the storage formation. These adaptive capabilities are particularly valuable in the unpredictable and dynamic environment of the deep ocean.

Monitoring and verification are central to the success of AI-controlled sequestration pilots, as they provide the evidence needed to demonstrate containment and assess environmental impacts. Autonomous sensor networks, integrated with AI algorithms, offer continuous, high-resolution data on key indicators such as pH, dissolved inorganic carbon, and seismic activity. These systems can detect anomalies that may indicate leakage or other issues, triggering alerts and guiding corrective actions. For example, if a sensor network detects a localized drop in pH near the injection site, an AI system might analyze the data to determine whether the change is due to natural variability or a potential breach in the reservoir.

Robotic platforms, such as autonomous underwater vehicles (AUVs) and remotely operated vehicles (ROVs), further enhance monitoring efforts by providing the ability to inspect storage sites and surrounding environments in detail. Equipped with cameras, sonar, and chemical sensors, these robots can conduct visual surveys, measure gas concentrations, and collect sediment samples to verify the stability and performance of storage systems. AI enables these platforms to operate autonomously, navigating complex seafloor terrains and prioritizing areas of interest based on real-time data.

Piloting AI-controlled sequestration projects also involves assessing their environmental and social implications. The deep ocean is a critical component of Earth's biosphere, supporting diverse ecosystems and providing essential services such as nutrient cycling and climate regulation. Ensuring that sequestration projects align with sustainability goals requires a thorough understanding of their interactions with marine ecosystems. AI-driven environmental models play a key role in this effort, simulating the potential impacts of CO_2 storage on biodiversity, habitat conditions, and biogeochemical processes. For example, a model might predict how changes in water chemistry caused by CO_2 injection could affect the behavior and survival of marine species, informing mitigation measures that minimize ecological disruption.

Stakeholder engagement is another critical aspect of piloting AI-controlled sequestration projects. Building trust and legitimacy among communities, policymakers, and industry stakeholders requires transparent communication about the goals, risks, and benefits of the project. Public consultations, advisory committees, and participatory workshops provide platforms for stakeholders to express their concerns, share their perspectives, and contribute to decision-making processes. For instance, coastal communities that depend on marine resources for their livelihoods might participate in discussions about site selection and monitoring strategies, ensuring that their interests are considered and respected.

One of the key advantages of piloting AI-controlled sequestration projects is the ability to generate insights that inform the scaling of these efforts to regional or global levels. Pilots provide a controlled environment for testing technologies, refining methodologies, and addressing challenges before they are deployed on a larger scale. For example, a pilot project might identify best practices for integrating AI with autonomous robotics, demonstrating how these technologies can work together to enhance monitoring and operational efficiency. These insights are invaluable for guiding future projects, reducing costs, and improving outcomes.

Economic considerations are also central to the design and evaluation of pilot projects. The deployment of AI-controlled sequestration systems involves significant upfront investments in infrastructure, technology, and expertise. However, the long-term benefits of these projects, such as avoided climate damages, job creation, and contributions to carbon markets, can outweigh the initial costs. Piloting projects provides an opportunity to evaluate their economic feasibility, testing different financing models, cost-sharing mechanisms, and revenue streams. For example, a pilot might explore how carbon credits generated through sequestration can offset operational expenses, making the project financially sustainable.

Despite their potential, piloting AI-controlled sequestration projects faces challenges that must be addressed to ensure their success. The availability and quality of data are critical, as accurate site selection, monitoring, and modeling depend on comprehensive and high-resolution datasets. Addressing this challenge requires investment in data collection technologies, such as seismic surveys, oceanographic sensors, and environmental sampling, as well as initiatives to standardize and integrate data across platforms.

Ethical and governance considerations are equally important. The use of AI to manage deep ocean systems raises questions about accountability, transparency, and the potential for unintended consequences. Pilots must be guided by robust governance frameworks that include mechanisms for oversight, ethical review, and stakeholder engagement. For instance, an independent advisory board might evaluate the environmental and social impacts of the project, ensuring that it aligns with shared values and regulatory standards.

Carbon offset strategies are an integral component of global efforts to mitigate climate change. These strategies allow individuals, corporations, and nations to compensate for their emissions by investing in projects that reduce or remove an equivalent amount of carbon dioxide (CO_2) from the atmosphere. As the urgency of addressing climate change intensifies, enhancing global carbon offset strategies through innovative approaches such as deep sea CO_2 sequestration presents a transformative opportunity. By integrating advanced technologies, robust governance, and global collaboration, stakeholders can align offset mechanisms with climate targets and ensure that they deliver tangible and lasting benefits.

Deep sea CO_2 sequestration has emerged as a powerful tool for enhancing carbon offset strategies, offering the potential to store vast quantities of carbon in geological formations or sediments beneath the ocean floor. Unlike traditional offsets, such as reforestation or soil carbon storage, deep sea sequestration provides long-term, scalable solutions capable of addressing the residual emissions from hard-to-abate sectors, such as heavy industry, aviation, and shipping. Incorporating this technology into global carbon offset frameworks requires addressing key challenges related to measurement, verification, cost-effectiveness, and equity.

The cornerstone of any carbon offset strategy is the ability to accurately measure and verify the emissions reductions or removals achieved by the project. For deep sea sequestration, this involves monitoring the injection, containment, and long-term behavior of CO_2 within storage reservoirs. Advanced technologies, such as artificial intelligence (AI), autonomous sensor networks, and geophysical imaging, play a critical role in ensuring the integrity of sequestration projects. AI-driven monitoring systems analyze real-time data from sensors to detect anomalies, assess risks, and verify the permanence of storage. For example, an AI model might analyze pressure and chemical data from injection sites to confirm that CO_2 remains securely contained and does not migrate toward the seafloor or adjacent ecosystems.

The integration of deep sea sequestration into carbon offset markets also hinges on the development of standardized methodologies for quantifying and reporting emissions reductions. These methodologies must account for the unique characteristics of ocean-based storage, including the potential for leakage, the interaction between injected CO_2 and surrounding sediments, and the environmental impacts of storage activities. For instance, frameworks developed by the Intergovernmental Panel on Climate Change (IPCC) or international carbon market organizations could include specific protocols for measuring the additionality, permanence, and co-benefits of deep sea sequestration projects.

Cost-effectiveness is another critical consideration for enhancing global carbon offset strategies. While deep sea sequestration offers significant potential for large-scale carbon removal, its initial costs are high due to the need for advanced infrastructure, such as pipelines, injection wells, and monitoring systems. However, as technology advances and economies of scale are achieved, these costs are expected to decline, making sequestration more competitive with other offset options. Financial incentives, such as carbon credits or subsidies, can further improve the economic viability of sequestration projects, encouraging investment and innovation.

Carbon markets provide a powerful mechanism for financing deep sea sequestration projects by creating economic value for emissions reductions or removals. In cap-and-trade systems, for example, companies that exceed their emissions caps can purchase credits generated by sequestration projects to offset their excess emissions. These credits represent a measurable and verified amount of carbon stored in the deep ocean, providing a financial incentive for project developers. Similarly, voluntary carbon markets enable corporations and individuals to offset their carbon footprints by investing in high-quality sequestration projects that align with their sustainability goals.

The integration of deep sea sequestration into carbon markets also offers opportunities for co-benefits that extend beyond climate mitigation. For instance, projects that incorporate habitat restoration or marine conservation measures can enhance biodiversity, support fisheries, and protect coastal communities from erosion and storm surges. These co-benefits increase the appeal of sequestration projects to buyers in voluntary markets who seek to support broader environmental and social objectives. For example, a sequestration project that restores seagrass meadows while storing CO_2 might attract companies looking to demonstrate leadership in both climate action and marine conservation.

Ethical and equity considerations are central to enhancing global carbon offset strategies through deep sea sequestration. Decisions about where and how to implement sequestration projects must ensure that the benefits and burdens of these efforts are distributed fairly. Developing countries and small island states, which are often the most vulnerable to climate change, should have access to the economic opportunities created by sequestration projects, such as jobs in project development, monitoring, and maintenance. Mechanisms for capacity building and technology transfer are essential for empowering these nations to participate in and benefit from sequestration efforts.

Stakeholder engagement is critical for addressing these ethical dimensions and building trust in carbon offset strategies. Transparent communication about the goals, risks, and benefits of sequestration projects fosters accountability and ensures that stakeholders have a voice in decision-making processes. For instance, public consultations and participatory workshops can provide a platform for communities, conservation groups, and industry representatives to share their perspectives and contribute to project design and governance. These processes not only enhance the legitimacy of sequestration projects but also align them with local needs and values.

The role of international cooperation in enhancing carbon offset strategies cannot be overstated. The global nature of climate change requires collaborative efforts to establish shared standards, align incentives, and coordinate actions. Multilateral agreements, such as the Paris Agreement, provide a framework for integrating sequestration projects into national and international climate commitments. Article 6 of the Paris Agreement, which addresses cooperative approaches and market mechanisms, creates opportunities for countries to collaborate on sequestration efforts by trading emissions reductions or removals. For example, a developed country with limited sequestration potential might invest in a deep sea storage project in a developing country, achieving cost-effective reductions while supporting sustainable development.

Robust governance frameworks are essential for ensuring the integrity and effectiveness of enhanced carbon offset strategies. These frameworks should include mechanisms for monitoring, reporting, and verifying the outcomes of sequestration projects, as well as safeguards to prevent double counting or over-crediting. For instance, an international registry of sequestration projects could track the ownership and impact of carbon credits, ensuring that they represent real and additional emissions reductions. Governance frameworks should also address potential risks, such as leakage or environmental impacts, by establishing protocols for monitoring and mitigation.

Artificial intelligence and digital technologies play a transformative role in supporting these governance efforts. AI systems provide the analytical and predictive capabilities needed to evaluate the performance of sequestration projects, assess their alignment with climate targets, and identify opportunities for improvement. For example, an AI-driven platform might analyze data from multiple sequestration sites to compare their cost-effectiveness, environmental impacts, and scalability, providing insights that inform policy decisions and market strategies.

Despite their promise, carbon offset strategies that rely on deep sea sequestration face challenges that must be addressed to realize their full potential. The scientific and technical uncertainties associated with deep ocean systems, the high upfront costs of infrastructure, and the need for robust monitoring and governance frameworks all pose significant hurdles. However, these challenges also present opportunities for innovation and collaboration. For instance, advances in material science, robotics, and geophysical modeling can reduce costs and improve the performance of sequestration systems, while interdisciplinary research can provide insights into the ecological and social impacts of storage projects.

Achieving a carbon-neutral future is one of humanity's most urgent and transformative challenges. It requires not only a dramatic reduction in greenhouse gas emissions but also the active removal of carbon dioxide (CO_2) from the atmosphere to balance residual emissions. Among the strategies to accomplish this, deep sea CO_2 sequestration offers a powerful and scalable solution, leveraging the vast storage capacity of the ocean to support global climate goals. The path to a carbon-neutral future involves the integration of advanced technologies, robust policies, and international collaboration, ensuring that CO_2 sequestration aligns with sustainability principles, economic realities, and societal values.

The first step toward scaling deep sea CO_2 sequestration is the widespread adoption of technologies that enable efficient and secure carbon capture and storage (CCS). CO_2 capture begins at the point of emission, whether from industrial facilities, power plants, or other sources. Advances in capture technologies, such as chemical absorption, membrane separation, and cryogenic processes, have significantly improved the efficiency and cost-effectiveness of removing CO_2 from emissions streams. The integration of artificial intelligence (AI) into these processes further enhances their performance, optimizing capture rates and reducing energy consumption.

Once captured, CO_2 must be transported to offshore storage sites, often located in deep ocean reservoirs such as saline aquifers or depleted oil and gas fields. The construction of pipelines and transport vessels capable of handling compressed or supercritical CO_2 is a critical component of the infrastructure needed for large-scale sequestration. These systems must be designed to operate reliably in harsh marine environments, with materials and engineering that ensure safety and durability. AI-driven logistics and optimization tools enable the efficient coordination of transport networks, minimizing costs and environmental impacts.

The injection of CO_2 into deep sea reservoirs is a technically complex process that requires precise control over parameters such as pressure, flow rate, and temperature to ensure safe and effective storage. Advanced monitoring systems, powered by AI and autonomous robotics, provide real-time data on reservoir conditions, allowing operators to adjust injection strategies dynamically. For example, if sensors detect changes in pressure or temperature that could compromise reservoir integrity, AI algorithms can recommend corrective actions, such as adjusting injection rates or redistributing CO_2 to alternative sites. These adaptive capabilities are essential for maintaining the stability and performance of storage systems over the long term.

Long-term monitoring and verification are central to the success of deep sea CO_2 sequestration, ensuring that stored carbon remains securely contained and that storage projects contribute to climate goals. Autonomous sensor networks and robotic platforms provide continuous data on key indicators such as CO_2 concentrations, water chemistry, and sediment stability. AI-powered analytics enable the detection of anomalies and the assessment of potential risks, supporting proactive management and mitigation efforts. For instance, an AI-driven monitoring system might identify an increase in dissolved CO_2 levels near a storage site, prompting further investigation and the deployment of containment measures if necessary.

The integration of deep sea CO_2 sequestration into national and international climate strategies is a critical step on the path to a carbon-neutral future. Governments play a key role in creating the policy frameworks and financial incentives needed to support sequestration projects. Carbon pricing mechanisms, such as taxes or cap-and-trade systems, provide a direct economic signal that rewards emissions reductions and removals. For example, a robust carbon pricing system that includes deep sea storage as an eligible mitigation option can encourage industries to invest in CCS technologies, reducing their emissions footprint while contributing to climate goals.

Subsidies and public funding for research, development, and deployment also play a vital role in advancing sequestration technologies. By lowering the financial barriers to entry, these incentives enable the scaling of sequestration projects and drive innovation in storage methods, monitoring systems, and infrastructure. For instance, government grants might support the development of AI-driven monitoring platforms or the construction of pipelines and injection wells in regions with high sequestration potential.

International cooperation and governance frameworks are essential for aligning deep sea sequestration efforts with global climate targets. The transboundary nature of the ocean and the global scale of climate change necessitate collaborative approaches that transcend national boundaries. Multilateral agreements, such as the Paris Agreement and the United Nations Convention on the Law of the Sea (UNCLOS), provide a foundation for coordinating sequestration efforts, establishing standards, and sharing knowledge. For example, international partnerships might focus on creating a global registry of CO_2 storage sites, documenting their performance and ensuring transparency and accountability.

Public engagement and stakeholder participation are critical for building trust and legitimacy in sequestration projects. The potential risks and benefits of deep sea storage must be communicated clearly and inclusively, addressing concerns and fostering a sense of shared ownership among communities, industries, and governments. Mechanisms for stakeholder engagement, such as public consultations, advisory boards, and participatory decision-making processes, provide valuable opportunities for dialogue and collaboration. For example, coastal communities that depend on marine ecosystems for their livelihoods might participate in discussions about the siting and monitoring of storage projects, ensuring that their perspectives are incorporated into project planning and governance.

The co-benefits of deep sea CO_2 sequestration further enhance its value as a cornerstone of a carbon-neutral future. In addition to reducing atmospheric CO_2 levels, sequestration projects can support biodiversity, protect ecosystems, and promote sustainable economic development. For instance, integrating habitat restoration or marine conservation measures into sequestration projects can enhance ecosystem services, such as fisheries productivity, water quality, and coastal protection. These co-benefits contribute to the resilience and well-being of both human and natural systems, creating synergies between climate mitigation and broader sustainability goals.

The role of innovation and interdisciplinary research in advancing deep sea sequestration cannot be overstated. Emerging technologies, such as quantum computing, bio-inspired robotics, and advanced materials, offer new possibilities for improving the efficiency, safety, and scalability of sequestration systems. Interdisciplinary research efforts that bring together oceanographers, engineers, data scientists, and policymakers provide the insights needed to address complex challenges and optimize solutions. For example, collaborative research might focus on understanding the long-term behavior of CO_2 in marine sediments, exploring how interactions between chemical, biological, and physical processes influence storage stability.

Despite its promise, the path to a carbon-neutral future through deep sea CO_2 sequestration is not without challenges. The high upfront costs of infrastructure, the need for long-term monitoring and governance, and the potential risks to marine ecosystems all pose significant hurdles. However, these challenges also present opportunities for innovation, collaboration, and leadership. For instance, advances in AI and robotics can reduce costs and improve the performance of sequestration systems, while partnerships between governments, industries, and research institutions can leverage resources and expertise to overcome barriers.

Ethical and equity considerations are central to the success of deep sea sequestration efforts. Decisions about how and where to implement sequestration projects must ensure that the benefits and burdens of these efforts are distributed fairly. Developing countries and marginalized communities, which are often disproportionately affected by climate change, should have access to the economic and environmental opportunities created by sequestration projects. Capacity building, technology transfer, and inclusive governance are essential for empowering these stakeholders to participate in and benefit from sequestration efforts.

10. Regulatory, Governance, and Ethical Considerations

10.1. International Legal Frameworks

The United Nations Convention on the Law of the Sea (UNCLOS) and the regulations established by the International Seabed Authority (ISA) form the cornerstone of international legal governance for activities in the deep ocean, particularly in areas beyond national jurisdiction. These frameworks are pivotal in regulating deep sea mining, balancing the exploitation of mineral resources with the need to protect marine ecosystems and ensuring equitable benefit-sharing among all nations. For stakeholders in the deep sea mining industry, understanding and navigating the provisions of UNCLOS and ISA regulations is not only a legal necessity but also a strategic imperative that influences project feasibility, investment decisions, and long-term sustainability.

UNCLOS, often referred to as the "Constitution of the Oceans," was adopted in 1982 as a comprehensive treaty to govern the use of the world's oceans. It defines the rights and responsibilities of states in their exploration and utilization of marine resources and establishes a legal framework for the management of areas beyond national jurisdiction, including the seabed and subsoil. A central tenet of UNCLOS is the designation of the international seabed area, the region beyond the limits of national exclusive economic zones (EEZs), as the "common heritage of mankind." This principle asserts that the seabed and its mineral resources belong to humanity as a whole, rather than to individual states or entities, and must be managed for the benefit of all, particularly developing countries.

The operationalization of this principle is entrusted to the ISA, an autonomous international organization established under UNCLOS with the mandate to regulate all activities related to the exploration and exploitation of mineral resources in the international seabed area. Headquartered in Kingston, Jamaica, the ISA serves as the central authority responsible for ensuring that resource extraction in this region aligns with the principles of equity, environmental protection, and sustainable development.

One of the ISA's primary functions is to issue licenses for mineral exploration and exploitation in the international seabed area. These licenses are granted to states or state-sponsored entities that demonstrate the technical and financial capacity to undertake deep sea mining operations responsibly. The licensing process involves rigorous requirements, including the submission of detailed plans of work, environmental impact assessments (EIAs), and financial commitments to ensure compliance with ISA regulations. The cost and complexity of obtaining and maintaining these licenses represent significant considerations for mining operators and investors, shaping the economic feasibility and timelines of projects.

Environmental protection is a core focus of the ISA's regulatory framework, reflecting the recognition of the deep ocean as a fragile and poorly understood ecosystem. ISA regulations mandate the adoption of precautionary approaches, requiring operators to identify and mitigate potential environmental impacts before commencing operations. This includes conducting baseline studies to establish the ecological characteristics of proposed mining sites, developing monitoring programs to track environmental changes, and implementing adaptive management strategies to address unforeseen impacts. Operators are also required to submit environmental management plans as part of their licensing applications, outlining measures to minimize harm to marine life and ensure the restoration of habitats affected by mining activities.

One of the most contentious aspects of the ISA's environmental regulations is the management of sediment plumes generated by mining operations. These plumes, caused by the disturbance of seafloor sediments during resource extraction, can spread over large areas, affecting water quality, benthic habitats, and pelagic ecosystems. The ISA has established guidelines for plume monitoring and mitigation, emphasizing the need for robust data collection and modeling to predict plume behavior and assess its ecological consequences. Compliance with these guidelines requires significant investment in environmental monitoring technologies, including high-resolution sensors, autonomous underwater vehicles (AUVs), and artificial intelligence (AI) tools for data analysis.

The ISA's benefit-sharing mechanism is another key regulatory influence on deep sea mining. Under UNCLOS, revenues generated from the exploitation of mineral resources in the international seabed area must be shared equitably among all member states, with a particular focus on supporting developing countries. This principle is operationalized through financial contributions by mining operators, which may include royalties, profit-sharing arrangements, or contributions to a common fund managed by the ISA. These payments are used to promote capacity-building initiatives, technology transfer, and socio-economic development in less developed nations. While the benefit-sharing mechanism fosters inclusivity and fairness, it also adds to the cost structure of mining projects, impacting their profitability and financial modeling.

In addition to its licensing and regulatory functions, the ISA plays a crucial role in advancing scientific research and fostering international cooperation in deep sea mining. The organization supports collaborative research initiatives to enhance the understanding of deep ocean ecosystems, resource distribution, and mining technologies. It also facilitates knowledge-sharing among member states, promoting the development and dissemination of best practices for sustainable resource management. These efforts contribute to the global knowledge base, enabling informed decision-making and the refinement of regulatory frameworks to address emerging challenges.

The regulatory framework established by UNCLOS and the ISA has not been without controversy. Critics argue that the pace of regulatory development has not kept up with the rapid advancements in mining technologies and the growing interest in deep sea resources. For example, the ISA has yet to finalize its exploitation regulations, creating uncertainty for operators and investors regarding the specific requirements and standards that will govern commercial mining activities. This regulatory ambiguity complicates project planning, investment decisions, and risk assessments, necessitating flexible strategies that account for a range of potential outcomes.

Another area of contention is the perceived trade-off between resource exploitation and environmental conservation. While the ISA's regulations emphasize the importance of minimizing environmental impacts, the potential consequences of large-scale mining operations on deep sea ecosystems remain poorly understood. The precautionary principle, which guides the ISA's approach to environmental management, is both a strength and a limitation, it provides a framework for avoiding harm in the face of uncertainty but also highlights the gaps in scientific knowledge that hinder comprehensive risk assessments.

The evolving nature of UNCLOS and ISA regulations reflects the broader challenges of governing activities in international waters, where jurisdictional boundaries are absent, and the interests of diverse stakeholders must be reconciled. Balancing the economic benefits of deep sea mining with the ethical, environmental, and social responsibilities enshrined in international law requires ongoing dialogue, collaboration, and innovation. For operators and investors, navigating this regulatory landscape demands not only compliance but also strategic engagement with the principles and processes that shape global governance.

Artificial intelligence and advanced analytics play a transformative role in navigating the complexities of UNCLOS and ISA regulations. AI-driven tools enable operators to model regulatory scenarios, evaluate compliance costs, and assess the financial implications of different regulatory outcomes. For example, machine learning algorithms can analyze historical licensing data, environmental assessments, and stakeholder feedback to identify trends and predict the likelihood of specific policy changes. These insights inform strategic planning, ensuring that projects are resilient to regulatory shifts and aligned with long-term objectives.

AI also enhances the efficiency and accuracy of compliance monitoring and reporting, a critical requirement under ISA regulations. Advanced sensor networks, autonomous monitoring platforms, and blockchain-based traceability systems enable real-time data collection and analysis, ensuring that operators meet environmental and operational standards. For instance, AI algorithms can analyze data from environmental monitoring systems to detect anomalies, assess compliance with sediment plume thresholds, and generate reports for regulatory authorities. By automating these processes, AI reduces administrative burdens, improves transparency, and strengthens stakeholder confidence in the industry's commitment to responsible practices.

The governance of maritime jurisdictions, particularly as it pertains to deep sea mining, is in a state of rapid evolution. Emerging policies reflect the growing recognition of the ocean's critical role in global ecosystems, the escalating demand for its resources, and the ethical imperatives of ensuring equitable access and sustainable use. These policies are being shaped by a complex interplay of national interests, regional cooperation, and international mandates, creating a dynamic regulatory landscape that directly impacts the viability and direction of deep sea mining projects. For industry stakeholders, understanding and engaging with these emerging policies is crucial to navigating the challenges and opportunities of operating in diverse and sometimes overlapping maritime jurisdictions.

Maritime jurisdictions are broadly categorized into areas within national jurisdiction, such as exclusive economic zones (EEZs), and areas beyond national jurisdiction, including the high seas and the international seabed area. Each category is governed by distinct legal frameworks, but both are increasingly influenced by emerging policies that address the environmental, social, and economic dimensions of ocean resource management. These policies are driven by advancements in science and technology, heightened awareness of environmental impacts, and the growing urgency to address global challenges such as climate change, biodiversity loss, and sustainable development.

One of the most significant trends in emerging policies is the strengthening of environmental safeguards within national maritime jurisdictions. Coastal states are increasingly adopting stringent regulations to protect marine ecosystems while pursuing the economic benefits of resource extraction within their EEZs. For example, Japan has developed a comprehensive legal framework for the exploration and exploitation of cobalt-rich ferromanganese crusts, emphasizing the use of advanced technologies to minimize environmental impacts. Similarly, Pacific island nations such as Kiribati and the Cook Islands, which hold jurisdiction over resource-rich areas in the Clarion-Clipperton Zone (CCZ), are implementing policies that balance economic aspirations with the preservation of biodiversity and ecosystem services.

These national policies often include requirements for environmental impact assessments (EIAs), baseline studies, and long-term monitoring programs, ensuring that resource extraction activities are informed by robust scientific data. In some cases, coastal states have established marine protected areas (MPAs) or biodiversity conservation zones within their EEZs, creating spatial constraints on mining operations to safeguard ecologically sensitive regions. For operators, these policies necessitate a deep understanding of local regulatory requirements and the integration of environmental considerations into project planning and execution.

In areas beyond national jurisdiction, emerging policies are being shaped by multilateral agreements and regional cooperation frameworks. The negotiations under the United Nations Convention on the Law of the Sea (UNCLOS) to establish a legally binding agreement on the conservation and sustainable use of marine biological diversity in areas beyond national jurisdiction (BBNJ) represent a landmark development. This agreement, commonly referred to as the BBNJ Treaty, aims to fill gaps in the existing governance framework by addressing issues such as marine genetic resources, environmental impact assessments, area-based management tools, and capacity-building for developing countries.

The BBNJ Treaty has significant implications for deep sea mining, particularly in its emphasis on protecting marine biodiversity and ensuring that activities in the high seas are conducted responsibly. The treaty's provisions on area-based management tools, such as marine spatial planning and the establishment of MPAs, could introduce new constraints on mining operations in international waters. At the same time, its focus on capacity-building and technology transfer aligns with the principles of equity and inclusivity, providing opportunities for collaboration and innovation.

Regional organizations and initiatives also play a critical role in shaping emerging policies for maritime jurisdictions. In the Pacific region, the Pacific Islands Forum has been at the forefront of advocating for a precautionary approach to deep sea mining, calling for a moratorium on commercial extraction until its environmental and socioeconomic impacts are better understood. The African Union's African Blue Economy Strategy similarly emphasizes the need for sustainable and inclusive governance of ocean resources, with a focus on supporting the interests of coastal and island states.

These regional policies reflect a growing recognition of the interconnectedness of ocean ecosystems and the need for coordinated governance to address transboundary challenges. For example, sediment plumes generated by mining activities in one jurisdiction can affect water quality and biodiversity in neighboring areas, necessitating regional cooperation on monitoring, mitigation, and impact assessment. Regional agreements and frameworks provide a platform for addressing these challenges, promoting dialogue, and aligning policies to ensure that the benefits of resource extraction are equitably distributed while minimizing cumulative impacts.

The role of international standards and certification schemes in shaping emerging policies cannot be overstated. Voluntary initiatives such as the Global Reporting Initiative (GRI) and the Responsible Mining Index (RMI) have established benchmarks for transparency, accountability, and sustainability in the mining sector, including deep sea mining. These standards encourage operators to adopt best practices, disclose their environmental and social performance, and engage with stakeholders in a meaningful and inclusive manner. For instance, adherence to the GRI standards requires companies to report on their greenhouse gas emissions, water usage, and biodiversity impacts, providing a basis for assessing their alignment with global sustainability goals.

Emerging policies in maritime jurisdictions are also influenced by technological advancements that enhance the capacity for monitoring, enforcement, and compliance. Remote sensing, satellite monitoring, and autonomous underwater vehicles (AUVs) equipped with advanced sensors provide real-time data on mining activities, enabling regulators to assess compliance with environmental standards and detect unauthorized operations. Artificial intelligence (AI) further enhances these capabilities by analyzing large datasets, identifying anomalies, and generating insights that inform decision-making. For example, AI algorithms can model the dispersion of sediment plumes, predict their ecological impacts, and recommend mitigation measures, supporting the implementation of evidence-based policies.

The integration of blockchain technology into governance frameworks represents another innovation with significant implications for emerging policies. Blockchain's decentralized and immutable ledger system enables the transparent and verifiable tracking of resource extraction, transportation, and processing, ensuring that operators adhere to regulatory requirements and ethical standards. For instance, blockchain-based traceability systems can document the origin, composition, and environmental credentials of extracted materials, providing assurance to regulators, consumers, and downstream industries.

Despite these advancements, the development and implementation of emerging policies face several challenges. One of the most pressing is the need for capacity-building in developing countries, many of which lack the technical expertise, financial resources, and institutional frameworks to effectively govern deep sea mining activities. Addressing this gap requires international cooperation, knowledge-sharing, and targeted investments in education, infrastructure, and technology transfer.

Another challenge is the potential for policy fragmentation and jurisdictional conflicts, particularly in areas where national, regional, and international regulations overlap or diverge. Aligning these policies requires dialogue, coordination, and a commitment to multilateralism, ensuring that governance frameworks are coherent, equitable, and enforceable. The role of international organizations such as the International Seabed Authority (ISA), the International Maritime Organization (IMO), and the United Nations Environment Program (UNEP) is critical in facilitating this alignment and promoting a unified vision for the sustainable use of ocean resources.

The governance of the deep ocean is fraught with a complex interplay of national and international claims, reflecting the competing priorities of states, industries, and global institutions. This tension arises from the dual nature of the ocean as a shared global commons and a domain where sovereign rights are asserted through exclusive economic zones (EEZs) and continental shelf extensions. For deep sea mining, this conflict is particularly pronounced, as the industry seeks to exploit valuable resources such as polymetallic nodules, cobalt-rich ferromanganese crusts, and seafloor massive sulfides (SMS), often in regions where national and international jurisdictions overlap or collide. Understanding and addressing these tensions is critical for ensuring equitable resource management, environmental stewardship, and long-term project viability.

The legal framework underpinning ocean governance is anchored in the United Nations Convention on the Law of the Sea (UNCLOS), which delineates maritime zones and defines the rights and responsibilities of states in their use of ocean resources. Within their EEZs, which extend 200 nautical miles from the coastline, coastal states enjoy sovereign rights over the exploration and exploitation of natural resources, including those on the seabed and in the subsoil. Beyond the EEZ, states can extend their jurisdiction over the continental shelf, up to 350 nautical miles or more, provided they submit scientific evidence to the Commission on the Limits of the Continental Shelf (CLCS). These extensions have significant implications for deep sea mining, as they expand national claims into areas previously considered part of the international seabed.

The international seabed area, referred to in UNCLOS as "the Area," lies beyond the limits of national jurisdiction and is governed by the principle of the "common heritage of mankind." This principle asserts that the resources of the Area must be managed for the benefit of all humanity, with particular consideration for the needs of developing countries. The International Seabed Authority (ISA) is the body responsible for regulating mineral exploration and exploitation in the Area, issuing licenses, and ensuring compliance with environmental and benefit-sharing obligations.

The tension between national and international claims emerges at the intersection of these jurisdictions, where the interests of coastal states, international institutions, and private operators converge and sometimes conflict. Coastal states, motivated by the economic potential of seabed resources within their EEZs and extended continental shelves, assert their sovereign rights to control access and set regulations. This often leads to a fragmentation of governance frameworks, as states adopt diverse policies and standards to manage activities in their waters. For instance, Papua New Guinea, Japan, and the Cook Islands have established national regulations for deep sea mining, each reflecting distinct priorities and approaches to environmental management and economic development.

In contrast, the ISA governs activities in the international seabed area under a unified regulatory framework designed to ensure equity, environmental protection, and sustainable development. This framework mandates that the benefits of resource exploitation in the Area be shared among all nations, particularly developing countries, through mechanisms such as royalties, capacity-building initiatives, and technology transfer. For operators, the dual requirements of complying with national regulations within EEZs and ISA regulations in the Area introduce significant complexity, increasing the administrative and financial burdens of project planning and execution.

The potential for jurisdictional overlap further exacerbates these tensions. In regions where extended continental shelves intersect with the international seabed area, questions arise about the scope and limits of national and international authority. Coastal states with extended continental shelf claims often assert greater control over resources in these areas, citing their sovereign rights under UNCLOS. However, such claims may conflict with the ISA's mandate to regulate activities in the Area and ensure that resources are managed as the common heritage of mankind. Resolving these disputes requires careful interpretation of UNCLOS provisions, diplomatic negotiation, and, in some cases, adjudication by international bodies such as the International Tribunal for the Law of the Sea (ITLOS).

Geopolitical factors further complicate the tension between national and international claims. The strategic importance of seabed resources, coupled with their potential to support economic growth and technological innovation, has heightened competition among states to secure access and control. For example, China's assertive approach to resource exploration in the South China Sea has raised concerns among neighboring countries and the international community about its compliance with international law. Similarly, the Arctic region has become a focal point for competing claims, as states such as Russia, Canada, and Denmark seek to extend their continental shelves to encompass resource-rich areas.

These geopolitical dynamics highlight the challenges of achieving equitable and cooperative governance in the deep ocean. The principle of the common heritage of mankind, while enshrined in UNCLOS, is often overshadowed by the economic and strategic interests of powerful states, raising questions about the inclusivity and effectiveness of international governance frameworks. For developing countries, which may lack the technical and financial capacity to exploit seabed resources independently, the tension between national and international claims can limit their access to the benefits of deep sea mining, exacerbating global inequalities.

The environmental implications of these tensions are equally significant. Fragmented governance frameworks and competing jurisdictional claims can hinder the development and enforcement of comprehensive environmental standards, increasing the risk of ecological degradation. For instance, mining activities conducted under divergent national regulations may fail to address transboundary impacts, such as sediment plumes or biodiversity loss, that extend beyond the boundaries of individual EEZs. Similarly, the lack of alignment between national and international environmental standards can create loopholes that undermine conservation efforts and exacerbate cumulative impacts.

Addressing the tension between national and international claims requires a multi-faceted approach that prioritizes collaboration, transparency, and innovation. Enhanced coordination between coastal states, the ISA, and other international bodies is essential for harmonizing regulatory frameworks and ensuring that activities in the deep ocean are managed equitably and sustainably. For example, the development of regional environmental management plans, supported by scientific research and stakeholder engagement, can provide a basis for aligning policies and addressing transboundary challenges.

Artificial intelligence and advanced data analytics offer valuable tools for navigating these tensions and supporting cooperative governance. AI-driven models can analyze jurisdictional boundaries, resource distribution, and environmental data to identify areas of overlap and potential conflict, informing negotiations and decision-making. For instance, a machine learning model might integrate geological and ecological data to assess the implications of overlapping continental shelf claims, providing insights into the trade-offs between resource exploitation and conservation.

Blockchain technology further enhances transparency and accountability in managing national and international claims. By creating a decentralized and immutable ledger of transactions, blockchain enables the verification of resource provenance, compliance with regulations, and benefit-sharing arrangements. For example, blockchain-based traceability systems can document the origin and ownership of extracted materials, ensuring that resources from disputed areas are managed in accordance with international law.

Despite the challenges, the tension between national and international claims also presents opportunities for innovation and progress in ocean governance. Collaborative initiatives, such as joint ventures between coastal states and international operators or partnerships between the ISA and regional organizations, can promote shared benefits and knowledge exchange. Similarly, the integration of traditional knowledge from indigenous and local communities into governance frameworks can enrich decision-making and foster inclusivity.

10.2. AI Ethics and Governance

The integration of artificial intelligence (AI) into deep sea mining operations has revolutionized the way resources are explored, extracted, and managed. AI-driven systems provide unparalleled efficiency and precision, enabling operators to navigate complex environments, optimize logistics, and maximize resource recovery. However, the growing reliance on AI also raises critical ethical and governance concerns, particularly regarding algorithmic fairness, bias, and transparency in resource allocation. These issues have far-reaching implications for the legitimacy and equity of the industry, influencing everything from operational decisions to global perceptions of the sector's ethical standing.

At its core, algorithmic fairness refers to the principle that AI systems should operate without unjust discrimination, ensuring that decisions are equitable and inclusive. In the context of deep sea mining, fairness takes on a multifaceted dimension, encompassing the equitable distribution of resources among stakeholders, the prioritization of environmentally and socially responsible practices, and the mitigation of biases that may disadvantage certain regions, communities, or ecosystems. Achieving algorithmic fairness is particularly challenging in a sector where resource allocation decisions are influenced by a wide array of technical, economic, and geopolitical factors, each carrying the potential for bias and inequity.

Bias in AI systems arises from the data and algorithms that underpin their operation. Machine learning models, which form the foundation of many AI-driven systems, are trained on historical data to recognize patterns and make predictions. If the training data reflects existing biases, such as overrepresentation of certain regions or underrepresentation of others, the AI system is likely to perpetuate these biases in its decisions. For instance, an AI model trained primarily on geophysical data from well-explored areas like the Pacific's Clarion-Clipperton Zone may prioritize similar regions for future exploration, overlooking less studied but potentially resource-rich areas. This bias not only limits the inclusivity of resource allocation but also risks reinforcing existing disparities in global access to seabed resources.

Transparency is critical for addressing these challenges and ensuring that AI systems are accountable to stakeholders. Transparent AI systems provide clear and accessible explanations for their decisions, enabling operators, regulators, and affected communities to understand how resource allocation choices are made. For example, an AI-driven model used to prioritize mining sites should clearly articulate the criteria it considers, such as resource grade, environmental sensitivity, and logistical feasibility, and how these criteria are weighted in its decision-making process. Transparency fosters trust and legitimacy, particularly in a sector where decisions have significant environmental and social implications.

One of the most pressing concerns in algorithmic fairness and transparency is the equitable allocation of seabed resources among nations, particularly in areas beyond national jurisdiction. Under the United Nations Convention on the Law of the Sea (UNCLOS), the international seabed area is designated as the "common heritage of mankind," with resources to be managed for the benefit of all, particularly developing countries. AI-driven resource allocation systems play a pivotal role in operationalizing this principle, determining which areas are prioritized for exploration and how the resulting benefits are distributed. Ensuring fairness in these decisions requires algorithms that account for the diverse needs and capabilities of different nations, promoting inclusivity and reducing global inequalities.

For example, an AI system designed to allocate exploration licenses in the international seabed area must consider a range of factors, including the technical and financial capacity of applicant states, the potential environmental impacts of their activities, and their alignment with global development goals. A fair algorithm would balance these considerations to ensure that licenses are distributed equitably, avoiding favoritism toward technologically advanced nations or private entities at the expense of developing countries. Achieving this balance requires not only technical expertise but also a commitment to ethical principles and stakeholder engagement.

The risk of bias in AI systems extends beyond resource allocation to encompass the prioritization of environmental and social objectives. For instance, an AI model used to optimize mining operations may prioritize economic efficiency over ecological preservation if its objectives are narrowly defined. This bias can lead to decisions that exacerbate environmental degradation or disproportionately impact vulnerable communities, undermining the sector's commitment to sustainability and social responsibility. To address this, AI systems must be designed with explicit safeguards that prioritize environmental and social considerations alongside economic goals. For example, an AI-driven site selection model could incorporate criteria such as biodiversity richness, habitat sensitivity, and proximity to cultural heritage sites, ensuring that these factors are weighted appropriately in its decisions.

Bias in AI systems also has implications for the governance and regulation of deep sea mining. Regulators rely on AI-driven models to evaluate compliance with environmental standards, assess the cumulative impacts of multiple operations, and monitor adherence to benefit-sharing agreements. If these models are biased or opaque, they may produce outcomes that undermine regulatory objectives or erode public trust. For example, a biased environmental impact assessment model might underestimate the impacts of sediment plumes in ecologically sensitive areas, leading to insufficient mitigation measures and long-term harm to marine ecosystems. Ensuring that regulatory AI systems are fair, transparent, and accountable is therefore essential for maintaining the integrity and legitimacy of governance frameworks.

Addressing algorithmic fairness and transparency in deep sea mining requires a multifaceted approach that combines technical innovation with ethical oversight. One key strategy is the development of explainable AI (XAI), which enhances the interpretability of machine learning models without sacrificing their predictive power. XAI techniques, such as feature importance analysis and visualization tools, enable stakeholders to understand how AI systems arrive at their decisions and identify potential biases in their operation. For example, an XAI framework could reveal that a site prioritization model places disproportionate weight on certain geological features, prompting a reassessment of its criteria and ensuring a more balanced approach.

Another important strategy is the use of diverse and representative datasets in training AI models. Bias in AI systems often stems from imbalances in the data used to train them. For deep sea mining, this means ensuring that datasets include a wide range of geographic regions, environmental conditions, and socioeconomic contexts. Collaborative data-sharing initiatives among nations, research institutions, and industry stakeholders can help build comprehensive and representative datasets that support fair and inclusive AI systems. For instance, a global database of seabed resource assessments, environmental baselines, and cultural heritage sites could provide the foundation for AI models that reflect the diversity and complexity of the deep ocean.

Stakeholder engagement is also critical for ensuring algorithmic fairness and transparency. Decisions about resource allocation, environmental management, and benefit-sharing affect a wide range of stakeholders, including coastal states, international organizations, indigenous communities, and environmental groups. Engaging these stakeholders in the design and oversight of AI systems helps ensure that their values, priorities, and concerns are reflected in the algorithms' objectives and criteria. For example, participatory workshops and consultation processes could involve stakeholders in defining the weight assigned to different factors in a resource allocation model, such as economic efficiency, environmental sensitivity, and social equity.

Ethical oversight mechanisms, such as independent audits and accountability frameworks, provide an additional layer of protection against bias and unfairness. Regular audits of AI systems can identify potential sources of bias, evaluate their alignment with ethical principles, and recommend improvements. Accountability frameworks, which define the roles and responsibilities of AI developers, operators, and regulators, ensure that all parties are held to high standards of transparency, fairness, and responsibility. For instance, an accountability framework for AI-driven mining operations could require operators to publish detailed explanations of their algorithms' criteria and performance, enabling independent verification and fostering public trust.

The integration of artificial intelligence (AI) into maritime operations, particularly in the domain of deep sea mining, has brought unprecedented capabilities for exploration, monitoring, and resource management. However, it has also introduced significant challenges related to data privacy and security. The maritime sector operates in a uniquely complex environment, where vast amounts of sensitive data are collected and exchanged across international borders, involving a diverse array of stakeholders, including private operators, governments, and global institutions. Ensuring the integrity, confidentiality, and security of this data is critical for the effective governance of maritime activities, the protection of proprietary information, and the prevention of cyber threats.

The foundation of data privacy and security in maritime domains lies in the nature of the data itself. Deep sea mining generates a wealth of information, encompassing geophysical surveys, environmental monitoring data, operational metrics, and proprietary algorithms. This data is essential for planning and executing mining activities, complying with regulatory requirements, and assessing environmental impacts. However, it is also highly sensitive, containing commercially valuable insights into resource distribution, technological capabilities, and operational strategies. In addition, environmental data often includes information about biodiversity, ecosystem health, and cultural heritage, raising ethical and legal questions about its ownership, accessibility, and use.

One of the primary challenges in ensuring data privacy and security is the decentralized and transnational nature of maritime operations. Deep sea mining projects often involve multiple stakeholders operating across different jurisdictions, each with its own legal and regulatory frameworks for data protection. For example, a project might include exploration conducted by a private operator registered in one country, environmental assessments overseen by international regulators, and logistical support provided by subcontractors from multiple nations. This fragmentation creates significant challenges for establishing consistent data privacy and security standards, as stakeholders must navigate conflicting regulations and varying levels of enforcement.

Cybersecurity is a critical concern in the maritime sector, particularly as AI systems and connected technologies become increasingly integral to operations. Autonomous underwater vehicles (AUVs), remotely operated vehicles (ROVs), and other smart systems rely on real-time data exchange and communication, making them vulnerable to cyberattacks. Hackers targeting these systems could disrupt operations, compromise sensitive data, or even gain control of critical infrastructure. For instance, a cyberattack on an AUV conducting resource mapping could result in the theft of proprietary geophysical data or the manipulation of its mission parameters, leading to financial losses and reputational damage.

The rise of cyber threats underscores the importance of robust data security measures in maritime domains. Encryption is a fundamental tool for protecting sensitive information, ensuring that data transmitted between systems is secure from interception or tampering. For example, encrypted communication protocols can safeguard the exchange of data between subsea vehicles and surface vessels, preventing unauthorized access to operational metrics or environmental monitoring results. In addition to encryption, authentication mechanisms such as multi-factor authentication and digital certificates help verify the identity of users and systems, reducing the risk of unauthorized access.

AI plays a dual role in cybersecurity, serving both as a target and a defense mechanism. On one hand, the reliance on AI systems for critical decision-making makes them attractive targets for cyberattacks, as compromising these systems can have far-reaching consequences. On the other hand, AI is a powerful tool for detecting and mitigating cyber threats. Machine learning algorithms can analyze network traffic patterns, identify anomalies, and flag potential intrusions in real-time. For instance, an AI-driven cybersecurity system might detect unusual data transmission activity from an AUV, indicating a potential breach, and initiate countermeasures to isolate the affected system and prevent further damage.

Data privacy concerns in maritime domains extend beyond cybersecurity to encompass issues of data ownership, access, and use. The collection and analysis of environmental data, in particular, raise questions about who owns the information and how it should be shared. Under the United Nations Convention on the Law of the Sea (UNCLOS), resources in the international seabed area are considered the "common heritage of mankind," suggesting that data related to these resources should be shared equitably among all nations. However, private operators often invest significant resources in collecting and processing this data, leading to tensions between proprietary interests and the principle of shared access.

Balancing these competing interests requires clear guidelines and governance frameworks for data management. Transparency and accountability are essential for ensuring that data is used responsibly and ethically. For example, data collected during environmental impact assessments (EIAs) should be made publicly available to facilitate independent verification and promote scientific research. At the same time, mechanisms should be in place to protect commercially sensitive information, ensuring that operators can maintain a competitive edge while fulfilling their obligations to stakeholders.

One approach to addressing data privacy and security concerns is the adoption of blockchain technology, which provides a decentralized and immutable ledger for recording data transactions. Blockchain can enhance transparency and accountability in data management while ensuring the security of sensitive information. For instance, a blockchain-based system could document the collection, processing, and sharing of environmental data, creating a verifiable record that demonstrates compliance with regulatory requirements. Operators could grant access to specific data entries while keeping proprietary information secure, enabling a balance between transparency and confidentiality.

International cooperation is also critical for addressing data privacy and security concerns in maritime domains. The global nature of the maritime sector demands harmonized standards and practices that transcend national boundaries. Initiatives such as the International Maritime Organization's (IMO) guidelines on maritime cybersecurity and the European Union's General Data Protection Regulation (GDPR) provide valuable frameworks for promoting data protection and security. However, achieving consistent implementation and enforcement requires collaboration among governments, industry stakeholders, and international organizations.

Ethical considerations play a central role in shaping data privacy and security policies. The use of AI and big data in maritime domains raises questions about the potential for misuse or unintended consequences. For example, data collected for environmental monitoring could be repurposed for unauthorized surveillance or exploited for geopolitical advantage. Ensuring that data is used in ways that align with ethical principles and respect human rights requires robust governance structures and oversight mechanisms. Stakeholder engagement is essential for building consensus around the ethical use of data and fostering trust in the systems that manage it.

Looking to the future, emerging technologies and innovations offer new opportunities for enhancing data privacy and security in maritime domains. Advances in quantum encryption, for example, promise to provide unprecedented levels of data security, making it virtually impossible for unauthorized parties to intercept or decipher communications. Similarly, the development of federated learning models allows AI systems to train on distributed datasets without centralizing sensitive information, reducing the risk of data breaches while enabling collaborative research and innovation.

The rapid integration of artificial intelligence (AI) into the domain of deep sea mining has unlocked immense potential for societal benefits, from advancing technological innovation to securing critical minerals essential for renewable energy and economic growth. However, this potential is inextricably tied to the ethical obligation of safeguarding the fragile marine ecosystems that lie beneath the ocean's surface. Balancing societal benefits with environmental stewardship is not merely a challenge of competing priorities; it is a profound test of humanity's ability to harness technological progress responsibly and equitably.

The societal benefits of deep sea mining, driven in large part by AI-enhanced capabilities, are both significant and urgent. The minerals targeted by deep sea mining, cobalt, nickel, manganese, and rare earth elements, are critical for manufacturing batteries, wind turbines, and other technologies central to the global transition to renewable energy. As the world strives to meet ambitious climate goals, the demand for these materials is projected to outstrip current terrestrial supplies, necessitating the exploration of alternative sources such as the ocean floor. Additionally, deep sea mining promises economic development opportunities, particularly for resource-poor nations with access to resource-rich marine areas. By leveraging AI to optimize exploration, extraction, and logistics, these nations can participate more equitably in the global economy, reducing dependency on traditional resource-exporting powers.

However, the pursuit of these societal benefits brings with it profound environmental risks. The deep ocean is a realm of extraordinary biodiversity and ecological significance, home to unique species and habitats adapted to extreme conditions. Mining activities, including the disturbance of seafloor sediments and the generation of sediment plumes, threaten these ecosystems in ways that are not yet fully understood. The potential for irreversible damage to benthic habitats, water column dynamics, and species interactions underscores the urgency of integrating environmental stewardship into every stage of deep sea mining operations.

AI serves as both a catalyst and a safeguard in balancing societal benefits with environmental stewardship. By enabling unprecedented precision and efficiency, AI reduces the operational footprint of mining activities, minimizing their environmental impact. For example, machine learning algorithms can analyze geological and geophysical data to identify high-grade resource deposits with pinpoint accuracy, reducing the need for extensive and disruptive exploration. Similarly, AI-driven monitoring systems provide real-time insights into the spread of sediment plumes, allowing operators to adjust extraction techniques dynamically to mitigate ecological harm.

The ethical deployment of AI requires a commitment to transparency and accountability, ensuring that decisions made by AI systems reflect a balance between economic objectives and ecological responsibilities. Transparency involves clear communication about how AI-driven models prioritize competing goals, such as resource recovery and habitat preservation. For instance, an AI system used for site selection must explain how it weighs factors such as mineral grade, biodiversity richness, and proximity to sensitive habitats. This transparency fosters trust among stakeholders, including regulators, conservation groups, and local communities, and ensures that decisions are subject to scrutiny and alignment with broader ethical principles.

Accountability extends to the design and governance of AI systems, requiring that they be programmed and managed in ways that reflect societal values and environmental priorities. This includes embedding safeguards that prioritize ecological protection, such as thresholds for acceptable levels of sediment disturbance or biodiversity loss. By integrating these safeguards into the algorithms themselves, operators can ensure that AI-driven systems act as stewards of the environment, aligning their actions with sustainability goals.

One of the most significant challenges in balancing societal benefits with environmental stewardship is the inherent uncertainty surrounding the impacts of deep sea mining. The deep ocean remains one of the least understood environments on Earth, and the long-term consequences of disturbing its ecosystems are difficult to predict. AI offers a powerful tool for navigating this uncertainty, enabling predictive modeling and scenario analysis that inform responsible decision-making. For example, AI algorithms can simulate the potential impacts of mining-induced sediment plumes on benthic organisms, providing insights into the spatial and temporal scales of disturbance. These models can guide the development of adaptive management strategies, ensuring that operations can be adjusted in real time to address unforeseen impacts.

Environmental monitoring is another critical area where AI bridges the gap between societal benefits and ecological stewardship. Continuous, autonomous sensor networks equipped with AI-driven analytics provide high-resolution data on key environmental parameters, such as water quality, biodiversity, and habitat conditions. By processing this data in real time, AI systems can detect anomalies, track recovery rates, and assess compliance with environmental regulations. For instance, an AI system might analyze eDNA samples collected by autonomous underwater vehicles (AUVs) to monitor changes in species composition near a mining site, flagging declines in biodiversity that require intervention. These monitoring capabilities not only ensure regulatory compliance but also demonstrate a tangible commitment to environmental stewardship, enhancing the industry's social license to operate.

The integration of societal benefits and environmental stewardship also necessitates a commitment to equity and inclusivity in decision-making. Deep sea mining has the potential to benefit a wide range of stakeholders, from resource-dependent industries and national economies to local communities and indigenous peoples. However, these benefits must be distributed fairly, and the voices of all stakeholders must be heard in shaping the governance and operation of mining projects. AI systems must be designed and managed in ways that reflect this diversity, ensuring that their objectives and criteria align with the values and priorities of affected populations. For example, participatory workshops involving local communities and conservation groups could help define the weight assigned to ecological and social factors in AI-driven decision models, ensuring that these systems promote inclusive and responsible outcomes.

Ethical frameworks for balancing societal benefits with environmental stewardship must also address the global implications of deep sea mining. The resources targeted by this industry are not only critical for individual nations but also essential for achieving global sustainability goals, such as the United Nations Sustainable Development Goals (SDGs). AI-driven systems play a pivotal role in aligning deep sea mining operations with these objectives, optimizing resource use while minimizing environmental and social costs. For instance, AI algorithms can model the cumulative impacts of multiple mining projects, identifying strategies to maximize resource recovery while preserving ecosystem services and biodiversity.

Collaboration and knowledge-sharing are essential for realizing the potential of AI to balance societal benefits with environmental stewardship. The development of global standards and best practices for AI deployment in deep sea mining requires input from a diverse array of stakeholders, including governments, industry leaders, scientists, and civil society organizations. International cooperation can facilitate the sharing of data, technologies, and expertise, enabling a more comprehensive and equitable approach to ocean resource management. For example, an international consortium could develop an open-access database of environmental baselines and mining impacts, providing the foundation for AI-driven models that reflect the full complexity of the deep ocean.

Despite its transformative potential, the use of AI in balancing societal benefits and environmental stewardship faces significant challenges. The reliability and robustness of AI models depend on the quality and representativeness of the data they process, yet deep ocean data remains sparse and unevenly distributed. Addressing this challenge requires investment in exploration and monitoring technologies, as well as collaborative efforts to build comprehensive and standardized datasets. Transparency and interpretability are also critical, as stakeholders must be able to understand and trust the decisions made by AI systems. Developing explainable AI (XAI) techniques that provide clear insights into the rationale behind model outputs is essential for building confidence and ensuring accountability.

10.3. Stakeholder Engagement and Public Perceptions

Community inclusion in resource management decisions is an essential aspect of ethical governance in deep sea mining, reflecting the growing recognition that sustainable development must be rooted in fairness, transparency, and respect for the voices of all affected stakeholders. Deep sea mining, while offering the promise of critical resources for renewable energy and global development, also raises significant concerns about environmental impacts, equitable benefit-sharing, and the long-term health of marine ecosystems. Communities, particularly those in coastal regions and small island nations, are often at the frontline of these impacts and must be meaningfully included in decision-making processes to ensure that their needs, values, and perspectives are addressed.

The principle of community inclusion acknowledges that those who are most directly affected by resource extraction should have a seat at the table in shaping its governance. For coastal and island communities, the ocean is not merely a distant economic opportunity but a vital source of cultural identity, sustenance, and livelihood. Fishing, tourism, and traditional practices are deeply intertwined with the health of marine ecosystems, making any disruption to these systems a matter of immediate and profound concern. Inclusive resource management recognizes this interconnectedness, ensuring that community members are active participants in decisions about how, where, and under what conditions deep sea mining takes place.

At its core, community inclusion is about creating platforms for dialogue, collaboration, and shared decision-making. These platforms must go beyond perfunctory consultations to genuinely engage communities as partners in the governance of marine resources. Effective community inclusion begins with the acknowledgment of local knowledge and expertise, which can provide invaluable insights into the ecological, social, and economic dynamics of specific regions. For example, traditional ecological knowledge held by indigenous communities often contains a nuanced understanding of marine habitats, species behavior, and environmental cycles, offering a complementary perspective to scientific assessments.

One of the most effective ways to incorporate community voices is through participatory resource management frameworks, where stakeholders collectively develop policies, strategies, and operational plans. These frameworks ensure that community members have a meaningful influence on decisions, from the selection of mining sites to the design of environmental monitoring programs. For instance, a participatory approach to site selection might involve workshops where community representatives, scientists, and policymakers assess the trade-offs between resource potential, ecological sensitivity, and cultural significance. By collaboratively defining criteria for decision-making, these frameworks foster trust and legitimacy while aligning resource management with the values and priorities of affected populations.

Community inclusion also extends to benefit-sharing, a critical component of equitable resource management. The extraction of deep sea minerals generates significant economic value, but the distribution of these benefits often disproportionately favors mining operators and distant markets, leaving local communities with minimal gains and disproportionate burdens. Inclusive governance ensures that communities receive a fair share of the benefits through mechanisms such as revenue-sharing agreements, local employment opportunities, and investments in infrastructure, education, and healthcare. For example, a portion of mining royalties might be allocated to a community development fund, supporting initiatives that enhance local resilience and well-being.

Transparency is a cornerstone of community inclusion, enabling stakeholders to make informed decisions and hold operators accountable. Transparent governance requires that information about mining activities, environmental impacts, and benefit-sharing arrangements be accessible, accurate, and understandable to all stakeholders. For example, operators might publish environmental impact assessments (EIAs) in local languages, accompanied by visual aids and summaries that convey key findings in a clear and concise manner. Public hearings and town hall meetings provide additional opportunities for community members to engage with the data, ask questions, and voice their concerns.

The integration of artificial intelligence (AI) into resource management offers new opportunities for enhancing community inclusion. AI-driven tools can facilitate data collection, analysis, and visualization, making complex information more accessible to non-expert audiences. For instance, AI algorithms can process large datasets from environmental monitoring systems, generating user-friendly dashboards that display real-time information about sediment plumes, water quality, and biodiversity near mining sites. These tools empower communities to monitor the impacts of mining activities independently, fostering accountability and enabling them to advocate for necessary interventions.

AI can also support participatory decision-making by modeling the potential outcomes of different management strategies. Scenario analysis tools powered by AI can simulate the environmental, economic, and social impacts of alternative approaches to resource extraction, providing communities with a clearer understanding of the trade-offs involved. For example, a scenario model might compare the implications of high-intensity versus low-intensity mining operations, illustrating how each approach affects resource recovery, ecosystem health, and local livelihoods. By presenting these scenarios in an interactive and visual format, AI tools enable communities to engage meaningfully with the decision-making process, ensuring that their preferences and priorities are reflected in final policies.

Despite its potential, community inclusion in deep sea mining faces significant challenges. Power imbalances between mining operators, governments, and local communities can undermine efforts to create equitable and inclusive governance frameworks. Communities often lack the technical expertise, financial resources, and political leverage to assert their interests effectively, leaving them vulnerable to exclusion or marginalization. Addressing these challenges requires targeted investments in capacity-building, education, and advocacy, ensuring that communities have the knowledge and tools to participate fully in resource management.

Another challenge is the diversity of interests and perspectives within communities themselves. Stakeholders are not monolithic; they include fishermen, tourism operators, indigenous groups, conservationists, and others, each with their own priorities and concerns. Inclusive governance must navigate these internal dynamics, fostering dialogue and consensus-building to reconcile competing interests. For example, a resource management plan might involve negotiations between fishing cooperatives seeking to protect spawning grounds and local governments pursuing economic diversification through mining revenues. Mediated discussions and collaborative problem-solving can help align these interests, creating solutions that balance economic development with ecological preservation.

Ethical considerations are central to community inclusion, particularly when it comes to respecting the rights and values of indigenous peoples. Many indigenous communities have deep spiritual and cultural connections to the ocean, viewing it as a sacred entity rather than a commodity. Inclusive governance must honor these perspectives, integrating traditional knowledge and respecting indigenous sovereignty in resource management decisions. For example, an indigenous council might be established to provide guidance on the cultural implications of mining activities, ensuring that decisions align with the community's values and traditions.

Collaboration and partnership are key to overcoming these challenges and advancing community inclusion. Governments, industry stakeholders, and international organizations have a responsibility to engage with communities in a spirit of mutual respect and shared purpose. Collaborative initiatives, such as joint management committees or community advisory boards, provide platforms for continuous dialogue, monitoring, and adaptation. These partnerships ensure that resource management is not only inclusive but also dynamic and responsive to evolving needs and circumstances.

The long-term success of deep sea mining depends on its ability to balance economic opportunities with social and environmental responsibilities. Community inclusion is not merely a box to be checked; it is a foundational principle that underpins the legitimacy, sustainability, and equity of the industry. By actively involving communities in resource management decisions, operators and policymakers can build trust, mitigate conflicts, and create outcomes that benefit all stakeholders. As the industry continues to evolve, its commitment to community inclusion will be a defining factor in its ability to achieve responsible and sustainable development in the deep ocean.

The use of artificial intelligence (AI) in deep sea mining has revolutionized decision-making processes, enabling unprecedented precision and efficiency in areas such as resource exploration, environmental monitoring, and logistical optimization. However, as AI becomes increasingly central to the industry's operations, the need for transparency and disclosure of AI-driven decisions has become a critical ethical and practical concern. Transparency ensures that stakeholders, including governments, local communities, regulators, and conservation groups, understand how AI systems influence decisions that have significant environmental, economic, and social implications. It fosters trust, accountability, and legitimacy, addressing concerns about potential biases, unintended consequences, and the perceived opacity of AI algorithms.

Best practices for transparency and disclosure of AI-driven decisions are grounded in principles of accessibility, accountability, and inclusivity. These practices ensure that AI systems operate in ways that align with societal values, regulatory requirements, and the expectations of diverse stakeholders. In the context of deep sea mining, transparency involves not only explaining how AI systems work but also providing meaningful insights into the criteria, data, and processes that inform their outputs. This chapter explores best practices for achieving transparency and disclosure in AI-driven decision-making, highlighting strategies for building trust, enhancing accountability, and promoting informed engagement.

A cornerstone of transparency in AI-driven decisions is the ability to explain how algorithms generate their outputs. Many AI systems, particularly those based on deep learning, operate as "black boxes," producing results without clear explanations of the logic or data that underpin them. This opacity can create uncertainty and skepticism among stakeholders, who may question the fairness, reliability, or validity of AI-driven decisions. To address this, the adoption of explainable AI (XAI) techniques is essential. XAI provides tools and methods for interpreting AI models, offering insights into the factors that influence their outputs.

For example, an AI system used to prioritize deep sea mining sites might incorporate multiple variables, such as mineral grade, environmental sensitivity, and logistical feasibility. An XAI framework could reveal how these variables are weighted in the decision-making process, enabling stakeholders to understand the trade-offs and assumptions embedded in the algorithm. Visualization tools, such as feature importance charts or decision trees, further enhance transparency by presenting complex data in an intuitive and accessible format. By making AI-driven decisions interpretable, XAI fosters trust and enables stakeholders to evaluate the system's alignment with their values and objectives.

Transparency also extends to the data used to train and inform AI systems. Data quality, representativeness, and potential biases are critical factors that influence the performance and fairness of AI algorithms. For instance, if an AI model used for environmental impact assessments is trained on incomplete or unbalanced data, it may fail to account for certain ecological variables or overemphasize others, leading to biased or inaccurate predictions. Best practices for transparency require operators to disclose the sources, characteristics, and limitations of the data underpinning their AI systems. This includes providing information about how data was collected, curated, and validated, as well as acknowledging any gaps or uncertainties that may affect the system's outputs.

In addition to data transparency, the documentation of algorithmic design and governance processes is essential for accountability. Operators should provide clear and detailed descriptions of how AI systems are developed, implemented, and monitored, including the roles and responsibilities of developers, operators, and oversight bodies. For example, a report on an AI-driven sediment plume monitoring system might outline the system's objectives, data inputs, modeling techniques, and validation protocols, as well as the procedures for detecting and addressing anomalies. This documentation not only enhances transparency but also serves as a reference for auditing, evaluation, and continuous improvement.

Stakeholder engagement is a critical component of best practices for transparency and disclosure. Decisions made by AI systems in deep sea mining have far-reaching implications for a diverse array of stakeholders, including local communities, environmental organizations, and industry partners. Ensuring that these stakeholders have access to information about AI-driven decisions and the opportunity to provide input is essential for building trust and legitimacy. Public consultations, participatory workshops, and advisory committees provide platforms for dialogue and collaboration, enabling stakeholders to express their concerns, priorities, and expectations.

For example, a public consultation on the deployment of an AI system for site selection might involve presentations by developers and operators, followed by discussions where stakeholders can ask questions, provide feedback, and propose criteria for the algorithm's design. By incorporating stakeholder perspectives into the development and implementation of AI systems, operators can ensure that their decisions reflect a broader range of values and interests, reducing the risk of conflict and enhancing the credibility of their operations.

The use of real-time and interactive tools further enhances transparency and stakeholder engagement. Dashboards and visualization platforms powered by AI provide stakeholders with dynamic access to information about mining activities, environmental conditions, and decision-making processes. For instance, a dashboard might display real-time data on sediment plume dispersion, water quality, and biodiversity near mining sites, along with explanations of how this data informs operational adjustments. Interactive features, such as scenario modeling or what-if analysis, allow stakeholders to explore the implications of different decisions and provide feedback on preferred approaches. These tools not only make complex information more accessible but also empower stakeholders to participate meaningfully in decision-making.

Ethical oversight and independent audits are additional pillars of transparency and accountability. Operators should establish mechanisms for evaluating the fairness, accuracy, and reliability of their AI systems, including third-party reviews and certifications. Independent audits assess whether AI-driven decisions comply with regulatory requirements, ethical standards, and stakeholder expectations, providing an objective basis for trust and accountability. For example, an audit of an AI system used for environmental compliance might evaluate its adherence to thresholds for sediment disturbance, its alignment with conservation priorities, and its responsiveness to stakeholder feedback. The findings of these audits should be disclosed publicly, demonstrating the operator's commitment to transparency and continuous improvement.

While transparency and disclosure are essential, they must be balanced with considerations of data privacy, security, and commercial confidentiality. Operators must navigate the tension between providing stakeholders with meaningful information and protecting sensitive data, such as proprietary algorithms or geophysical surveys. Best practices for transparency include developing clear policies and protocols for data sharing, specifying what information will be disclosed, to whom, and under what conditions. For example, environmental monitoring data might be made publicly available in aggregated or anonymized form, ensuring that stakeholders can evaluate ecological impacts without compromising the operator's competitive position.

The role of international standards and frameworks in promoting transparency and disclosure cannot be overstated. Initiatives such as the United Nations Guiding Principles on Business and Human Rights, the Global Reporting Initiative (GRI), and the Responsible Mining Index (RMI) provide benchmarks for ethical and transparent practices in resource extraction, including the use of AI. Adopting these standards demonstrates a commitment to accountability and aligns operations with global expectations for responsible business conduct. For instance, adherence to the GRI's reporting standards might involve publishing detailed disclosures about how AI systems are used to assess environmental impacts, manage risks, and engage stakeholders.

The advancement of deep sea mining presents an unprecedented opportunity to harness the ocean's vast mineral wealth to address global demands for critical resources. However, this opportunity is accompanied by significant ethical obligations, particularly the responsibility to minimize ecological impact in some of the Earth's most fragile and poorly understood ecosystems. Ethical guidelines for minimal ecological impact serve as a critical framework for guiding the industry, balancing the pursuit of economic benefits with the need to protect marine biodiversity, ecosystem services, and the health of the planet's oceans. These guidelines reflect not only a moral imperative but also a pragmatic approach to securing the long-term viability and legitimacy of deep sea mining operations.

The principle of minimal ecological impact begins with a precautionary approach, which mandates that operators prioritize the avoidance of harm in the face of scientific uncertainty. The deep ocean, with its unique and often endemic species, intricate food webs, and slow ecological recovery rates, is particularly vulnerable to disturbance. Ethical guidelines emphasize that where there is uncertainty about the potential impacts of mining activities, caution must prevail, and operations should proceed only if comprehensive measures are in place to prevent significant harm. This precautionary principle aligns with international environmental law, including the provisions of the United Nations Convention on the Law of the Sea (UNCLOS), and is a cornerstone of sustainable ocean governance.

Environmental impact assessments (EIAs) are a central component of ethical guidelines for minimal ecological impact, serving as the foundation for informed decision-making. These assessments must go beyond regulatory compliance to provide a robust and transparent analysis of the potential effects of mining activities on marine ecosystems. Ethical guidelines call for EIAs to be comprehensive, science-based, and inclusive of diverse perspectives, including those of local communities, indigenous peoples, and conservation experts. For instance, an EIA for a polymetallic nodule mining project might include baseline studies of benthic habitats, water column dynamics, and species distributions, coupled with predictive modeling of sediment plumes and long-term ecosystem recovery scenarios. The findings of the EIA should be publicly disclosed, ensuring that stakeholders have access to the information needed to evaluate the project's alignment with environmental and ethical standards.

One of the most significant challenges in minimizing ecological impact is the management of sediment plumes, which are generated when mining activities disturb seafloor sediments. These plumes can spread over large areas, smothering benthic habitats, reducing water quality, and altering the availability of light and nutrients in the water column. Ethical guidelines emphasize the importance of monitoring, predicting, and mitigating plume behavior to minimize its ecological consequences. Advanced technologies, such as high-resolution sensors, autonomous underwater vehicles (AUVs), and artificial intelligence (AI)-driven modeling, play a critical role in achieving this goal. For example, AI algorithms can analyze real-time data on currents, sediment composition, and plume dispersion, enabling operators to adjust mining techniques dynamically to reduce plume spread and intensity.

Another key aspect of ethical guidelines is the protection of biodiversity, particularly in areas of high ecological value or vulnerability. Deep sea ecosystems are characterized by high levels of endemism, meaning that many species found in these environments exist nowhere else on Earth. Mining activities pose a significant risk to these species, not only through direct habitat destruction but also by altering the chemical and physical properties of their surroundings. Ethical guidelines call for the identification and avoidance of ecologically sensitive areas, such as hydrothermal vent fields, seamounts, and cold-water coral habitats. Marine spatial planning, informed by scientific research and stakeholder input, provides a valuable tool for balancing resource extraction with biodiversity conservation, ensuring that mining activities are restricted to areas where the risks to ecosystems are minimal.

The concept of "no net loss" of biodiversity is increasingly recognized as a standard for responsible resource management, and it features prominently in ethical guidelines for deep sea mining. This principle requires operators to offset any biodiversity losses resulting from their activities by contributing to the restoration or protection of comparable habitats elsewhere. For example, a mining company might invest in the restoration of degraded coral reefs or the establishment of marine protected areas (MPAs) as part of its commitment to achieving no net loss. Ethical guidelines emphasize that offsets should be measurable, verifiable, and designed to deliver long-term ecological benefits, avoiding the risk of tokenistic or inadequate measures.

Ethical guidelines also address the need for continuous environmental monitoring and adaptive management. Deep sea mining operates in a dynamic and poorly understood environment, where the impacts of human activities may not be immediately apparent. Continuous monitoring, supported by autonomous sensor networks and AI-driven analytics, enables operators to track changes in environmental conditions and ecosystem health in real time. For instance, a monitoring system might detect changes in water chemistry, such as decreases in dissolved oxygen or increases in turbidity, signaling potential disturbances to marine life. Adaptive management frameworks ensure that this data is used to inform decision-making, allowing operators to modify their practices, reduce their impacts, and respond proactively to emerging risks.

Transparency and accountability are fundamental to the ethical governance of deep sea mining and are integral to minimizing ecological impact. Ethical guidelines call for operators to disclose information about their environmental performance, including the results of monitoring programs, compliance with mitigation measures, and progress toward sustainability targets. This transparency fosters trust among stakeholders and provides a basis for independent oversight and evaluation. For example, a mining company might publish annual sustainability reports detailing its efforts to minimize sediment disturbance, protect biodiversity, and contribute to marine conservation initiatives.

Stakeholder engagement is another critical element of ethical guidelines, ensuring that decisions about deep sea mining reflect the values and priorities of those most affected by its impacts. Local communities, indigenous peoples, and environmental organizations often possess unique knowledge and perspectives that can inform the design of more sustainable and equitable mining practices. For instance, an indigenous community with a deep cultural connection to the ocean might advocate for the protection of specific marine areas, while a conservation group might provide expertise on the recovery rates of benthic species. Ethical guidelines emphasize the importance of meaningful engagement with these stakeholders, creating opportunities for dialogue, collaboration, and shared decision-making.

The integration of ethical guidelines with international standards and best practices is essential for promoting consistency and accountability across the deep sea mining industry. Initiatives such as the International Seabed Authority's (ISA) Mining Code, the Global Reporting Initiative (GRI), and the International Finance Corporation's (IFC) Performance Standards provide valuable frameworks for aligning operations with global sustainability goals. Ethical guidelines encourage operators to adopt these standards and participate in certification schemes, such as the Responsible Mining Index (RMI), to demonstrate their commitment to responsible practices.

Despite their importance, the implementation of ethical guidelines for minimal ecological impact faces significant challenges. The deep ocean is one of the least explored and understood environments on Earth, and scientific knowledge about its ecosystems, processes, and recovery capacities remains limited. Addressing this knowledge gap requires sustained investment in research, exploration, and monitoring, as well as collaboration among governments, industry stakeholders, and the scientific community. Ethical guidelines also require a balance between ecological protection and economic feasibility, ensuring that sustainability measures are both effective and practical for operators to implement.

11. Human-AI Collaboration and Skillsets Required

11.1. Interdisciplinary Teams and Skill Gaps

The complexities of deep sea mining demand a level of interdisciplinary collaboration rarely encountered in traditional industries. The successful execution of deep sea projects hinges on the seamless integration of expertise from a diverse array of disciplines, including geology, oceanography, robotics, data science, and AI ethics. Each of these fields plays a critical role in addressing the technical, environmental, and ethical challenges posed by operating in one of the planet's most extreme and enigmatic environments. Building and managing interdisciplinary teams capable of navigating these challenges is not only a logistical necessity but also a strategic imperative for achieving innovation, sustainability, and long-term success.

Geologists are at the forefront of identifying and characterizing the mineral resources that drive deep sea mining. Their expertise in understanding the composition, formation, and distribution of seabed deposits such as polymetallic nodules, seafloor massive sulfides (SMS), and cobalt-rich ferromanganese crusts is foundational to every stage of resource exploration and extraction. Geologists analyze geophysical survey data, core samples, and seafloor imagery to assess the quality and quantity of resources in a given area, providing the critical insights needed to prioritize exploration targets and design efficient extraction plans. Their work often involves collaboration with data scientists and roboticists to process complex datasets and deploy advanced sensing technologies, ensuring that resource assessments are accurate, comprehensive, and aligned with operational objectives.

Oceanographers bring a complementary perspective to the interdisciplinary team, focusing on the physical, chemical, and biological processes that define the deep ocean environment. Their expertise is essential for understanding the dynamic systems that influence resource distribution and impact mining activities, including ocean currents, sediment transport, and ecosystem interactions. For instance, oceanographers play a pivotal role in modeling the behavior of sediment plumes generated by mining operations, assessing their potential effects on water quality, benthic habitats, and pelagic ecosystems. By providing detailed insights into the environmental conditions and vulnerabilities of target areas, oceanographers help ensure that resource extraction is conducted in a manner that minimizes ecological harm and adheres to regulatory standards.

Roboticists are central to the technological innovation that underpins deep sea mining, designing and deploying the autonomous systems that make it possible to operate in the extreme conditions of the deep ocean. These engineers develop remotely operated vehicles (ROVs), autonomous underwater vehicles (AUVs), and specialized extraction tools that perform tasks ranging from seafloor mapping and sample collection to nodule harvesting and environmental monitoring. Roboticists collaborate closely with geologists and oceanographers to integrate advanced sensors, cameras, and manipulators into their designs, ensuring that these systems can navigate complex terrains, adapt to changing conditions, and collect high-quality data. Their work also involves leveraging the capabilities of artificial intelligence (AI) to enhance the autonomy, precision, and efficiency of robotic systems, enabling them to perform complex operations with minimal human intervention.

Data scientists play a transformative role in processing and analyzing the vast amounts of data generated by deep sea mining activities. These professionals develop machine learning algorithms and statistical models to extract actionable insights from geophysical surveys, environmental monitoring systems, and operational metrics. For example, data scientists might create predictive models that identify high-value resource deposits based on patterns in sonar and geochemical data, or algorithms that optimize the routes of AUVs to maximize survey coverage while minimizing energy consumption. Their expertise is also critical for integrating diverse datasets into cohesive frameworks, enabling interdisciplinary teams to make informed decisions that account for both resource potential and environmental constraints.

AI ethicists add an essential dimension to the interdisciplinary team, ensuring that the deployment of AI in deep sea mining is guided by ethical principles and aligned with societal values. These specialists address issues such as algorithmic fairness, transparency, and accountability, providing the oversight needed to prevent biases, mitigate unintended consequences, and build trust among stakeholders. For instance, an AI ethicist might evaluate the criteria used by an AI system to prioritize mining sites, ensuring that decisions are equitable and consider the interests of diverse stakeholders, including local communities and environmental organizations. Their work often involves collaborating with data scientists, regulators, and industry leaders to establish governance frameworks that balance the benefits of AI with its ethical risks, promoting responsible innovation and long-term sustainability.

The integration of these disciplines into cohesive interdisciplinary teams requires careful coordination and a shared commitment to collaboration. Each field brings its own methodologies, terminologies, and perspectives, which must be harmonized to achieve common goals. For example, a geologist's focus on resource characterization must align with an oceanographer's concerns about ecosystem impacts, while a roboticist's technical innovations must be informed by the ethical considerations raised by an AI ethicist. Establishing clear communication channels, fostering mutual respect, and promoting cross-disciplinary learning are essential for overcoming these challenges and unlocking the full potential of the team.

Skill gaps pose a significant challenge to the formation and effectiveness of interdisciplinary teams in deep sea mining. The rapid pace of technological advancement and the unique demands of the deep ocean environment require a workforce with specialized knowledge and skills that are often in short supply. For instance, the development and operation of autonomous systems for deep sea exploration demand expertise in robotics, AI, and marine engineering, a combination that is not commonly found in existing training programs. Similarly, the integration of environmental and ethical considerations into mining operations requires professionals who are well-versed in both scientific research and regulatory frameworks, a rare and highly valuable skill set.

Addressing these skill gaps necessitates targeted investments in education, training, and professional development. Academic institutions, industry stakeholders, and government agencies have a shared responsibility to create programs that equip the next generation of professionals with the interdisciplinary skills needed for deep sea mining. For example, universities could establish joint degree programs in marine science and robotics, or offer specialized courses on AI ethics in resource management. Industry-sponsored apprenticeships and internships provide valuable opportunities for hands-on learning, enabling students and early-career professionals to gain practical experience in cutting-edge technologies and collaborative problem-solving.

Professional development programs for current practitioners are equally important for bridging skill gaps and fostering interdisciplinary collaboration. Workshops, conferences, and online training platforms provide opportunities for geologists, oceanographers, roboticists, data scientists, and AI ethicists to deepen their understanding of each other's fields and explore new approaches to shared challenges. For instance, a workshop on sediment plume modeling might bring together oceanographers, data scientists, and roboticists to develop integrated solutions for monitoring and mitigating plume impacts, enhancing both technical and collaborative capacities.

The diversity of expertise within interdisciplinary teams is a source of strength, enabling deep sea mining operations to address the complex and interconnected challenges of resource extraction, environmental protection, and ethical governance. However, it also requires intentional efforts to cultivate a culture of collaboration and mutual respect. Team leaders play a critical role in fostering this culture, creating an environment where diverse perspectives are valued and contributions are recognized. Leadership that emphasizes inclusivity, transparency, and adaptability ensures that interdisciplinary teams can work effectively toward their shared goals.

As the field of deep sea mining evolves, the demand for a highly skilled, interdisciplinary workforce becomes increasingly urgent. The challenges of operating in one of the planet's most extreme environments require expertise that spans geology, oceanography, robotics, artificial intelligence (AI), ethics, and more. To address these needs, universities and professional training programs must design curricula that prepare the next generation of professionals with the technical, environmental, and ethical knowledge necessary to navigate the complexities of deep sea mining. These programs must go beyond traditional disciplinary boundaries, fostering a culture of collaboration and innovation while addressing critical skill gaps that exist in the current workforce.

A foundational principle in the design of curricula for deep sea mining is the recognition of its inherently interdisciplinary nature. Unlike conventional industries, which may rely on siloed expertise, deep sea mining demands an integrated approach where scientists, engineers, ethicists, and policymakers work together to solve complex problems. For example, understanding the formation and distribution of polymetallic nodules requires geological expertise, while assessing their ecological impacts demands input from oceanographers. Similarly, designing and deploying autonomous underwater vehicles (AUVs) for resource extraction involves robotics and data science, while ensuring these technologies align with societal values necessitates the guidance of AI ethicists.

To address these diverse demands, university programs must offer interdisciplinary degrees and specialized tracks that combine multiple fields of study. For instance, a master's program in "Marine Resource Technology and Sustainability" could integrate courses in marine geology, environmental science, robotics, and AI ethics. Such programs should emphasize both theoretical knowledge and practical skills, providing students with a holistic understanding of deep sea mining operations and their implications. For example, a course on environmental monitoring might cover the principles of sediment plume dynamics, the use of sensor networks for data collection, and the application of AI algorithms for real-time analysis.

Hands-on learning is a critical component of effective curricula, enabling students to apply their knowledge to real-world challenges. Universities should establish partnerships with industry stakeholders, research institutions, and regulatory agencies to provide opportunities for fieldwork, internships, and collaborative projects. For example, a student might participate in an internship with a mining company, working alongside oceanographers and roboticists to develop a monitoring system for benthic habitats. Similarly, a capstone project could involve designing an AI-driven decision-support tool for site selection, incorporating considerations of resource potential, environmental sensitivity, and stakeholder priorities.

In addition to interdisciplinary degrees, universities should offer certificate programs and short courses tailored to specific aspects of deep sea mining. These programs can serve as entry points for professionals from related fields, such as terrestrial mining, renewable energy, or AI development, who wish to transition into the deep sea mining sector. For instance, a certificate in "Marine Robotics and Autonomous Systems" might cover the design, operation, and maintenance of AUVs and remotely operated vehicles (ROVs), providing hands-on experience with cutting-edge technologies. Similarly, a course on "AI Ethics in Resource Management" could explore the ethical considerations of deploying AI in sensitive marine environments, offering tools for evaluating algorithmic fairness, transparency, and accountability.

Professional training programs play a complementary role in addressing skill gaps, providing opportunities for lifelong learning and career development. These programs are particularly valuable for mid-career professionals who seek to enhance their expertise or adapt to new challenges in the industry. For example, a geologist with experience in terrestrial mining might benefit from a training program on marine resource exploration, which could include modules on geophysical surveying techniques, core sampling methods, and the geological characteristics of the seafloor. Similarly, an engineer specializing in robotics might pursue training on the design and deployment of subsea systems, focusing on the unique technical requirements of operating under high pressure and low visibility.

The incorporation of ethics and sustainability into curricula and training programs is essential for ensuring that the workforce is prepared to navigate the broader implications of deep sea mining. Courses on environmental ethics, corporate social responsibility, and international law should be integral components of educational offerings, fostering a sense of accountability and stewardship among professionals. For instance, a module on "Environmental Impact Assessments for Deep Sea Mining" might cover the principles of the precautionary approach, the design of monitoring programs, and the integration of traditional ecological knowledge into decision-making. By emphasizing the ethical dimensions of their work, these programs help ensure that professionals are equipped to balance economic development with the protection of marine ecosystems and the interests of affected communities.

To foster a global perspective, curricula should include content on international governance frameworks, such as the United Nations Convention on the Law of the Sea (UNCLOS) and the International Seabed Authority (ISA). Understanding these frameworks is critical for navigating the regulatory complexities of deep sea mining and ensuring compliance with international standards. For example, a course on "Global Ocean Governance" might examine the principles of the "common heritage of mankind," the role of the ISA in licensing and benefit-sharing, and the implications of regional agreements on biodiversity conservation. Case studies and simulations can provide valuable context, allowing students to explore real-world scenarios and develop strategies for addressing regulatory challenges.

Collaboration with international institutions and organizations is key to ensuring that curricula and training programs reflect the latest developments and best practices in deep sea mining. Partnerships with entities such as the ISA, the International Maritime Organization (IMO), and the Global Ocean Observing System (GOOS) can provide access to expertise, resources, and networks that enhance the quality and relevance of educational offerings. For example, a university might collaborate with the ISA to develop a training program on benefit-sharing mechanisms, incorporating insights from global policy discussions and stakeholder consultations. Similarly, a partnership with GOOS could support the integration of cutting-edge oceanographic data and technologies into coursework and research projects.

The use of technology in education is another important consideration for curriculum design, particularly in the context of deep sea mining. Virtual reality (VR) and simulation-based tools offer immersive learning experiences, enabling students to explore underwater environments, operate robotic systems, and analyze geophysical data in realistic settings. For instance, a VR simulation might allow students to navigate an AUV through a seafloor terrain, collecting samples and avoiding obstacles, while receiving real-time feedback on their performance. Online learning platforms and digital resources further enhance accessibility, allowing professionals from diverse locations and backgrounds to participate in training programs and stay updated on industry trends.

Despite the opportunities, curriculum design for deep sea mining faces challenges that must be addressed to ensure its effectiveness. One of the primary challenges is the scarcity of faculty and trainers with expertise in this emerging field. Addressing this gap requires investment in faculty development, as well as the recruitment of industry professionals and researchers who can bring practical insights to the classroom. Another challenge is the need for interdisciplinary integration, which requires breaking down traditional academic silos and fostering collaboration across departments and disciplines. Universities must create structures and incentives that support cross-disciplinary initiatives, such as joint appointments, interdisciplinary research centers, and team-taught courses.

11.2. Cognitive Systems and Decision-Making

In the domain of deep sea mining, cognitive systems that enable intelligent decision-making are transforming how operations are planned, executed, and monitored. These systems rely heavily on artificial intelligence (AI) and data analytics to process vast amounts of information, predict outcomes, and optimize processes in one of the Earth's most challenging environments. Central to the design of these cognitive systems is the balance between human oversight, often referred to as "human-in-the-loop" systems, and fully autonomous pipelines that operate with minimal human intervention. Each approach has distinct advantages, limitations, and implications for operational efficiency, safety, and ethical governance, making their careful integration essential for the success and sustainability of deep sea mining.

Human-in-the-loop systems are built on the principle that human expertise and judgment remain essential for critical decision-making, particularly in complex or uncertain scenarios. These systems combine AI's computational power with human insight, ensuring that operators retain control over key aspects of operations while benefiting from the speed and precision of automated processes. In deep sea mining, human-in-the-loop systems are widely used in areas such as resource exploration, environmental monitoring, and risk assessment, where the stakes are high and the variables numerous.

For instance, a human-in-the-loop system might be deployed during seafloor mapping, where AI algorithms analyze sonar and geophysical data to identify potential resource deposits. While the AI system can process large datasets and highlight areas of interest with remarkable efficiency, human geologists review the results to validate the findings, interpret geological patterns, and account for contextual factors that the AI may not fully understand. This collaboration ensures that decisions about exploration targets are informed by both advanced analytics and domain expertise, reducing the risk of errors and enhancing the reliability of the results.

Human-in-the-loop systems also play a critical role in managing ethical and environmental considerations, ensuring that decisions align with societal values and regulatory requirements. For example, an AI-driven sediment plume monitoring system might detect deviations from baseline conditions, flagging areas where turbidity levels exceed acceptable thresholds. Human operators can then evaluate the context, considering factors such as ecological sensitivity, operational constraints, and stakeholder concerns, before determining the appropriate course of action. By keeping humans in the loop, these systems allow for nuanced decision-making that balances economic objectives with ethical responsibilities.

Despite their strengths, human-in-the-loop systems face challenges related to scalability, consistency, and real-time responsiveness. The involvement of human operators can limit the system's ability to process information and make decisions at the speed required for certain deep sea mining operations. For example, in situations where immediate adjustments are needed to prevent equipment damage or mitigate environmental impacts, the latency introduced by human review could lead to suboptimal outcomes. Additionally, the reliance on human expertise introduces variability, as decisions may be influenced by individual biases, experience levels, or subjective interpretations.

Fully autonomous pipelines, by contrast, operate without direct human involvement, relying entirely on AI algorithms and automated systems to make decisions and execute tasks. These pipelines are designed to maximize efficiency, consistency, and scalability, making them well-suited to repetitive or time-critical operations where human oversight may not add significant value. In deep sea mining, fully autonomous pipelines are increasingly used for tasks such as resource extraction, equipment maintenance, and data analysis, where precision and speed are paramount.

An example of a fully autonomous pipeline is the operation of subsea robots for nodule collection. Equipped with advanced sensors, cameras, and AI-driven control systems, these robots navigate the seafloor, identify and collect nodules, and transport them to surface vessels with minimal human intervention. The autonomy of these systems allows them to operate continuously, adapting to changing conditions and optimizing their performance in real time. For instance, an autonomous robot might adjust its route to avoid obstacles, modify its suction power to minimize sediment disturbance, or recalibrate its position based on updated sonar data, all without requiring input from human operators.

Fully autonomous pipelines are particularly valuable in areas where environmental or operational conditions are too hazardous or unpredictable for direct human involvement. For example, the extreme pressures, low temperatures, and darkness of the deep ocean make it impractical for humans to oversee every aspect of subsea operations in real time. Autonomous systems, by contrast, are designed to withstand these conditions and operate reliably over extended periods, ensuring the continuity of operations and reducing the risks associated with human error or fatigue.

However, the deployment of fully autonomous pipelines raises important ethical and practical questions, particularly regarding accountability and transparency. Without human oversight, decisions made by autonomous systems may lack the contextual understanding or ethical considerations that humans can provide. For example, an autonomous system tasked with optimizing resource recovery might prioritize efficiency over ecological sensitivity, inadvertently causing harm to vulnerable habitats or species. Ensuring that fully autonomous pipelines operate responsibly requires the integration of ethical principles into their algorithms, as well as robust governance frameworks to monitor and evaluate their performance.

The choice between human-in-the-loop systems and fully autonomous pipelines is not binary; rather, it depends on the specific requirements, risks, and objectives of each operation. In many cases, the optimal approach involves a hybrid model that combines elements of both systems, leveraging their respective strengths while mitigating their limitations. For example, a hybrid system might use fully autonomous pipelines for routine tasks such as data collection and processing, while reserving human-in-the-loop systems for critical decision points where human judgment adds significant value. This approach ensures that operations remain efficient, adaptive, and aligned with ethical standards, even in the most challenging environments.

The integration of human-in-the-loop systems and fully autonomous pipelines also necessitates investment in training and capacity-building to equip professionals with the skills needed to interact with and oversee these systems effectively. Operators, engineers, and managers must be able to understand the capabilities and limitations of AI-driven systems, interpret their outputs, and intervene when necessary. For example, a training program for AUV operators might include modules on AI algorithms, sensor calibration, and ethical decision-making, ensuring that they can supervise autonomous pipelines with confidence and competence.

Transparency and accountability are critical for ensuring the legitimacy and acceptance of both human-in-the-loop systems and fully autonomous pipelines. Stakeholders, including regulators, environmental organizations, and local communities, must have access to information about how these systems operate, what criteria they use to make decisions, and how their impacts are monitored and evaluated. For instance, an operator might publish detailed reports on the performance of an autonomous sediment monitoring system, including data on compliance with environmental thresholds, the accuracy of predictions, and the frequency of human interventions. This transparency fosters trust and enables informed dialogue about the benefits and risks of using AI in deep sea mining.

The integration of artificial intelligence (AI) into deep sea mining operations has brought about remarkable advancements in efficiency, precision, and adaptability. However, these benefits are only realized when human operators can trust and effectively supervise the AI tools at their disposal. The design of interfaces and user experiences (UX) plays a pivotal role in bridging the gap between sophisticated AI systems and the humans who rely on them to make informed decisions. Effective interfaces must not only communicate complex information clearly and intuitively but also foster trust, accountability, and seamless collaboration between humans and machines.

Trust is a cornerstone of successful human-AI interaction. In deep sea mining, where decisions often have significant environmental, economic, and ethical consequences, operators must have confidence in the accuracy, reliability, and fairness of AI tools. Achieving this trust begins with transparency—ensuring that AI systems provide clear explanations of their processes, criteria, and outputs. For example, an AI-driven site selection tool might identify areas with high resource potential based on geological, environmental, and logistical data. To foster trust, the interface must display the factors that influenced the decision, such as mineral grade, ecological sensitivity, and proximity to infrastructure, along with a breakdown of how these factors were weighted in the algorithm.

Explainable AI (XAI) techniques are instrumental in achieving this level of transparency. XAI methods, such as feature importance analysis and decision path visualization, allow operators to understand how AI systems generate their outputs. For instance, a sediment plume monitoring system powered by AI might display a heatmap of plume dispersion patterns, annotated with the key variables driving the predictions, such as current velocity and sediment composition. By making the AI's reasoning visible and interpretable, XAI enhances the operator's ability to evaluate the system's conclusions, identify potential biases, and make informed decisions.

The design of interfaces must also prioritize clarity and accessibility, ensuring that complex information is presented in a way that is easy to understand and actionable. Deep sea mining generates vast amounts of data, from geophysical surveys and environmental monitoring to operational metrics and market analytics. Without thoughtful design, the sheer volume and complexity of this data can overwhelm operators, leading to errors, inefficiencies, or mistrust in the AI system. Intuitive interfaces that prioritize usability and streamline workflows are essential for enabling operators to focus on critical tasks.

For example, a dashboard designed for managing autonomous underwater vehicles (AUVs) might include real-time visualizations of vehicle locations, sensor readings, and operational statuses, accompanied by alerts for anomalies or maintenance needs. Interactive elements, such as zoomable maps or drill-down menus, allow operators to explore data at varying levels of detail, providing both high-level overviews and granular insights. Color-coded indicators and progress bars can highlight key metrics, such as the percentage of a survey area completed or the remaining battery life of an AUV, enabling quick assessments and timely interventions.

Another critical aspect of user experience is the ability to customize interfaces to meet the diverse needs of operators with varying levels of expertise and responsibilities. In deep sea mining, the workforce may include geologists, oceanographers, roboticists, and environmental managers, each with distinct priorities and workflows. A geologist might focus on resource distribution maps, while an environmental manager might monitor compliance with sediment plume thresholds. Interfaces that allow users to configure their dashboards, select relevant data streams, and set personalized alerts enhance efficiency and ensure that each stakeholder has access to the information they need to make informed decisions.

Collaboration is another essential consideration in designing AI interfaces for deep sea mining. Many decisions involve input from multiple stakeholders, requiring interfaces that facilitate communication and coordination. For example, an interface for site selection might include features for annotating maps, sharing comments, and proposing alternative scenarios, enabling team members to collaborate on decisions in real time. These collaborative tools ensure that decisions are not made in isolation but are informed by diverse perspectives and expertise, enhancing both the quality and legitimacy of the outcomes.

To support effective supervision, interfaces must also include mechanisms for manual overrides and escalation, allowing operators to intervene when necessary. While AI systems excel at processing data and identifying patterns, they may lack the contextual understanding or ethical judgment required for certain decisions. For example, an AI system tasked with optimizing resource extraction might recommend intensifying operations in a high-value area without accounting for its cultural or ecological significance. In such cases, the interface should provide a clear and accessible pathway for operators to override the AI's recommendations, ensuring that human judgment remains central to critical decisions.

Building trust in AI systems also requires addressing concerns about fairness, bias, and accountability. Operators must be confident that the AI is making decisions that are equitable and aligned with regulatory and ethical standards. Interfaces can support this confidence by displaying information about the system's fairness metrics, such as the distribution of resource allocation across regions or the consistency of environmental impact assessments. For example, a dashboard might include a fairness audit panel that highlights any disparities in how the AI's recommendations affect different stakeholder groups or ecological zones, prompting further review and adjustments if necessary.

The use of feedback loops is another powerful tool for fostering trust and improving the effectiveness of AI systems. Interfaces should allow operators to provide feedback on the system's outputs, such as flagging errors, suggesting improvements, or confirming successful outcomes. This feedback can be used to refine the AI model, ensuring that it learns from real-world experiences and evolves to better meet the needs of its users. For instance, if an operator identifies a false positive in a sediment plume alert, the feedback can be incorporated into the model's training data, reducing the likelihood of similar errors in the future.

Ethical considerations are central to the design of AI interfaces in deep sea mining, particularly in the context of stakeholder engagement and public perceptions. Transparency and inclusivity are key to building trust not only among operators but also among external stakeholders, such as regulators, conservation groups, and local communities. Interfaces should include features for generating and sharing reports, visualizations, and summaries that communicate the AI system's decisions and impacts to a broader audience. For example, a public-facing dashboard might display information about the environmental performance of a mining operation, such as compliance with sediment disturbance thresholds or progress toward biodiversity restoration goals. By making this information accessible and understandable, operators can demonstrate their commitment to accountability and foster trust among stakeholders.

Designing AI tools that humans can trust and effectively supervise also requires addressing the challenges of training and capacity-building. Operators must be equipped with the skills and knowledge needed to interpret AI outputs, evaluate their implications, and make informed decisions. Training programs should include modules on AI fundamentals, data analysis, and ethical decision-making, ensuring that operators are confident in their ability to interact with and oversee AI systems. For example, a training course might include hands-on exercises where participants use AI-driven dashboards to analyze sediment plume data, identify potential issues, and propose mitigation strategies.

11.3. Organizational Transformation

The adoption of advanced technologies in traditional mining and maritime companies is not merely a technical challenge but a profound organizational transformation that reshapes workflows, business models, and corporate cultures. As the demands of the global economy evolve, driven by the transition to renewable energy, the need for critical minerals, and the pursuit of sustainability, companies in these sectors must integrate cutting-edge innovations such as artificial intelligence (AI), robotics, and advanced data analytics into their operations. For traditional organizations with established practices and legacy systems, this transformation is both an opportunity and a challenge, requiring carefully designed strategies to ensure that technology adoption aligns with long-term goals and delivers tangible value.

The successful adoption of technology in traditional mining and maritime companies begins with a clear and compelling vision for transformation. This vision articulates how new technologies will enable the organization to achieve its objectives, whether by improving efficiency, reducing environmental impacts, enhancing safety, or expanding market opportunities. For example, a mining company might envision leveraging AI to optimize resource extraction, minimize waste, and comply with stringent environmental regulations. Similarly, a maritime company might seek to use autonomous vessels and predictive analytics to streamline logistics, reduce fuel consumption, and improve operational reliability. This vision serves as a unifying framework that guides decision-making, secures stakeholder buy-in, and aligns the organization's efforts toward a common purpose.

Leadership plays a pivotal role in driving technology adoption, setting the tone for the organization's commitment to innovation and fostering a culture that embraces change. Executives and managers must champion the adoption of new technologies, demonstrating their benefits and addressing concerns about disruption or obsolescence. For instance, a CEO might highlight the role of AI-driven environmental monitoring in enhancing the company's sustainability credentials, positioning it as a competitive advantage in an increasingly eco-conscious market. By articulating the strategic value of technology and actively supporting its implementation, leaders can inspire confidence and enthusiasm among employees, creating a foundation for successful transformation.

One of the most significant challenges in technology adoption is overcoming resistance to change, which often stems from fear of the unknown, skepticism about the value of new tools, or attachment to established practices. Addressing these concerns requires a combination of education, communication, and empowerment. Employees at all levels of the organization must be provided with the knowledge and skills needed to understand and utilize new technologies effectively. For example, a training program on AI and data analytics might include workshops on interpreting predictive models, hands-on exercises with decision-support dashboards, and discussions about the ethical implications of algorithmic decision-making. These programs not only build technical competencies but also foster a sense of ownership and agency, helping employees see themselves as active participants in the organization's transformation.

Communication is equally critical for addressing resistance and building trust. Transparent and consistent messaging about the rationale, benefits, and expected outcomes of technology adoption helps dispel uncertainty and align stakeholders around a shared vision. For instance, a mining company might use newsletters, town hall meetings, and intranet portals to share success stories from pilot projects, provide updates on implementation progress, and solicit feedback from employees. These communications reinforce the message that technology adoption is a collaborative effort that values input from all parts of the organization.

Pilot projects are a valuable strategy for demonstrating the feasibility and benefits of new technologies while minimizing risks. By starting with small-scale implementations in specific areas, companies can test and refine their approaches, gather data on performance and outcomes, and build confidence among stakeholders. For example, a maritime company might pilot an AI-driven route optimization system on a single shipping route, comparing its efficiency and cost savings to traditional methods before scaling up to the entire fleet. The insights gained from pilot projects inform broader rollouts, ensuring that the organization is prepared to address challenges and capitalize on opportunities as it scales its technology adoption efforts.

Integration is another critical aspect of technology adoption, particularly in organizations with complex operations and legacy systems. New technologies must be seamlessly incorporated into existing workflows, infrastructure, and business processes, ensuring that they complement rather than disrupt the organization's operations. For instance, a mining company adopting autonomous vehicles for resource extraction must integrate these systems with its logistics, maintenance, and environmental monitoring platforms, creating a cohesive ecosystem that supports end-to-end efficiency. Achieving this integration requires collaboration across departments, as well as partnerships with technology providers who can customize solutions to meet the organization's unique needs.

Collaboration extends beyond internal teams to include external stakeholders, such as technology vendors, academic institutions, and industry consortia. Partnerships with these entities provide access to expertise, resources, and innovation that enhance the organization's capabilities and accelerate technology adoption. For example, a maritime company might collaborate with a university's robotics lab to develop autonomous vessels tailored to its operational requirements, or participate in an industry consortium focused on advancing sustainability standards for shipping. These collaborations foster a culture of continuous learning and innovation, ensuring that the organization remains at the forefront of technological advancements.

The adoption of advanced technologies also requires a focus on sustainability and ethics, particularly in industries such as mining and maritime that have significant environmental and social impacts. Companies must ensure that their technology strategies align with global sustainability goals, such as the United Nations Sustainable Development Goals (SDGs), and demonstrate a commitment to responsible practices. For instance, a mining company adopting AI-driven exploration tools might prioritize technologies that minimize habitat disturbance and reduce greenhouse gas emissions, aligning its operations with regulatory requirements and stakeholder expectations. Similarly, a maritime company might implement blockchain-based traceability systems to ensure that its supply chain meets ethical and environmental standards.

Measurement and evaluation are essential for assessing the success of technology adoption initiatives and identifying areas for improvement. Companies must establish clear metrics and benchmarks to track progress, evaluate performance, and quantify the benefits of new technologies. For example, a mining company might measure the impact of AI-driven resource optimization on cost savings, resource recovery rates, and environmental compliance. These metrics provide valuable insights that inform future investments and enable the organization to demonstrate the value of its technology adoption efforts to stakeholders.

Finally, technology adoption in traditional mining and maritime companies is not a one-time effort but an ongoing process of innovation and adaptation. As technologies continue to evolve and new challenges emerge, organizations must remain agile, continuously evaluating and refining their strategies to stay competitive and responsive. This requires a culture of curiosity, experimentation, and openness to change, where employees are encouraged to explore new ideas, take calculated risks, and embrace lifelong learning.

The integration of artificial intelligence (AI) into organizational workflows and decision-making processes represents one of the most transformative opportunities for traditional mining and maritime companies. AI has the potential to enhance operational efficiency, improve resource utilization, and minimize environmental impacts, but its successful adoption requires a structured approach that aligns technological capabilities with organizational readiness. Maturity models and readiness assessments provide critical frameworks for evaluating an organization's current state, identifying gaps, and charting a pathway for integrating AI effectively and sustainably. These tools ensure that AI adoption is not just a technological upgrade but a holistic transformation that enhances the organization's strategic goals and operational resilience.

Maturity models are structured frameworks that assess an organization's progression through various stages of AI adoption, from initial experimentation to full integration and optimization. These models typically consist of several levels, each representing a greater degree of capability, sophistication, and alignment with organizational objectives. For mining and maritime companies, a maturity model might begin with basic awareness and experimentation, advance to pilot projects and targeted implementations, and culminate in enterprise-wide integration where AI systems are embedded across all critical processes.

At the foundational level of the maturity model, organizations typically exhibit limited awareness of AI's potential and rely on manual or legacy systems for decision-making and operations. In this stage, there may be isolated efforts to explore AI technologies, such as experimenting with predictive analytics for equipment maintenance or using machine learning to analyze geophysical survey data. These early efforts are often uncoordinated and lack the strategic direction needed for scaling AI initiatives. Readiness assessments at this stage focus on building foundational capabilities, such as fostering organizational awareness, identifying potential use cases, and securing executive support.

The transition to the next stage of maturity involves piloting AI applications in specific areas to test their feasibility and impact. For example, a mining company might deploy an AI-driven sediment plume monitoring system in a single project to evaluate its accuracy and operational benefits. Similarly, a maritime company might pilot an AI-based route optimization tool for one shipping route to assess its potential for reducing fuel consumption and emissions. These pilots provide valuable insights into the technical, operational, and organizational challenges of AI implementation, laying the groundwork for broader adoption.

As organizations progress through the maturity model, they begin to integrate AI into core processes, moving from isolated applications to interconnected systems that support end-to-end decision-making and operations. For instance, a mining company might use AI to optimize the entire resource extraction lifecycle, from site selection and exploration to extraction and logistics. Similarly, a maritime company might implement AI-driven systems across its fleet, integrating navigation, maintenance, and cargo management into a cohesive ecosystem. At this stage, readiness assessments focus on scaling successful pilots, addressing infrastructure and data integration challenges, and fostering cross-departmental collaboration.

The highest level of AI maturity is characterized by enterprise-wide integration, where AI is fully embedded into the organization's strategic and operational fabric. At this stage, AI systems operate autonomously or semi-autonomously, providing real-time insights, automating routine tasks, and supporting complex decision-making processes. For example, an AI-driven digital twin might simulate the operations of a mining site or shipping terminal, enabling operators to optimize performance, predict maintenance needs, and respond dynamically to changing conditions. Achieving this level of maturity requires a deep alignment between AI initiatives and the organization's goals, as well as a robust governance framework to ensure transparency, accountability, and ethical compliance.

Readiness assessments are essential tools for guiding organizations through the stages of the maturity model, providing a systematic evaluation of their capabilities, resources, and challenges. These assessments typically examine several key dimensions, including leadership, culture, technology, data, skills, and governance. For each dimension, the assessment identifies strengths, weaknesses, and areas for improvement, enabling organizations to prioritize actions and allocate resources effectively.

Leadership and culture are critical dimensions of readiness, as they shape the organization's attitude toward innovation and change. A readiness assessment might evaluate whether executives and managers have a clear vision for AI adoption, whether they actively support AI initiatives, and whether the organization fosters a culture of curiosity, experimentation, and learning. For example, an assessment might reveal that while senior leaders are enthusiastic about AI, middle managers are hesitant due to concerns about job displacement or the complexity of new technologies. Addressing these concerns through targeted communication, training, and engagement strategies is essential for building a unified commitment to AI integration.

The technology dimension focuses on the organization's infrastructure, systems, and tools for supporting AI initiatives. A readiness assessment might examine the compatibility of legacy systems with AI technologies, the availability of high-quality data for training AI models, and the scalability of computing resources for processing large datasets. For instance, an assessment might highlight that a maritime company's existing fleet management software lacks the real-time data capabilities needed for AI-driven route optimization. Addressing this gap might involve upgrading hardware, adopting cloud-based solutions, or partnering with technology providers to develop customized tools.

Data is the lifeblood of AI, and its quality, accessibility, and governance are critical factors in readiness assessments. Organizations must ensure that their data is accurate, comprehensive, and representative of the contexts in which AI systems will operate. For example, a mining company deploying an AI system for site selection must have access to high-resolution geological, geophysical, and environmental data, along with robust protocols for managing and securing this data. Readiness assessments might identify gaps in data availability or highlight the need for data standardization and integration across departments or external partners.

Skills and talent are another critical dimension of readiness, as AI adoption requires a workforce with the technical expertise and interdisciplinary knowledge to develop, implement, and oversee AI systems. A readiness assessment might evaluate whether the organization has sufficient data scientists, AI engineers, and domain specialists, as well as whether employees across all levels are equipped to interact with and benefit from AI tools. For example, an assessment might reveal a need for training programs to help operations managers understand how to interpret and act on AI-driven insights, or for partnerships with academic institutions to build a pipeline of skilled graduates.

Governance is the final dimension of readiness, encompassing the policies, standards, and frameworks that ensure AI systems are used responsibly and effectively. A readiness assessment might examine whether the organization has clear guidelines for algorithmic transparency, data privacy, and ethical compliance, as well as mechanisms for monitoring and evaluating the performance of AI systems. For instance, an assessment might recommend the establishment of an AI ethics committee to oversee the deployment of AI tools in environmentally sensitive areas, ensuring that decisions align with regulatory requirements and societal values.

The insights gained from readiness assessments inform the development of tailored strategies for AI integration, ensuring that organizations can navigate the complexities of the maturity model effectively. For example, a mining company at the pilot stage might focus on building technical infrastructure and fostering cross-departmental collaboration, while a maritime company nearing enterprise-wide integration might prioritize governance frameworks and stakeholder engagement. These strategies must be dynamic and iterative, allowing organizations to adapt to new challenges and opportunities as they progress through the maturity model.

The successful integration of innovation into traditional mining and maritime companies requires governance structures that not only foster technological progress but also mitigate the risks associated with complex operations and emerging technologies. In the context of deep sea mining, this balance is particularly critical. The industry's reliance on advanced systems such as artificial intelligence (AI), autonomous robotics, and real-time environmental monitoring introduces unprecedented opportunities for efficiency and precision, but it also amplifies risks related to ethical considerations, environmental impacts, regulatory compliance, and operational failures. Governance structures must be designed to encourage innovation while ensuring that risks are identified, assessed, and managed effectively, safeguarding the long-term viability of the organization and its social license to operate.

Governance structures in this context begin with a clear articulation of the organization's values, priorities, and objectives, providing a framework for evaluating and guiding innovation initiatives. These values must reflect not only the company's business goals but also its commitments to sustainability, transparency, and ethical responsibility. For example, a deep sea mining company might adopt a governance charter that prioritizes minimizing ecological impacts, promoting equitable benefit-sharing, and adhering to international regulatory frameworks such as the United Nations Convention on the Law of the Sea (UNCLOS). This charter serves as a touchstone for decision-making, ensuring that innovation aligns with the organization's overarching mission and values.

At the operational level, governance structures should incorporate risk management processes that are integrated into the innovation lifecycle, from ideation and development to deployment and evaluation. These processes must be dynamic and iterative, capable of adapting to new information and evolving challenges. For instance, a mining company developing an AI-driven site selection tool might conduct an initial risk assessment to identify potential biases in the algorithm, such as the overrepresentation of well-explored areas or the underweighting of environmental criteria. Mitigation measures, such as diversifying the training dataset and incorporating fairness metrics into the algorithm, are then implemented and continuously monitored to ensure that the tool operates as intended.

Transparency is a cornerstone of effective governance structures, fostering trust and accountability among stakeholders. For innovation initiatives, this requires clear and consistent communication about the goals, processes, and outcomes of new technologies. For example, a maritime company deploying autonomous vessels for resource transportation might publish detailed reports on the safety, efficiency, and environmental performance of the vessels, including information about how risks such as cyberattacks or equipment failures are managed. Transparency not only builds confidence in the organization's commitment to responsible innovation but also provides a basis for informed dialogue with regulators, communities, and industry partners.

Stakeholder engagement is another critical element of governance structures that balance innovation with risk management. The decisions made during the development and implementation of new technologies often have far-reaching implications for a diverse array of stakeholders, including local communities, environmental organizations, regulators, and investors. Governance structures should include mechanisms for soliciting input and feedback from these stakeholders, ensuring that their perspectives are reflected in the organization's innovation strategies. For instance, a mining company might establish a stakeholder advisory board to review and provide recommendations on the development of a new sediment plume monitoring system, incorporating insights from scientists, conservationists, and community representatives.

The role of leadership in governance structures cannot be overstated. Executives and managers must champion innovation while modeling a commitment to ethical responsibility and risk management. Leadership is responsible for fostering a culture that values both creativity and accountability, encouraging employees to explore new ideas while maintaining rigorous standards for safety, compliance, and sustainability. For example, a CEO might emphasize the importance of adhering to environmental thresholds in the development of resource extraction technologies, ensuring that teams prioritize ecological considerations alongside operational objectives. By setting the tone for the organization's approach to innovation, leaders create an environment where risks are acknowledged and addressed constructively, rather than ignored or minimized.

Governance structures must also address the technical and operational complexities of integrating new technologies into existing workflows. This requires robust oversight mechanisms that ensure the reliability, safety, and performance of innovative systems. For example, a mining company adopting AI-driven robotics for nodule collection might establish a dedicated innovation oversight committee to evaluate the deployment of these systems. The committee would be tasked with reviewing the technical specifications, conducting safety audits, and assessing the systems' compliance with regulatory and ethical standards. Regular reporting and performance evaluations ensure that the oversight process is both thorough and transparent, providing a clear record of how risks are managed and mitigated.

Regulatory compliance is a fundamental consideration for governance structures in the deep sea mining industry, where operations are subject to a complex web of international, national, and regional regulations. Governance structures must include processes for ensuring that innovation initiatives meet these regulatory requirements, from obtaining licenses and permits to adhering to environmental monitoring and reporting standards. For instance, a maritime company deploying blockchain-based traceability systems might need to ensure that the system complies with data privacy laws, such as the European Union's General Data Protection Regulation (GDPR), while also addressing the transparency requirements of international agreements like the International Seabed Authority's (ISA) Mining Code.

The integration of artificial intelligence and advanced data analytics into governance structures offers new opportunities for enhancing risk management and decision-making. AI-driven tools can analyze large and complex datasets to identify emerging risks, assess the effectiveness of mitigation measures, and optimize resource allocation. For example, an AI system might analyze historical data on equipment failures and environmental incidents to predict potential risks associated with a new mining project. These insights inform proactive measures, such as targeted maintenance schedules, enhanced training programs, or the adoption of alternative technologies, ensuring that risks are managed before they escalate.

Ethical oversight is another critical component of governance structures that balance innovation with risk management. As organizations adopt advanced technologies, they must ensure that these systems operate in ways that align with ethical principles and societal values. This requires the establishment of dedicated ethics committees or advisory boards to review the development and deployment of new technologies, evaluating their impacts on issues such as fairness, transparency, and environmental sustainability. For example, an ethics committee might review an AI-driven environmental monitoring system to ensure that its algorithms are unbiased, its data collection methods respect privacy, and its outputs are accessible to all stakeholders. The committee's recommendations provide a framework for aligning the system's design and operation with the organization's ethical commitments.

Continuous learning and improvement are essential for the effectiveness of governance structures, ensuring that they remain responsive to new challenges and opportunities. This requires the establishment of feedback loops that enable the organization to evaluate the outcomes of innovation initiatives, learn from successes and failures, and refine its governance processes accordingly. For instance, a maritime company might conduct post-implementation reviews of autonomous vessel deployments, analyzing data on safety incidents, operational performance, and stakeholder feedback to identify areas for improvement. These reviews inform updates to governance policies, training programs, and technical standards, ensuring that the organization continuously enhances its approach to innovation and risk management.

Collaboration with external partners, such as academic institutions, industry consortia, and regulatory bodies, further strengthens governance structures by providing access to expertise, resources, and best practices. For example, a mining company might partner with a university's environmental science department to develop and validate models for predicting the impacts of sediment plumes, or join an industry consortium focused on advancing sustainability standards for AI-driven technologies. These collaborations ensure that the organization's governance structures are informed by the latest research and aligned with broader efforts to promote responsible innovation.

12. Future-Forward Technologies and Theoretical Explorations

12.1. Quantum Computing and Next-Generation AI

The exploration and exploitation of the deep ocean present some of the most complex challenges in resource management, requiring precise modeling of dynamic and interconnected systems. Among these challenges, the accurate representation of seafloor conditions, encompassing geological, hydrodynamic, and ecological processes, stands out as a formidable task. Traditional computational methods, while effective in certain domains, often struggle to capture the intricate multiphysics interactions that define seafloor environments, particularly in the context of deep sea mining. Quantum computing, paired with next-generation artificial intelligence (AI), has the potential to revolutionize these efforts by enabling complex optimization and modeling at scales and speeds previously unimaginable.

Multiphysics modeling involves the simulation of systems governed by multiple physical processes, such as fluid dynamics, heat transfer, geomechanics, and chemical interactions. For deep sea mining, these processes are crucial to understanding the behavior of sediment plumes, assessing the stability of seafloor substrates, and predicting the impacts of resource extraction on marine ecosystems. For instance, the dispersion of sediment plumes generated during mining operations is influenced by a combination of factors, including ocean currents, sediment composition, and water density gradients. Capturing these interactions in a coherent and predictive model requires immense computational resources and advanced algorithms capable of handling non-linearities and high-dimensional datasets.

Quantum computing represents a paradigm shift in computational power and efficiency, leveraging the principles of quantum mechanics to process information in fundamentally new ways. Unlike classical computers, which use binary bits (0s and 1s) to perform calculations, quantum computers use quantum bits, or qubits, which can exist in multiple states simultaneously through the phenomenon of superposition. This capability allows quantum computers to perform parallel computations on an exponentially larger scale, making them particularly well-suited for solving optimization problems and simulating complex systems.

In the context of multiphysics modeling for seafloor conditions, quantum computing offers the potential to address several key limitations of classical approaches. One of the most significant challenges in traditional modeling is the computational cost of solving coupled equations that describe the interactions between different physical processes. For example, simulating the behavior of sediment plumes requires solving the Navier-Stokes equations for fluid dynamics, the advection-diffusion equations for particle transport, and additional equations for chemical interactions and ecological impacts. These calculations are computationally intensive, often requiring significant simplifications or approximations that reduce the accuracy and reliability of the results.

Quantum algorithms, such as the quantum phase estimation and variational quantum eigensolver, can dramatically accelerate these computations by exploiting the inherent parallelism of quantum systems. These algorithms enable quantum computers to solve partial differential equations and linear systems with greater speed and efficiency than classical methods, reducing the need for approximations and enabling more detailed and accurate simulations. For example, a quantum-enabled model of sediment plume dynamics could account for finer spatial and temporal resolutions, capturing the intricate interactions between ocean currents, particle settling velocities, and biological responses.

Another critical advantage of quantum computing in multiphysics modeling is its ability to handle uncertainty and probabilistic phenomena, which are inherent in seafloor conditions. Deep ocean environments are characterized by variability and unpredictability, influenced by factors such as fluctuating currents, tectonic activity, and the presence of diverse geological features. Traditional models often struggle to incorporate this uncertainty, relying on deterministic assumptions that may not accurately reflect real-world conditions. Quantum computing, through its use of qubits and superposition, naturally represents probabilities and uncertainties, allowing for the development of probabilistic models that capture the full range of possible outcomes.

For instance, a quantum model of seafloor stability could account for the probabilistic nature of sediment cohesion and pore pressure, providing more accurate predictions of how mining activities might affect slope stability and the risk of submarine landslides. Similarly, a quantum-enabled ecological model could simulate the probabilistic interactions between species and their environment, predicting the impacts of sediment plumes on biodiversity with greater confidence and granularity.

The integration of next-generation AI with quantum computing further enhances its potential for complex optimization and multiphysics modeling. AI algorithms, particularly those based on machine learning and neural networks, excel at identifying patterns and relationships within large and complex datasets. When combined with the computational power of quantum systems, AI can analyze and interpret the vast amounts of data generated by seafloor monitoring technologies, such as high-resolution sonar, environmental DNA (eDNA) analysis, and real-time sensor networks.

For example, a quantum-AI hybrid system could process data from thousands of sensors deployed across a mining site, identifying patterns in sediment movement, water quality, and ecological responses. This system could then use these insights to optimize mining operations, such as adjusting extraction techniques to minimize sediment plume dispersion or timing activities to align with periods of lower ecological sensitivity. By leveraging the strengths of both quantum computing and AI, these systems enable real-time decision-making that is both data-driven and adaptive, enhancing the sustainability and efficiency of deep sea mining.

The potential applications of quantum computing and next-generation AI in multiphysics modeling extend beyond operational optimization to include risk assessment and scenario planning. Deep sea mining is inherently uncertain, with significant risks related to environmental impacts, regulatory compliance, and technical failures. Quantum-enabled models can simulate a wide range of scenarios, incorporating variables such as resource distribution, operational constraints, and regulatory thresholds. These simulations provide a probabilistic understanding of risks and trade-offs, allowing stakeholders to evaluate the implications of different strategies and select the most resilient and responsible options.

For instance, a quantum model of sediment plume behavior could simulate how different mining techniques and environmental conditions affect plume dispersion, identifying scenarios where impacts exceed acceptable thresholds. This information can guide the design of mitigation measures, such as optimizing the placement of sediment containment systems or modifying extraction schedules to reduce peak impacts. Similarly, a quantum-AI system could model the cumulative impacts of multiple mining operations within a region, providing insights into how activities interact and informing regional management plans that balance resource extraction with ecosystem conservation.

Despite its transformative potential, the application of quantum computing in deep sea mining remains in its early stages, with significant challenges to overcome. Quantum hardware is still in development, with current systems limited by issues such as qubit coherence, error rates, and scalability. Additionally, the integration of quantum computing with existing modeling frameworks and data infrastructures requires advances in software, algorithms, and interdisciplinary collaboration. Addressing these challenges will require sustained investment in research and development, as well as partnerships between industry, academia, and government to accelerate the adoption of quantum technologies.

Ethical considerations are also critical in the deployment of quantum-enabled models, particularly in the context of deep sea mining's environmental and social impacts. The enhanced predictive power of these models must be accompanied by transparency, accountability, and stakeholder engagement, ensuring that their outputs are used to support equitable and sustainable decision-making. For example, the results of quantum simulations should be shared with regulators, conservation groups, and local communities, providing a basis for informed dialogue and collaborative problem-solving.

The exploration of mineral resources in the deep ocean represents one of the most demanding computational challenges in modern science and industry. The sheer scale, complexity, and uncertainty of seabed environments require advanced algorithms capable of processing vast datasets and simulating intricate geological and environmental systems. Early-stage research into quantum algorithms offers a transformative approach to these challenges, leveraging the principles of quantum mechanics to achieve breakthroughs in resource exploration. By unlocking new capabilities for optimization, pattern recognition, and probabilistic modeling, quantum computing has the potential to revolutionize the way deep sea mining companies identify and assess resource deposits.

At the heart of resource exploration is the need to analyze and interpret data from a variety of sources, including high-resolution sonar, geophysical surveys, core samples, and environmental monitoring systems. These datasets are not only large but also highly multidimensional, capturing complex relationships between geological formations, sediment layers, mineral compositions, and hydrodynamic processes. Traditional computational methods, while powerful, are often limited by the exponential growth of data complexity and the time required to process it. Quantum algorithms, designed to operate on qubits rather than classical bits, offer a paradigm shift in computational efficiency, enabling the rapid analysis of large-scale datasets with unprecedented precision.

One of the most promising areas of early-stage research on quantum algorithms is in the optimization of resource exploration strategies. Deep sea mining operations often face the challenge of determining the most efficient and effective way to explore vast and poorly mapped seabed regions. This involves decisions about where to deploy survey equipment, how to allocate resources, and how to sequence exploration activities to maximize the likelihood of identifying high-value deposits. Quantum optimization algorithms, such as the quantum approximate optimization algorithm (QAOA) and variational quantum eigensolvers, are particularly well-suited to addressing these challenges.

For example, a quantum optimization algorithm could be used to develop an exploration plan that minimizes the time and cost of surveying while maximizing data coverage and resource discovery potential. By processing data from previous surveys and incorporating constraints such as equipment availability and environmental regulations, the algorithm can identify optimal survey routes, prioritize target areas, and adapt to changing conditions in real time. This capability is especially valuable in the deep ocean, where operational costs are high, and the window for conducting surveys is often limited by weather, currents, and other environmental factors.

Another area of focus for early-stage quantum research is pattern recognition, a critical component of resource exploration. Identifying resource-rich deposits in the deep ocean requires the ability to detect subtle patterns in geophysical and geochemical data, distinguishing promising areas from background noise and false positives. Quantum machine learning algorithms, such as quantum support vector machines and quantum neural networks, offer the potential to enhance pattern recognition by processing data in high-dimensional quantum states. These algorithms can identify complex correlations and nonlinear relationships that are difficult or impossible to capture using classical methods.

For instance, a quantum machine learning model might analyze sonar and seismic data to identify geological features associated with polymetallic nodules, such as flat abyssal plains with low sedimentation rates. By incorporating additional data, such as magnetic anomalies or geochemical signatures, the model can refine its predictions and generate a ranked list of exploration targets. These insights enable operators to focus their efforts on the most promising areas, reducing the need for extensive and costly surveys while increasing the likelihood of resource discovery.

Quantum algorithms also hold promise for improving the accuracy and reliability of probabilistic modeling in resource exploration. The deep ocean is characterized by significant uncertainty, with limited data on the distribution of resources, the behavior of sediments, and the interactions between geological and ecological systems. Probabilistic models are essential for capturing this uncertainty, enabling stakeholders to evaluate risks, trade-offs, and potential outcomes. Quantum computing's inherent ability to represent and manipulate probabilities makes it an ideal tool for advancing this field.

For example, a quantum algorithm could be used to simulate the probabilistic distribution of polymetallic nodules across a target region, incorporating data on sediment thickness, ocean currents, and tectonic history. By running simulations that account for different scenarios and uncertainties, the algorithm provides a range of possible outcomes, helping operators assess the likelihood of resource presence and the potential risks of exploration. This probabilistic approach not only informs decision-making but also supports adaptive exploration strategies that can respond dynamically to new information.

One of the most exciting aspects of early-stage research on quantum algorithms for resource exploration is the potential for hybrid systems that combine quantum computing with classical and AI-driven approaches. These hybrid systems leverage the strengths of each technology, using quantum algorithms to tackle the most computationally intensive tasks while relying on classical and AI methods for preprocessing, integration, and interpretation. For example, a hybrid system might use classical AI to preprocess raw sonar data, extracting key features and reducing noise, before feeding the data into a quantum optimization algorithm for exploration planning. Similarly, a quantum-AI hybrid could combine the predictive power of quantum models with the interpretability and visualization capabilities of classical machine learning, creating a comprehensive framework for resource exploration.

Despite its promise, early-stage research on quantum algorithms for resource exploration faces significant challenges. Quantum hardware is still in its infancy, with current systems limited by issues such as qubit coherence, error rates, and scalability. Developing quantum algorithms that can operate effectively on these nascent platforms requires advances in both hardware and software, as well as interdisciplinary collaboration between physicists, computer scientists, and domain experts in geology and oceanography. Additionally, integrating quantum computing into existing exploration workflows and infrastructures presents technical and logistical challenges, necessitating the development of compatible interfaces, data pipelines, and training programs.

Ethical considerations are also central to the deployment of quantum algorithms in resource exploration. As these algorithms enhance the precision and efficiency of exploration, they also raise questions about their environmental and social impacts. For example, the ability to identify high-value deposits more quickly and accurately could accelerate the pace of deep sea mining, potentially increasing the risks to fragile marine ecosystems. Ensuring that quantum-enabled exploration is aligned with principles of sustainability and equity requires robust governance frameworks, transparency, and stakeholder engagement. These frameworks must address issues such as data privacy, algorithmic fairness, and the equitable sharing of benefits derived from resource discoveries.

The future of quantum computing and next-generation AI in resource exploration will be shaped by ongoing advancements in hardware, algorithms, and interdisciplinary collaboration. Partnerships between industry, academia, and government will play a critical role in accelerating progress and translating research breakthroughs into practical applications. For example, collaborations between quantum computing firms and deep sea mining companies could lead to the development of tailored algorithms for specific exploration challenges, such as modeling the behavior of SMS deposits or predicting the impacts of sediment disturbances. Similarly, partnerships with academic institutions and research consortia can provide access to expertise, data, and experimental platforms that enhance the quality and scope of quantum research.

12.2. Bio-Inspired Computing and Soft Robotics

The deep ocean, with its intricate ecosystems and extraordinary adaptations, serves as an unparalleled source of inspiration for developing novel artificial intelligence (AI) paradigms. These bio-inspired approaches, modeled after the behaviors, processes, and structures observed in biological systems, offer transformative potential for addressing the unique challenges of deep sea mining. By mimicking the strategies employed by marine organisms to navigate, communicate, and adapt to extreme environments, AI systems can achieve new levels of efficiency, resilience, and sustainability. These innovations not only enhance the capabilities of robotic systems but also redefine the principles of decision-making, optimization, and interaction in one of the most demanding frontiers on Earth.

Biological ecosystems operate as complex, self-organizing systems where individual agents interact dynamically with their environment and one another to achieve collective goals. This emergent behavior, characterized by adaptability, robustness, and efficiency, provides a compelling framework for designing AI systems that can navigate the uncertainties and complexities of the deep ocean. For example, swarm intelligence, a concept derived from the collective behavior of social organisms such as fish schools or ant colonies, has inspired algorithms for coordinating fleets of autonomous underwater vehicles (AUVs). In a deep sea mining context, these algorithms enable AUVs to collaborate on tasks such as resource mapping, sediment monitoring, and nodule collection, distributing workloads dynamically and responding to changing conditions in real time.

One of the key advantages of bio-inspired AI paradigms is their ability to optimize operations in decentralized and data-scarce environments. Unlike traditional algorithms that rely on centralized control and comprehensive datasets, bio-inspired systems emulate the decentralized decision-making processes observed in nature, where individual agents operate with limited information and rely on local interactions to achieve global objectives. For instance, a bio-inspired navigation system for AUVs might mimic the way fish use local cues such as water currents and visual landmarks to navigate complex environments. By incorporating similar principles, AI systems can adapt to dynamic ocean conditions, avoid obstacles, and optimize their trajectories without requiring constant communication with a central control unit.

The adaptability of bio-inspired AI is particularly valuable for managing the environmental and operational uncertainties inherent in deep sea mining. Marine organisms have evolved over millions of years to survive in extreme and unpredictable conditions, developing strategies for resilience and self-repair. Mimicking these adaptations, bio-inspired AI systems can adjust their behavior to account for unexpected changes in seafloor topography, equipment malfunctions, or shifts in environmental conditions. For example, an AI-driven sediment plume monitoring system might draw inspiration from the foraging behavior of marine organisms, deploying sensor-equipped drones in a pattern that maximizes data collection while minimizing redundancy and energy use.

Another compelling aspect of bio-inspired AI is its potential to enhance energy efficiency, a critical consideration for deep sea operations where resources are often constrained. Marine organisms are masters of energy conservation, employing streamlined movements, efficient hunting strategies, and adaptive behaviors to survive in resource-limited environments. Similarly, AI systems can incorporate bio-inspired optimization techniques to reduce energy consumption and extend the operational lifespan of robotic platforms. For instance, an AUV designed for long-duration missions might use algorithms inspired by the energy-efficient swimming patterns of jellyfish, adjusting its propulsion to match current conditions and minimize drag.

Communication and coordination are central themes in bio-inspired AI, reflecting the ways in which biological ecosystems achieve complex tasks through collective effort. Marine organisms use a variety of mechanisms, such as chemical signaling, bioluminescence, and acoustic communication, to share information and synchronize their actions. These mechanisms inspire AI paradigms for multi-agent systems, where fleets of robots or drones work together to achieve shared objectives. For example, a swarm of robotic crawlers tasked with nodule collection might use bio-inspired algorithms to coordinate their movements, avoiding collisions, optimizing coverage, and ensuring that resources are distributed evenly among the group.

Bio-inspired AI also offers innovative solutions for environmental monitoring and conservation, leveraging the principles of symbiosis, adaptation, and regeneration observed in nature. For instance, certain marine species form symbiotic relationships with other organisms, benefiting from mutual cooperation while maintaining ecological balance. Similarly, bio-inspired AI systems can be designed to interact harmoniously with their environment, minimizing disturbance and even contributing to ecosystem restoration. A robotic system inspired by coral polyps, for example, might be used to repair damaged reefs or stabilize sediments in areas affected by mining activities, aligning resource extraction with ecological stewardship.

In addition to their operational advantages, bio-inspired AI paradigms contribute to the ethical and philosophical discourse surrounding deep sea mining. By drawing inspiration from nature, these systems embody a more integrated and holistic approach to technology, emphasizing harmony and sustainability rather than dominance and exploitation. This perspective aligns with the principles of environmental ethics and the precautionary approach, fostering a mindset of respect and responsibility toward marine ecosystems.

The development of bio-inspired AI paradigms requires a multidisciplinary approach, bringing together expertise in biology, robotics, data science, and oceanography. Researchers study the behaviors and adaptations of marine organisms, translating these insights into algorithms and models that guide the design of AI systems. For example, a team might analyze the movement patterns of squid or octopuses to develop a soft robotic manipulator capable of handling fragile seafloor specimens with precision and care. Similarly, insights into the collective decision-making processes of coral reef ecosystems could inform the development of AI systems for managing multi-objective trade-offs in mining operations.

Despite their promise, bio-inspired AI paradigms face significant challenges in their development and application. Translating biological principles into computational algorithms is a complex task that requires not only a deep understanding of biological systems but also the ability to abstract and generalize their mechanisms. Additionally, the integration of bio-inspired AI into existing deep sea mining workflows and infrastructures demands significant investment in research, development, and testing. Addressing these challenges requires collaboration between academia, industry, and government, as well as sustained funding for interdisciplinary research initiatives.

Ethical considerations are also central to the deployment of bio-inspired AI in deep sea mining. While these systems offer the potential to reduce environmental impacts and enhance sustainability, their design and operation must be guided by transparent and inclusive governance frameworks. For example, stakeholders must have confidence that bio-inspired AI systems used for resource allocation or environmental monitoring are free from biases, aligned with regulatory standards, and accountable to public scrutiny. Ensuring that these systems operate ethically and responsibly requires ongoing dialogue among developers, operators, regulators, and affected communities.

The future of bio-inspired AI in deep sea mining holds immense promise, driven by advances in computing power, sensor technology, and interdisciplinary collaboration. As researchers continue to explore the behaviors and adaptations of marine organisms, new paradigms for AI systems will emerge, enabling more efficient, resilient, and sustainable operations. These innovations have the potential to transform not only the field of deep sea mining but also the broader landscape of marine technology and environmental management, setting a new standard for how humanity interacts with the ocean.

The application of soft robotics in deep sea operations marks a significant leap forward in the pursuit of sustainable and minimally invasive resource extraction. Traditional robotic systems, while highly effective for heavy-duty tasks, are often rigid, bulky, and designed primarily for robustness rather than sensitivity to their surroundings. In contrast, soft robotics draws inspiration from the pliability, adaptability, and resilience of biological organisms, offering a new paradigm for interacting with delicate marine ecosystems. By prioritizing flexibility, compliance, and environmental harmony, soft robotic systems have the potential to transform deep sea mining, enabling resource extraction that minimizes ecological disturbance while enhancing operational precision and efficiency.

The deep ocean is a realm of unparalleled biodiversity, characterized by fragile ecosystems and species that are uniquely adapted to its extreme conditions. From hydrothermal vent communities to abyssal plains, these environments host organisms that are highly sensitive to physical disruptions, such as sediment disturbances, mechanical impacts, and habitat fragmentation. The introduction of traditional, rigid robotic systems into such environments can pose significant risks, inadvertently causing harm to habitats or displacing vulnerable species. Soft robotics addresses these challenges by mimicking the gentle and adaptive interactions observed in nature, ensuring that robotic systems can operate safely and responsibly within sensitive marine settings.

Soft robots are constructed from compliant materials, such as silicone, elastomers, and hydrogels, that enable them to bend, stretch, and deform without causing damage to their surroundings. These materials are often combined with advanced actuation systems, such as pneumatic, hydraulic, or electroactive mechanisms, which allow the robots to perform complex movements with precision and control. For example, a soft robotic manipulator designed for sampling seafloor sediments might incorporate tentacle-like appendages that conform to the shape of the sediment, applying just enough force to collect a sample without disturbing nearby organisms or structures. This level of dexterity and gentleness is critical for operating in environments where even minor disruptions can have cascading ecological effects.

One of the most promising applications of soft robotics in deep sea operations is in the collection of polymetallic nodules, which are scattered across the seafloor and often embedded within layers of sediment. Traditional collection methods, such as suction dredging or mechanical harvesting, can generate significant sediment plumes, smothering benthic habitats and disrupting marine ecosystems over wide areas. Soft robotic systems offer a less invasive alternative, using biomimetic designs and adaptive mechanisms to gently extract nodules while minimizing sediment disturbance. For instance, a soft robotic gripper inspired by the tentacles of an octopus might envelop a nodule with a flexible membrane, lifting it from the seafloor with minimal disruption to the surrounding sediment.

The adaptability of soft robotics extends beyond resource collection to encompass environmental monitoring and restoration. Monitoring is a critical component of responsible deep sea mining, providing the data needed to assess and mitigate ecological impacts. Soft robotic systems are uniquely suited for tasks such as collecting samples, measuring water quality, and observing marine life, as their compliant designs allow them to interact safely with delicate structures and organisms. For example, a soft robotic drone equipped with chemical sensors and cameras could navigate through coral habitats, collecting data on water chemistry and biodiversity without risking physical damage to the reef.

Restoration represents another frontier for soft robotics, particularly in mitigating the impacts of mining activities on benthic habitats. Marine organisms such as sponges and corals often play a key role in stabilizing sediments, maintaining ecosystem balance, and supporting biodiversity. When these organisms are displaced or damaged, the recovery of the ecosystem can be severely hindered. Soft robotic systems offer innovative solutions for habitat restoration, such as deploying artificial substrates, reseeding damaged areas with coral fragments, or stabilizing sediments with bio-inspired anchors. These systems mimic natural processes, leveraging the flexibility and adaptability of soft robotics to integrate seamlessly into the environment and promote recovery.

The design and operation of soft robotics are heavily informed by the principles of biomimicry, which seeks to emulate the strategies and adaptations of living organisms. Marine animals such as jellyfish, sea stars, and cephalopods serve as models for the development of robotic systems that are not only flexible and efficient but also capable of interacting harmoniously with their surroundings. For instance, a robotic jellyfish might use undulating movements to navigate through water, generating minimal turbulence and avoiding harm to nearby organisms. Similarly, a robotic sea star might use its multi-armed design to grip and manipulate objects on the seafloor with precision and care.

Advances in soft robotics are also driving innovations in material science, leading to the development of new materials that combine strength, durability, and compliance. For example, hydrogels that mimic the mechanical properties of biological tissues enable the creation of robots that can withstand the high pressures and low temperatures of the deep ocean while maintaining their flexibility and functionality. These materials are often coupled with sensors and actuators that allow the robots to respond dynamically to environmental conditions, such as adjusting their grip strength based on the texture of a surface or altering their shape to navigate through narrow crevices.

The integration of artificial intelligence (AI) further enhances the capabilities of soft robotics, enabling real-time decision-making, learning, and adaptation. AI algorithms process data from sensors embedded in the robots, such as pressure, temperature, and chemical sensors, to inform their movements and actions. For example, a soft robotic arm used for sediment sampling might use AI to analyze sensor data and adjust its approach based on the composition and stability of the sediment, ensuring that the sample is collected efficiently and without disturbance. Similarly, AI-driven navigation systems enable soft robots to autonomously explore and map seafloor habitats, identifying areas of ecological significance and avoiding sensitive zones.

Despite their potential, the deployment of soft robotics in deep sea mining faces several challenges, including the need for reliable power sources, robust communication systems, and scalable manufacturing processes. Operating in the deep ocean requires systems that can function autonomously for extended periods, often without direct human intervention or access to surface support. Addressing these challenges requires advances in energy storage, wireless communication, and the development of modular designs that allow for rapid repair and reconfiguration.

Ethical considerations are central to the development and deployment of soft robotics, particularly in the context of deep sea mining's environmental impacts. Ensuring that these systems are used responsibly requires transparent governance frameworks, stakeholder engagement, and adherence to international standards for environmental protection. For example, the deployment of soft robotics for nodule collection should be guided by comprehensive environmental impact assessments (EIAs) and monitoring programs that evaluate their performance and ensure compliance with ecological thresholds.

The future of soft robotics in deep sea mining holds immense promise, driven by ongoing advancements in materials, design, and AI integration. These innovations have the potential to transform the industry, enabling resource extraction that aligns with principles of sustainability and environmental stewardship. By prioritizing minimal ecological disturbance, soft robotics not only enhances operational efficiency but also fosters a more harmonious relationship between technology and the natural world.

12.3. Futuristic Scenarios and Roadmaps

As humanity grapples with the twin imperatives of economic development and environmental sustainability, the role of the oceans as a source of critical resources and ecosystem services becomes increasingly central to our future. Over the next 20 to 50 years, visionary models of ocean resource utilization suggest that advances in technology, governance, and scientific understanding will fundamentally reshape the way we interact with marine environments. These models envision a future where the oceans are managed as interconnected systems, balancing the extraction of valuable resources with the preservation of biodiversity, climate regulation, and cultural heritage. At the heart of this transformation are innovative approaches to deep sea mining, renewable energy, and bioprospecting, all of which promise to unlock the potential of the ocean in ways that are both sustainable and equitable.

One of the most transformative aspects of future ocean resource utilization is the integration of artificial intelligence (AI) and advanced robotics into every stage of marine operations. Over the next two decades, AI systems will become increasingly sophisticated, capable of processing vast amounts of data in real time to optimize resource extraction, monitor environmental impacts, and ensure compliance with international standards. In the context of deep sea mining, this means deploying fleets of autonomous underwater vehicles (AUVs) and robotic systems that can map the seafloor with millimeter precision, identify resource-rich deposits, and extract minerals with minimal ecological disturbance. These systems will operate as part of interconnected networks, sharing data and coordinating activities to maximize efficiency and minimize duplication.

Looking 50 years into the future, the concept of fully autonomous "ocean cities" dedicated to resource utilization may become a reality. These self-sustaining hubs, located in international waters, would serve as platforms for a range of activities, including deep sea mining, aquaculture, and renewable energy generation. Powered by advanced energy systems, such as floating solar farms and wave energy converters, these hubs would operate with minimal human presence, relying on AI and robotics to perform tasks ranging from resource extraction to waste recycling. Data collected by these hubs would feed into global ocean monitoring networks, providing real-time insights into marine health and supporting adaptive management strategies.

One of the most intriguing possibilities for future ocean resource utilization lies in the convergence of bioprospecting and biotechnology. The deep ocean is home to an extraordinary diversity of life, much of which remains undiscovered and poorly understood. Over the next 20 to 50 years, advances in genomics, synthetic biology, and AI will enable researchers to explore the biochemical and genetic diversity of marine organisms at an unprecedented scale. This bioprospecting effort will uncover new enzymes, metabolites, and biomaterials with applications in medicine, agriculture, and industry. For example, heat-resistant enzymes derived from hydrothermal vent organisms could be used to develop more efficient industrial processes, while bioluminescent proteins might find applications in medical imaging and environmental sensing.

As the potential of bioprospecting becomes more apparent, the need for ethical and equitable governance will become increasingly critical. Visionary models of ocean resource utilization emphasize the importance of benefit-sharing frameworks that ensure the discoveries and profits derived from marine biodiversity are distributed fairly, particularly to developing countries and indigenous communities with traditional knowledge of marine ecosystems. Over the next half-century, international agreements such as the United Nations Convention on the Law of the Sea (UNCLOS) and the emerging Biodiversity Beyond National Jurisdiction (BBNJ) treaty will likely play a central role in shaping these frameworks, fostering collaboration and inclusivity in the exploration and use of marine resources.

In addition to mineral and biological resources, the future of ocean utilization will be defined by the expansion of renewable energy infrastructure. The vast energy potential of the oceans, ranging from offshore wind and wave energy to ocean thermal energy conversion, offers a pathway to decarbonizing the global economy and addressing the climate crisis. By 2050, the oceans could become a major source of clean energy, with integrated energy grids connecting offshore wind farms, tidal turbines, and floating solar arrays to coastal and island communities. These developments will not only reduce reliance on fossil fuels but also create new opportunities for economic growth and energy security in regions that currently lack access to reliable power.

In this vision of the future, the governance of ocean resources will undergo a fundamental transformation, moving toward integrated and adaptive management frameworks that recognize the interconnectedness of marine systems. Over the next 20 to 50 years, international and regional bodies will adopt more holistic approaches to ocean governance, integrating data from satellite monitoring, autonomous sensors, and AI-driven analytics to make informed decisions about resource use, conservation, and climate adaptation. These frameworks will prioritize ecosystem-based management, balancing resource extraction with the preservation of ecosystem services such as carbon sequestration, nutrient cycling, and biodiversity maintenance.

Public engagement and stakeholder inclusion will also play a central role in shaping the future of ocean resource utilization. Visionary models emphasize the importance of transparent and participatory governance, where decisions about the oceans are informed by the perspectives of diverse stakeholders, including local communities, environmental organizations, industry representatives, and scientists. Over the next five decades, the development of digital platforms and interactive tools will enable more inclusive decision-making processes, allowing stakeholders to visualize the impacts of different scenarios and contribute to the design of sustainable management strategies.

The ethical dimensions of ocean resource utilization will become increasingly prominent in the coming decades, driven by growing awareness of the need to balance human development with environmental stewardship. Future models will place a strong emphasis on the precautionary principle, ensuring that activities in the ocean are conducted responsibly and with minimal risk to marine ecosystems. This will require robust environmental impact assessments (EIAs), continuous monitoring, and the adoption of best practices for reducing ecological footprints. For example, advanced soft robotics and bio-inspired technologies will be used to conduct minimally invasive resource extraction, while AI-driven systems will optimize operations to minimize energy use and waste.

Over the next 50 years, the concept of "blue carbon" will play a key role in the vision for ocean resource utilization. Recognizing the oceans as a critical component of the global carbon cycle, future models will prioritize the protection and restoration of marine habitats that act as carbon sinks, such as seagrass meadows, mangroves, and deep-sea sediments. These efforts will be supported by international agreements and market-based mechanisms, such as carbon credits and offsets, that incentivize the preservation of blue carbon ecosystems as part of global climate strategies.

Technological innovation will also drive new approaches to waste management and pollution control in the oceans. By 2050, closed-loop systems and circular economy principles will be fully integrated into ocean resource utilization, ensuring that materials and energy are reused and recycled to the greatest extent possible. Autonomous systems will be deployed to collect and process marine debris, turning waste into valuable resources and reducing the environmental impact of human activities. These technologies will be complemented by stricter regulations on marine pollution and the development of alternative materials that reduce the reliance on plastics and other non-biodegradable substances.

The future of ocean resource utilization is poised to be shaped by transformational innovations that redefine how we interact with the marine environment. Among these, the advent of 3D printing for subsea structures and the development of automated subsea factories stand out as groundbreaking advancements with the potential to revolutionize industries such as deep sea mining, aquaculture, and renewable energy. These technologies promise to enhance operational efficiency, minimize environmental impacts, and enable entirely new ways of utilizing the ocean's resources while aligning with principles of sustainability and resilience.

One of the most compelling applications of 3D printing in the ocean is the construction of subsea structures that support resource extraction, habitat restoration, and infrastructure development. Traditional methods for deploying subsea infrastructure, such as pipelines, platforms, and foundations, are often constrained by logistical challenges, high costs, and environmental risks.

Transporting large, pre-fabricated components to remote ocean locations requires significant energy and resources, while the installation process can disrupt marine ecosystems. 3D printing offers an alternative by enabling the in situ fabrication of customized structures directly on the seafloor, reducing the need for extensive transportation and minimizing ecological disturbance.

Using 3D printing technologies, subsea structures can be designed with unprecedented precision and adaptability, tailoring their shapes, sizes, and materials to specific environmental conditions and operational requirements. For example, a 3D printer mounted on an autonomous underwater vehicle (AUV) could construct a coral-like lattice structure designed to stabilize sediments, provide habitats for marine life, and support resource extraction equipment. These structures could be fabricated using bio-inspired materials or composites that mimic the properties of natural substrates, ensuring compatibility with the surrounding ecosystem and promoting biodiversity.

The potential of 3D printing extends beyond infrastructure to include the creation of protective barriers, containment systems, and artificial reefs. For instance, a 3D-printed barrier could be used to contain sediment plumes generated during deep sea mining, preventing their spread to ecologically sensitive areas. Similarly, artificial reefs constructed using 3D printing could serve as offsets for habitats impacted by mining activities, providing a platform for coral growth and supporting the recovery of benthic communities. These applications demonstrate how 3D printing can be leveraged to align industrial activities with environmental stewardship, creating solutions that balance resource extraction with conservation.

The development of automated subsea factories represents another transformational innovation with profound implications for the future of ocean resource utilization. These factories, envisioned as fully autonomous systems located on the seafloor, would integrate advanced robotics, AI, and modular design to perform a wide range of functions, from resource processing to waste recycling. By bringing manufacturing and processing operations closer to the site of resource extraction, automated subsea factories eliminate the need for extensive transportation, reducing costs, energy use, and greenhouse gas emissions.

A key feature of these factories is their ability to operate continuously and independently, leveraging renewable energy sources such as underwater turbines or floating solar panels to power their systems. For example, a subsea factory designed for polymetallic nodule processing might include robotic arms for sorting and crushing nodules, reactors for extracting valuable metals, and waste management systems for safely returning non-target materials to the seafloor. These operations would be monitored and controlled by AI-driven systems capable of adapting to changing conditions, optimizing efficiency, and ensuring compliance with environmental regulations.

The modular nature of automated subsea factories allows them to be customized and scaled to meet specific operational needs. Modules can be added, removed, or reconfigured as required, enabling the factory to adapt to different resource types, environmental conditions, or production targets. For instance, a factory operating in an area rich in cobalt might prioritize modules for cobalt extraction and processing, while another factory in a region with significant manganese deposits might focus on manganese-specific workflows. This flexibility ensures that subsea factories can respond dynamically to market demands and resource availability, enhancing their economic viability and long-term sustainability.

In addition to resource extraction and processing, automated subsea factories could play a critical role in environmental monitoring and restoration. Equipped with advanced sensors and robotics, these factories could collect data on water quality, sediment composition, and biodiversity, feeding this information into global monitoring networks. They could also deploy systems for habitat restoration, such as planting seagrass, reseeding coral reefs, or stabilizing sediments in areas affected by mining or other industrial activities. By integrating environmental functions into their design, subsea factories align industrial operations with the broader goals of ocean health and resilience.

The deployment of 3D printing and automated subsea factories also opens up new possibilities for supporting renewable energy infrastructure in the ocean. For example, 3D printing could be used to fabricate customized components for floating wind turbines, tidal energy systems, or wave energy converters directly at sea, reducing the logistical challenges associated with transporting large structures from onshore facilities. Subsea factories could complement these efforts by serving as hubs for assembling, maintaining, and recycling renewable energy systems, creating a closed-loop approach that minimizes waste and maximizes resource efficiency.

Despite their promise, the realization of 3D printing for subsea structures and automated subsea factories faces several challenges. The harsh conditions of the deep ocean, including high pressures, low temperatures, and corrosive seawater, demand materials and technologies that can withstand extreme environments while maintaining their functionality and durability. Addressing these challenges requires ongoing research and development in areas such as material science, robotics, and energy systems, as well as interdisciplinary collaboration among industry, academia, and government.

The integration of these technologies into existing regulatory and governance frameworks also presents a significant hurdle. Ensuring that 3D printing and subsea factories are deployed responsibly requires clear guidelines and standards that address issues such as environmental impacts, safety, and accountability. For example, the creation of 3D-printed structures must be informed by rigorous environmental impact assessments (EIAs) to ensure that they do not disrupt critical habitats or alter natural processes. Similarly, the operation of subsea factories must be monitored to prevent pollution, overextraction, or other unintended consequences.

Ethical considerations are central to the development and deployment of these innovations, particularly in the context of their environmental and social impacts. The potential to accelerate resource extraction and industrialization in the ocean raises questions about the long-term consequences for marine ecosystems and the communities that depend on them. Ensuring that these technologies are used in ways that promote sustainability, equity, and transparency requires robust governance frameworks, stakeholder engagement, and a commitment to the principles of the precautionary approach.

Looking to the future, the adoption of 3D printing and automated subsea factories will be driven by advancements in AI, robotics, and renewable energy, as well as by the growing recognition of the need for sustainable and resilient approaches to ocean resource utilization. These innovations have the potential to redefine what is possible in the deep ocean, creating opportunities for economic development, environmental restoration, and technological leadership. By prioritizing sustainability and ethical responsibility, the industry can harness these technologies to build a future where the oceans are managed as a shared resource that benefits all of humanity.

The trajectory of ocean resource utilization over the next 20 to 50 years will be shaped not only by planned innovations and strategic advancements but also by wildcards, unpredictable or high-impact events and trends that could dramatically alter the landscape of deep sea mining and marine resource management. Among these wildcards, climate change and the potential rise of alternate space mining stand out as transformative forces with profound implications for the extraction and governance of sea-based resources. Understanding and preparing for these wildcards is critical for ensuring the resilience and sustainability of ocean industries in an increasingly uncertain world.

Climate change represents one of the most significant and far-reaching wildcards, influencing the physical, chemical, and biological systems of the ocean. Rising sea temperatures, ocean acidification, deoxygenation, and shifting currents are already altering marine ecosystems, with cascading effects on biodiversity, resource distribution, and ecosystem services. For deep sea mining, these changes pose both challenges and opportunities, reshaping the availability of resources and the conditions under which extraction occurs.

One of the primary impacts of climate change on sea-based resources is its effect on the distribution and quality of mineral deposits. The movement of tectonic plates, the upwelling of nutrient-rich waters, and the deposition of sediments have all contributed to the formation of valuable seabed resources such as polymetallic nodules, cobalt-rich ferromanganese crusts, and seafloor massive sulfides (SMS). Climate-driven changes in ocean currents, sediment transport, and temperature gradients could disrupt these processes, altering the locations and compositions of resource deposits. For example, changes in sedimentation rates due to shifting river inputs or melting ice caps might accelerate or inhibit the growth of polymetallic nodules in certain regions, complicating exploration efforts and economic forecasts.

Climate change also heightens the risks associated with deep sea mining by exacerbating the vulnerability of marine ecosystems. Many of the species and habitats found in the deep ocean are adapted to stable environmental conditions, making them particularly sensitive to disturbances. Rising temperatures and acidification could weaken these ecosystems' resilience, increasing the likelihood of irreversible damage from mining activities. For instance, corals and other calcifying organisms that play critical roles in benthic habitats are already threatened by declining pH levels, and the additional stress of sediment plumes or habitat destruction could push these systems past tipping points. Ethical and regulatory frameworks for deep sea mining will need to account for these compounded risks, emphasizing precautionary approaches and adaptive management.

On the operational side, climate change could disrupt the logistics and infrastructure of deep sea mining by increasing the frequency and severity of extreme weather events. Hurricanes, cyclones, and turbulent seas pose significant risks to offshore operations, including the deployment of autonomous underwater vehicles (AUVs), the stability of surface platforms, and the integrity of transport systems. Companies will need to invest in resilient technologies, such as storm-resistant vessels and modular, rapidly deployable infrastructure, to ensure continuity and safety in the face of climatic volatility. Advanced weather prediction models powered by artificial intelligence (AI) will also play a critical role in anticipating and mitigating the impacts of extreme weather on mining operations.

In contrast to the risks posed by climate change, the emergence of alternate space mining represents a wildcard that could fundamentally reshape the economic and strategic calculus of sea-based resource extraction. Space mining, or the extraction of resources from celestial bodies such as asteroids and the Moon, has moved from the realm of science fiction to a realistic prospect, driven by advances in space exploration technologies and private sector investment. Companies like Planetary Resources and Deep Space Industries, along with national space agencies, are exploring the potential of harvesting metals, water, and other resources from space to support both terrestrial industries and off-world colonization.

The implications of space mining for sea-based resources are multifaceted. On one hand, the availability of resources from space could reduce the demand for deep sea mining, particularly for critical minerals like cobalt, nickel, and platinum-group metals. Asteroids, for instance, are known to contain high concentrations of these elements, and their extraction in space could provide a viable alternative to mining in fragile marine ecosystems. This shift could alleviate some of the environmental pressures associated with deep sea mining, allowing for more stringent conservation efforts and the preservation of biodiversity hotspots.

On the other hand, the rise of space mining could also create competitive pressures that accelerate technological innovation and drive down costs in the ocean mining sector. As companies and governments invest in the infrastructure and expertise needed for space resource extraction, many of these advancements, such as modular robotics, autonomous systems, and AI-driven optimization, could be adapted for use in marine environments. This cross-pollination of technologies could enhance the efficiency and sustainability of sea-based operations, enabling deeper and more precise exploration of the ocean's resources.

The geopolitical implications of space mining as a wildcard are equally significant, influencing the governance and equity of resource utilization on Earth and beyond. If space mining becomes a dominant source of critical minerals, questions will arise about the distribution of its benefits and the potential for resource monopolies to exacerbate global inequalities. For sea-based resources, this dynamic could lead to shifts in market demand, regulatory priorities, and international cooperation, as nations reassess their reliance on ocean mining in the context of broader resource strategies. The development of legal frameworks, such as those under the United Nations Outer Space Treaty and the emerging Biodiversity Beyond National Jurisdiction (BBNJ) treaty, will play a critical role in balancing these interests and ensuring that both space and ocean resources are managed as shared assets for humanity.

The interaction between climate change and space mining as wildcards further complicates the future of sea-based resource utilization. For instance, the increasing impacts of climate change could intensify the urgency of finding alternative resource sources, accelerating investments in space mining technologies. Conversely, the successful development of space mining could provide new opportunities for addressing climate challenges, such as by sourcing rare earth elements needed for renewable energy technologies without further depleting Earth's ecosystems. These interconnected dynamics underscore the importance of adopting holistic and adaptive approaches to resource management that account for the interplay between terrestrial, marine, and extraterrestrial systems.

Preparing for these wildcards requires a combination of foresight, innovation, and collaboration. Governments, industry leaders, and researchers must invest in scenario planning and modeling to explore the potential impacts of climate change and space mining on sea-based resources. Advanced tools, such as AI-driven simulations and digital twins, can provide valuable insights into how these wildcards might interact with existing trends, enabling stakeholders to develop proactive strategies that enhance resilience and adaptability.

Ethical considerations must also guide the response to these wildcards, ensuring that decisions about resource utilization reflect principles of equity, sustainability, and intergenerational responsibility. For example, the development of space mining should not come at the expense of further marginalizing communities that depend on marine resources for their livelihoods. Similarly, strategies for mitigating the impacts of climate change on deep sea mining should prioritize the protection of vulnerable ecosystems and the equitable sharing of benefits derived from resource extraction.

13. Synthesis

13.1. Revisiting Key Findings

The exploration and development of deep sea resources, particularly through the lens of artificial intelligence (AI) and advanced technologies, offer transformative opportunities and formidable challenges. The journey toward unlocking the potential of the ocean's critical mineral wealth has illuminated key innovations that promise to revolutionize resource extraction and management while also highlighting significant hurdles and ethical questions that must be addressed. This chapter revisits the main findings, synthesizing the advancements in technology, the obstacles to progress, and the moral imperatives that frame this ambitious endeavor.

The integration of AI into deep sea mining stands out as one of the most groundbreaking innovations, reshaping how resources are discovered, extracted, and managed. AI algorithms excel at processing vast amounts of data from diverse sources, such as seismic surveys, sonar imaging, and environmental monitoring systems. These capabilities enable operators to identify high-value mineral deposits with unprecedented precision, reducing the time and cost associated with traditional exploration methods. For example, machine learning models can analyze geophysical data to pinpoint the locations of polymetallic nodules or cobalt-rich crusts, optimizing exploration efforts while minimizing environmental disturbance.

In addition to its role in exploration, AI has proven transformative in the realm of extraction. Autonomous underwater vehicles (AUVs) and remotely operated vehicles (ROVs), powered by AI-driven control systems, perform complex tasks such as resource collection and environmental monitoring with minimal human intervention. These robotic platforms navigate the challenging terrains of the deep ocean, adapting dynamically to changes in their surroundings. Their precision reduces waste, enhances efficiency, and ensures that operations are conducted with minimal disruption to marine ecosystems. Moreover, AI-driven predictive maintenance systems analyze data from mining equipment to identify potential failures before they occur, reducing downtime and extending the lifespan of critical assets.

Environmental monitoring has also been revolutionized by AI and autonomous systems. Networks of sensors collect continuous data on water quality, sediment movement, and biodiversity, providing a real-time view of the environmental impacts of mining activities. AI algorithms process this data to detect anomalies, assess risks, and recommend mitigation measures, enabling operators to align their activities with principles of sustainability and regulatory compliance. These innovations ensure that deep sea mining not only meets operational goals but also upholds environmental stewardship, setting a global benchmark for responsible resource development.

Despite these advancements, the path forward is fraught with challenges that test the limits of technology, governance, and public trust. The extreme conditions of the deep ocean, characterized by high pressures, low temperatures, and corrosive seawater, create technical and logistical obstacles that demand continuous innovation. Developing materials and systems capable of withstanding these conditions is both costly and time-intensive, requiring substantial investment in research and development. Moreover, the high upfront costs of deploying AI-driven technologies and autonomous systems pose financial barriers for operators, particularly in the early stages of adoption.

The fragmented and complex regulatory landscape further complicates the development of deep sea mining. Operators must navigate a web of overlapping jurisdictions, international agreements, and environmental standards, which can delay project timelines and increase costs. For example, the lack of comprehensive governance frameworks for international waters leaves gaps in oversight and accountability, creating uncertainty for stakeholders. Streamlining regulatory processes, harmonizing international standards, and fostering collaboration among nations are essential for addressing these challenges and providing the clarity needed to advance deep sea mining responsibly.

Public perception and stakeholder engagement present additional challenges. Deep sea mining often faces skepticism from environmental organizations, coastal communities, and the general public, who express concerns about its potential impacts on marine ecosystems and biodiversity. Building trust and legitimacy requires transparent communication about the benefits, risks, and trade-offs of mining activities, as well as meaningful engagement with stakeholders. Mechanisms such as public consultations, advisory committees, and participatory workshops are critical for ensuring that diverse perspectives are considered and that governance frameworks reflect shared values.

The ethical considerations surrounding deep sea mining are profound, encompassing issues of environmental justice, intergenerational responsibility, and equitable resource distribution. The ocean is a shared global resource, and its development carries implications for all of humanity. Decisions about where and how to mine must balance the need for critical minerals with the imperative to protect vulnerable ecosystems and respect the rights of affected communities. For instance, ensuring that resource development benefits marginalized populations and developing nations requires a commitment to capacity building, knowledge sharing, and inclusive decision-making.

Intergenerational responsibility is another key ethical dimension. The impacts of deep sea mining, both positive and negative, extend far beyond the present, shaping the ecological and economic landscapes of future generations. Ensuring that mining activities align with principles of sustainability and long-term resilience is critical for preserving the health and integrity of the ocean. For example, rigorous environmental impact assessments, adaptive management strategies, and the establishment of marine protected areas can safeguard biodiversity and ecosystem services for decades to come.

The equitable distribution of resources and benefits is central to the ethical governance of deep sea mining. Developing nations and small island states, which are often disproportionately affected by climate change and resource depletion, must have access to the economic opportunities created by mining activities. This requires mechanisms for technology transfer, financial support, and capacity building, ensuring that all nations can participate in and benefit from global resource development. For instance, the United States could lead initiatives that provide funding and technical assistance to help developing countries establish deep sea mining operations and integrate them into international supply chains.

The interplay between innovation, challenges, and ethics defines the future of deep sea mining, offering both opportunities and imperatives for action. The integration of AI and advanced technologies provides a roadmap for addressing technical and operational challenges while enhancing efficiency and sustainability. However, realizing this potential requires a commitment to responsible governance, collaborative frameworks, and continuous dialogue among stakeholders.

Policy recommendations play a vital role in shaping the trajectory of deep sea mining, ensuring that its development aligns with national and global priorities. For example, investing in research and development accelerates the deployment of innovative technologies, while public-private partnerships leverage the resources and expertise of diverse stakeholders. Streamlining regulatory processes reduces barriers to entry and provides operators with the clarity needed to integrate AI and autonomous systems. Meanwhile, international collaboration fosters the alignment of governance frameworks with shared values, promoting transparency, accountability, and inclusivity.

The integration of artificial intelligence (AI) into deep sea mining represents one of the most compelling examples of how technological advancements can drive sustainability in resource extraction. At its core, the symbiosis between AI and sustainable deep sea mining is founded on a shared objective: balancing the economic imperatives of extracting critical minerals with the ethical and environmental responsibility of preserving marine ecosystems. AI offers transformative tools that enable operators to optimize efficiency, minimize ecological impacts, and adhere to regulatory standards, making it an indispensable partner in the quest for sustainable resource development.

Deep sea mining is inherently challenging due to the extreme and remote conditions of the ocean floor. High pressures, low temperatures, limited visibility, and fragile ecosystems create an environment where traditional methods of exploration and extraction are often impractical or environmentally damaging. AI addresses these challenges by providing the analytical and operational capabilities needed to navigate this complexity with precision and adaptability. By leveraging machine learning algorithms, real-time data processing, and autonomous systems, AI enhances every stage of the mining process, from identifying high-value deposits to monitoring environmental impacts.

One of the most significant contributions of AI to sustainable deep sea mining is its ability to optimize exploration efforts. The search for critical minerals such as cobalt, nickel, manganese, and rare earth elements requires detailed mapping and assessment of seafloor resources. Traditional exploration methods often involve extensive physical surveys and manual data interpretation, which can be time-consuming, costly, and environmentally intrusive. AI streamlines this process by analyzing vast amounts of data from seismic surveys, sonar imaging, geochemical samples, and satellite observations. These algorithms identify patterns and correlations that indicate the presence of valuable deposits, allowing operators to prioritize exploration efforts and reduce unnecessary disturbance to the ocean floor.

For example, AI-driven geophysical models can process data from sonar scans to map the distribution of polymetallic nodules across abyssal plains or cobalt-rich ferromanganese crusts on seamounts. These models provide a high-resolution understanding of resource locations, enabling operators to focus their efforts on areas with the greatest economic potential and the least ecological sensitivity. This precision reduces the environmental footprint of exploration activities, aligning with the principles of sustainability.

The symbiosis between AI and sustainable mining extends beyond exploration to include the extraction process itself. Autonomous underwater vehicles (AUVs) and remotely operated vehicles (ROVs), equipped with AI-driven control systems, perform tasks such as collecting nodules, cutting crusts, and monitoring extraction sites. These robotic platforms operate in challenging deep ocean environments, using AI to navigate complex terrains, avoid obstacles, and execute extraction tasks with accuracy and efficiency. By minimizing human intervention and optimizing resource recovery, AI reduces the energy consumption, waste, and environmental disturbance associated with traditional extraction methods.

Real-time monitoring and adaptive management are critical components of sustainable deep sea mining, and AI plays a central role in enabling these capabilities. Autonomous sensor networks deployed around extraction sites collect continuous data on environmental indicators, such as water quality, sediment plumes, and marine biodiversity. AI algorithms process this data in real time, detecting anomalies, assessing impacts, and recommending operational adjustments. For example, if sensors detect an increase in turbidity caused by sediment disturbance, an AI system might suggest modifying extraction techniques or deploying sediment containment measures to mitigate the impact. These adaptive responses ensure that operations remain within regulatory thresholds and minimize harm to marine ecosystems.

AI-driven predictive modeling further enhances sustainability by enabling operators to anticipate and mitigate potential impacts before they occur. These models simulate the interactions between extraction activities and environmental systems, providing insights into the long-term effects of mining on biodiversity, sediment composition, and biogeochemical cycles. For instance, an AI model might predict how changes in sediment structure caused by mining could affect benthic organisms, informing the design of mitigation strategies that protect vulnerable species. By integrating these predictions into planning and decision-making, operators can align their activities with principles of environmental stewardship and long-term resilience.

The use of AI in regulatory compliance underscores its symbiotic relationship with sustainable mining. Governments and international organizations impose stringent regulations to ensure that resource extraction is conducted responsibly and with minimal environmental impact. AI-driven systems enhance compliance efforts by automating data collection, analysis, and reporting, reducing the administrative burden on operators and ensuring that regulatory requirements are met efficiently. For example, an AI-powered monitoring system can generate real-time reports on emissions, water usage, and habitat conditions, providing transparent and verifiable evidence of adherence to environmental standards. These capabilities not only reduce the risk of fines and legal disputes but also build trust with stakeholders and the public.

The environmental benefits of AI-driven sustainability efforts are complemented by their economic advantages, creating a virtuous cycle that supports the long-term viability of deep sea mining. By optimizing operations and reducing waste, AI enhances efficiency and profitability, making sustainable practices economically attractive. For instance, an AI system that identifies high-density resource areas for extraction minimizes energy and material use, reducing operational costs while maximizing output. These efficiencies create a compelling business case for integrating AI into deep sea mining, demonstrating that sustainability and profitability are not mutually exclusive but mutually reinforcing.

The role of AI in fostering transparency and accountability further highlights its contribution to sustainable mining. Transparent communication about the benefits, risks, and trade-offs of mining activities is essential for building trust with stakeholders and ensuring that governance frameworks reflect shared values. AI-driven data analytics and visualization tools enable operators to share real-time information about their activities and impacts, fostering dialogue and collaboration with regulators, communities, and environmental organizations. For example, a public-facing dashboard that displays data on sediment plumes, biodiversity changes, and compliance metrics allows stakeholders to monitor operations and provide feedback, creating a sense of shared ownership and accountability.

The integration of AI with other emerging technologies amplifies its potential to support sustainable deep sea mining. For example, combining AI with blockchain technology can create secure and transparent records of resource extraction, ensuring that minerals are sourced responsibly and in compliance with international standards. Similarly, integrating AI with digital twins, virtual replicas of physical systems, allows operators to simulate and optimize mining operations in a controlled environment, reducing risks and improving outcomes. These synergies highlight the transformative potential of technology to drive sustainability and innovation in resource extraction.

Despite its promise, the symbiosis between AI and sustainable mining is not without challenges. The development and deployment of AI-driven systems require significant investment in research, infrastructure, and training, as well as the resolution of ethical and governance issues. Ensuring that AI systems operate transparently and accountably is critical for addressing concerns about data security, decision-making biases, and unintended consequences. Additionally, the adoption of AI in deep sea mining must be accompanied by robust environmental safeguards and adaptive governance frameworks that prioritize sustainability and inclusivity.

Public policy and government support are essential for fostering the symbiosis between AI and sustainable mining. Initiatives such as research funding, tax incentives, and public-private partnerships can accelerate the development and deployment of AI technologies in resource extraction. For example, government grants could support the creation of AI-driven environmental monitoring systems, while regulatory frameworks could require the use of AI-powered tools to ensure compliance with sustainability standards. These policies not only promote innovation but also ensure that the benefits of AI are distributed equitably and aligned with broader societal goals.

13.2. Limitations, Gaps, and Future Research

The promise of deep sea mining as a solution to the global demand for critical minerals is accompanied by a range of scientific and engineering challenges that remain unresolved. These challenges span the spectrum of technological feasibility, environmental impact, and operational logistics, revealing significant gaps in our understanding and capabilities. Addressing these questions is essential for ensuring that deep sea mining can be conducted responsibly, sustainably, and profitably. A critical assessment of these unresolved issues highlights the need for ongoing research, innovation, and interdisciplinary collaboration.

One of the most fundamental scientific questions concerns the geological distribution and formation of mineral deposits in the deep ocean. While significant progress has been made in mapping seafloor resources such as polymetallic nodules, cobalt-rich ferromanganese crusts, and seafloor massive sulfides, much of the ocean floor remains unexplored. The processes that govern the formation and distribution of these deposits are complex and not fully understood, involving interactions between geological, chemical, and biological systems. For example, the mechanisms that control the growth rates of polymetallic nodules and their spatial variability are not well characterized, complicating efforts to estimate resource availability and design efficient extraction strategies.

Geophysical and geochemical modeling offer promising tools for addressing these knowledge gaps, but their effectiveness is limited by the quality and quantity of available data. High-resolution mapping technologies, such as autonomous underwater vehicles (AUVs) equipped with advanced sonar and imaging systems, provide valuable insights into the morphology and composition of seafloor deposits. However, the vastness of the deep ocean and the logistical challenges of data collection create barriers to comprehensive mapping efforts. Enhancing data acquisition and integration capabilities is critical for building accurate models of resource distribution and guiding exploration efforts.

Another unresolved question pertains to the long-term stability and behavior of geological formations used for CO_2 storage or impacted by mining activities. Understanding the structural integrity of these formations is essential for ensuring the safety and efficacy of carbon sequestration projects and minimizing the environmental impacts of mining. For instance, the potential for caprock fractures or sediment displacement to create pathways for CO_2 leakage or disrupt benthic habitats remains poorly understood. Advanced geotechnical studies, coupled with high-resolution seismic imaging and AI-driven predictive modeling, are needed to assess these risks and inform the design of containment and extraction systems.

The environmental impacts of deep sea mining represent another critical area of scientific uncertainty. The deep ocean is one of the least understood ecosystems on Earth, hosting unique and often fragile species that are adapted to extreme conditions. The potential effects of mining activities, such as habitat destruction, sediment plumes, and changes in chemical composition, are not fully characterized, raising concerns about biodiversity loss and ecosystem disruption. For example, the dispersal of sediment plumes generated during mining operations could smother benthic organisms and alter nutrient dynamics over large areas, with cascading effects on marine food webs.

Environmental monitoring and impact assessment tools have advanced significantly, but they are not yet capable of providing a comprehensive understanding of the ecological consequences of deep sea mining. Autonomous sensor networks, environmental DNA (eDNA) analysis, and AI-driven ecological models offer valuable insights into species distributions and ecosystem dynamics, but their resolution and coverage are limited by logistical and technical constraints. Expanding the use of these tools and integrating them with long-term monitoring programs is essential for addressing knowledge gaps and ensuring that environmental risks are managed effectively.

The engineering challenges of deep sea mining are equally formidable, particularly in the design and deployment of extraction systems capable of operating in extreme environments. The high pressures, low temperatures, and corrosive seawater of the deep ocean create conditions that test the limits of materials science and mechanical engineering. For example, developing robotic platforms and mining equipment that can withstand these conditions while maintaining precision and reliability is a major technical hurdle. The durability and performance of components such as robotic arms, cutting tools, and sensors must be validated through rigorous testing and iterative design.

Energy efficiency is another significant engineering challenge. Deep sea mining operations require substantial energy inputs for powering autonomous systems, transporting materials, and maintaining communication links with surface vessels. The reliance on traditional energy sources not only increases operational costs but also contributes to the carbon footprint of mining activities. Advances in renewable energy technologies, such as underwater turbines or solar-powered systems, offer potential solutions, but their integration with deep sea mining infrastructure remains an area of active research. Developing energy-efficient systems that balance operational demands with sustainability goals is a priority for the industry.

Operational logistics and scalability also present unresolved questions for deep sea mining. Coordinating the deployment of extraction systems, transporting materials to processing facilities, and managing supply chains require sophisticated logistical planning and real-time decision-making. AI-driven optimization algorithms can enhance these processes, but their effectiveness depends on the availability of high-quality data and the interoperability of different systems. For instance, integrating data from multiple autonomous platforms, sensor networks, and operational systems into a unified framework is a complex challenge that requires advancements in data sharing and communication technologies.

The social and ethical dimensions of deep sea mining add another layer of complexity to these unresolved questions. The potential benefits of resource development, such as economic growth and technological innovation, must be weighed against the risks of environmental harm, social disruption, and inequitable resource distribution. Ensuring that the benefits of deep sea mining are shared equitably among stakeholders, including developing nations and indigenous communities, requires a commitment to ethical governance and inclusive decision-making. For example, mechanisms for capacity building and knowledge transfer can empower underrepresented stakeholders to participate in and benefit from resource development initiatives.

Public perception and stakeholder engagement are critical for addressing the ethical and social challenges of deep sea mining. The lack of transparency and accountability in some mining operations has fueled skepticism and opposition from environmental organizations, coastal communities, and the general public. Building trust and legitimacy requires open communication about the risks and benefits of mining activities, as well as meaningful engagement with diverse stakeholders. Mechanisms such as public consultations, advisory committees, and participatory workshops provide platforms for dialogue and collaboration, fostering a sense of shared ownership and accountability.

Despite these challenges, the unresolved questions in deep sea mining represent opportunities for innovation, discovery, and collaboration. Addressing these knowledge gaps requires interdisciplinary research that brings together scientists, engineers, policymakers, and industry leaders to develop holistic solutions. Advances in AI, robotics, materials science, and environmental monitoring provide powerful tools for tackling these challenges, while public-private partnerships and international collaboration enhance the resources and expertise available for research and development.

The field of deep sea mining is rich with opportunities for groundbreaking research, offering numerous areas of inquiry that demand the intellectual rigor and innovation of PhD-level studies. The intersection of cutting-edge technologies, complex environmental systems, and pressing ethical considerations creates a fertile ground for advancing scientific knowledge and addressing practical challenges. This chapter proposes key areas for future doctoral research, emphasizing the need for interdisciplinary approaches and a commitment to sustainability, technological progress, and equitable resource management.

One of the most pressing areas for future research is the geological and geophysical characterization of deep ocean mineral deposits. Despite significant advancements in mapping and exploration technologies, much of the seafloor remains unexplored, and the processes governing the formation and distribution of critical minerals are not fully understood. Doctoral research in this area could focus on developing high-resolution models of resource-rich regions, incorporating data from seismic surveys, sonar imaging, and core sampling. For example, a PhD study might investigate the mechanisms that influence the growth rates and spatial variability of polymetallic nodules, shedding light on their economic potential and guiding sustainable extraction strategies.

The integration of artificial intelligence (AI) and machine learning into resource exploration is another promising avenue for doctoral research. AI-driven algorithms have demonstrated exceptional capabilities in processing large datasets and identifying patterns, but their application in deep sea mining is still in its infancy. Future research could explore the development of AI models that enhance the accuracy and efficiency of exploration efforts. For instance, a PhD candidate might design an AI system that predicts the locations of cobalt-rich ferromanganese crusts based on geochemical and geophysical indicators, enabling operators to target high-value deposits while minimizing environmental disturbance.

Environmental impact assessment and mitigation strategies represent a critical area for PhD-level research, particularly given the unique and poorly understood nature of deep ocean ecosystems. The potential effects of mining activities, such as sediment plumes, habitat destruction, and changes in chemical composition, raise concerns about biodiversity loss and ecosystem disruption. Doctoral studies could investigate the ecological responses of specific marine species to mining-induced changes, providing data that inform conservation measures and adaptive management strategies. For example, a PhD project might examine the impact of sediment plumes on filter-feeding organisms, identifying thresholds for turbidity and sediment deposition that minimize harm.

Long-term monitoring and environmental restoration are equally important areas for future research, as they ensure that mining activities align with principles of sustainability and ecosystem health. PhD candidates could focus on developing autonomous sensor networks and AI-driven analytics that provide continuous data on key environmental indicators. These systems could track changes in water quality, sediment composition, and biodiversity, enabling operators to implement proactive and adaptive measures. Additionally, doctoral research could explore techniques for restoring mined areas, such as reseeding coral reefs or stabilizing sediments with bioengineered materials, contributing to the resilience and recovery of marine ecosystems.

The design and optimization of extraction technologies also offer rich opportunities for doctoral research. The extreme conditions of the deep ocean, high pressures, low temperatures, and corrosive environments, pose significant challenges for mining equipment and infrastructure. Future studies could focus on developing materials and mechanical systems that withstand these conditions while maintaining precision and efficiency. For instance, a PhD project might investigate the use of advanced composites or nanomaterials for constructing robotic arms and cutting tools, enhancing their durability and performance. Similarly, research could explore bio-inspired robotics that mimic the movements and adaptability of marine organisms, enabling more sensitive and sustainable extraction methods.

Energy efficiency and renewable energy integration in deep sea mining operations are critical areas for addressing the dual challenges of cost and sustainability. The high energy demands of mining activities, particularly in remote and inaccessible locations, necessitate innovative solutions that reduce reliance on traditional energy sources. Doctoral research could investigate the feasibility of integrating underwater turbines, wave energy converters, or solar-powered systems into mining infrastructure. For example, a PhD study might evaluate the potential for harnessing ocean currents to power autonomous extraction platforms, providing a renewable and reliable energy source that aligns with climate goals.

Supply chain optimization and logistics represent another vital area for PhD-level research, particularly as deep sea mining scales to meet global demand for critical minerals. Coordinating the movement of materials, equipment, and personnel in complex and dynamic environments requires sophisticated planning and real-time decision-making. Future studies could focus on developing AI-driven optimization algorithms that enhance the efficiency of supply chains while reducing costs and environmental impacts. For instance, a PhD candidate might design a logistics platform that integrates data from autonomous vehicles, shipping routes, and processing facilities, enabling seamless coordination and resource allocation.

The social and ethical dimensions of deep sea mining present compelling topics for doctoral research, particularly in the context of equitable resource distribution and community engagement. The extraction of marine resources raises questions about who benefits from resource development and who bears its costs, highlighting the need for inclusive and participatory governance frameworks. PhD-level studies could explore mechanisms for ensuring that the benefits of mining are shared equitably among stakeholders, including developing nations and indigenous communities. For example, a doctoral project might investigate the role of capacity building and knowledge transfer in empowering underrepresented groups to participate in and benefit from mining activities.

Public perception and communication are also critical areas for future research, as they influence the social license to operate and the legitimacy of mining activities. Doctoral studies could examine how different stakeholders perceive the risks and benefits of deep sea mining, identifying strategies for building trust and fostering dialogue. For instance, a PhD project might analyze the effectiveness of public consultations, advisory committees, and participatory workshops in addressing community concerns and incorporating diverse perspectives into decision-making.

Policy and governance frameworks represent another rich field for doctoral research, particularly given the evolving nature of international agreements and standards for deep sea mining. The transboundary impacts of mining activities and the global demand for critical minerals necessitate cooperative approaches to governance that balance economic, environmental, and social considerations. Future studies could investigate the development of innovative policy instruments that promote transparency, accountability, and sustainability in resource extraction. For example, a PhD candidate might analyze the feasibility of creating a global registry for deep sea mining operations, documenting their location, scale, and environmental impacts to enhance oversight and public trust.

The potential for technological synergies in deep sea mining offers an exciting avenue for interdisciplinary research, combining expertise in AI, robotics, environmental science, and engineering. Doctoral projects could explore how emerging technologies such as blockchain, digital twins, and quantum computing can be integrated into mining operations to enhance efficiency, sustainability, and governance. For instance, a PhD study might investigate the use of blockchain to create secure and transparent records of resource extraction, ensuring traceability and compliance with ethical and environmental standards.

Despite its promise, the pursuit of these research areas requires addressing significant challenges, including the high costs and logistical complexities of conducting deep ocean studies. Collaborative research initiatives, supported by public-private partnerships and international funding, are essential for overcoming these barriers and advancing knowledge. Additionally, the interdisciplinary nature of these topics necessitates collaboration among scientists, engineers, policymakers, and industry leaders to develop holistic solutions that address the multifaceted challenges of deep sea mining.

13.3. Actionable Recommendations

The successful development and integration of deep sea mining into global resource strategies require a cohesive and well-coordinated roadmap that brings together policymakers, industry leaders, and research institutions. Each stakeholder plays a distinct yet interconnected role in shaping the future of this industry, addressing the technological, environmental, and societal challenges it presents. A roadmap provides the strategic framework needed to align these efforts, ensuring that deep sea mining achieves its potential as a source of critical minerals while adhering to principles of sustainability, transparency, and ethical governance.

For policymakers, the primary focus should be on establishing a regulatory and policy framework that balances innovation with responsibility. Clear and predictable regulations are essential for providing the certainty needed to attract investment and foster innovation while safeguarding environmental and social interests. Policymakers must prioritize the creation of comprehensive governance structures that address the unique challenges of deep sea mining, including its transboundary impacts and the need for long-term monitoring.

One critical step for policymakers is to streamline permitting processes, reducing delays and uncertainties that discourage investment and hinder technological adoption. This involves developing standardized templates for environmental impact assessments (EIAs) and creating fast-track pathways for projects that demonstrate strong commitments to sustainability and compliance. For example, policies could incentivize the use of artificial intelligence (AI) and autonomous monitoring systems, which enhance transparency and efficiency in environmental oversight. By embedding innovation into regulatory frameworks, policymakers can accelerate the deployment of advanced technologies that minimize the ecological footprint of mining activities.

Policymakers also have a crucial role in promoting international collaboration and alignment. The global nature of critical mineral demand and the shared responsibility for ocean stewardship necessitate cooperative approaches to governance. The United States, in particular, has the opportunity to lead efforts to establish international standards for transparency, accountability, and sustainability in deep sea mining. This could involve advocating for a global registry of mining activities, which documents their location, scale, and environmental impacts, providing a foundation for collective oversight and informed decision-making.

For industry leaders, the roadmap emphasizes the integration of cutting-edge technologies to optimize operations and align with sustainability goals. Companies engaged in deep sea mining must invest in research and development (R&D) to address the technical and logistical challenges of operating in extreme environments. Artificial intelligence, robotics, and advanced monitoring systems are at the forefront of this effort, offering solutions that enhance efficiency, reduce costs, and mitigate environmental impacts.

For example, industry leaders can adopt AI-driven systems for exploration and extraction, which process data from seismic surveys, sonar imaging, and environmental sensors to identify high-value deposits and execute operations with precision. These technologies not only improve resource recovery rates but also minimize disturbances to marine ecosystems, aligning with regulatory and societal expectations. Additionally, predictive maintenance systems powered by AI can analyze data from mining equipment to identify potential failures before they occur, reducing downtime and extending the lifespan of critical assets.

Collaboration within the industry is also essential for advancing the field. Mining companies, technology providers, and environmental consultancies must work together to develop and deploy innovative solutions that address shared challenges. Industry consortia, for example, can pool resources and expertise to fund demonstration projects, validate new technologies, and establish best practices for sustainable operations. By fostering a culture of collaboration, industry leaders can accelerate progress and build a strong foundation for long-term success.

Research institutions play a pivotal role in advancing knowledge and innovation in deep sea mining. Their contributions span fundamental and applied research, workforce development, and the dissemination of scientific findings that inform policy and practice. A key priority for research institutions is to address the significant knowledge gaps in our understanding of deep ocean ecosystems and the potential impacts of mining activities. This includes conducting long-term studies on biodiversity, sediment dynamics, and biogeochemical cycles, which provide the data needed to develop evidence-based mitigation strategies.

Research institutions should also focus on the development of new technologies and methodologies that enhance the sustainability of deep sea mining. For instance, engineers and scientists can work on designing bio-inspired robotics that mimic the movements and adaptability of marine organisms, enabling more sensitive and sustainable extraction methods. Similarly, researchers can explore the use of renewable energy systems, such as underwater turbines or wave energy converters, to power mining operations and reduce their carbon footprint.

The role of research institutions in training the next generation of professionals is equally critical. Deep sea mining requires a multidisciplinary workforce with expertise in areas such as AI, robotics, oceanography, environmental science, and policy analysis. Universities and training centers should develop specialized programs that equip students with the skills and knowledge needed to navigate the complexities of this industry. These programs can include hands-on experience with advanced technologies, interdisciplinary coursework, and partnerships with industry stakeholders to ensure that graduates are well-prepared to contribute to the field.

Collaboration between policymakers, industry leaders, and research institutions is the cornerstone of the roadmap for deep sea mining. Public-private partnerships provide a mechanism for aligning efforts and leveraging resources across sectors. For example, a partnership between a federal research agency, a mining company, and a university could support the development of an AI-driven environmental monitoring system, combining public funding with private sector innovation and academic expertise. These partnerships not only accelerate technological progress but also ensure that research and development activities align with national and global priorities.

Funding and investment are critical components of the roadmap, as the high costs of deep sea mining require substantial financial support to advance technologies and infrastructure. Policymakers can provide grants, tax incentives, and loan guarantees to de-risk investment and encourage private sector participation. Industry leaders can complement these efforts by allocating resources to R&D and demonstration projects that validate new approaches and reduce uncertainties. Additionally, research institutions can secure funding from government agencies, philanthropic organizations, and international consortia to support long-term studies and capacity building.

Public engagement and stakeholder participation are integral to the success of the roadmap. Transparent communication about the benefits, risks, and trade-offs of deep sea mining ensures that diverse perspectives are considered and that governance frameworks reflect the values and priorities of all stakeholders. Mechanisms such as public consultations, advisory committees, and participatory workshops provide platforms for dialogue and collaboration, fostering trust and legitimacy in decision-making processes. For instance, coastal communities, environmental organizations, and industry representatives could contribute to discussions about the design and implementation of mitigation measures, ensuring that their insights inform project planning and execution.

The roadmap also emphasizes the importance of ethical considerations in shaping the future of deep sea mining. Decisions about resource development must balance economic growth with environmental stewardship and social equity. This includes ensuring that the benefits of mining are distributed fairly among stakeholders, including developing nations and marginalized communities. For example, capacity building and knowledge transfer initiatives can empower underrepresented groups to participate in and benefit from mining activities, promoting inclusivity and shared prosperity.

The integration of artificial intelligence (AI) into deep sea mining presents a transformative opportunity to revolutionize resource extraction while minimizing its environmental footprint. However, the application of AI in this context must be guided by principles of responsibility, sustainability, and ethical governance to ensure that innovation serves the greater good. Fostering responsible, AI-driven innovation in deep sea mining requires a multifaceted approach that combines technological advancement, robust policy frameworks, stakeholder engagement, and continuous research and development.

One of the primary strategies for fostering responsible AI-driven innovation in deep sea mining is the establishment of clear and enforceable governance frameworks. Governments and international organizations must create regulatory structures that provide guidance on the use of AI in exploration, extraction, and environmental monitoring. These frameworks should prioritize transparency, accountability, and sustainability, ensuring that AI technologies are deployed in a manner that aligns with societal values and environmental goals. For instance, regulations could mandate the use of AI-driven monitoring systems to track the impacts of mining activities in real time, providing data that informs adaptive management strategies and supports compliance with environmental standards.

In addition to setting regulatory standards, policymakers can incentivize the adoption of AI technologies that enhance sustainability and efficiency. Financial incentives, such as grants, tax credits, and subsidies, can encourage companies to invest in the development and deployment of AI-driven systems. For example, a government program might provide funding for mining operators to integrate AI algorithms that optimize resource recovery while minimizing energy consumption and waste. These incentives not only accelerate the adoption of innovative technologies but also demonstrate a commitment to responsible resource development.

Public-private partnerships play a critical role in advancing responsible AI-driven innovation. By bringing together the resources and expertise of government agencies, industry leaders, and research institutions, these partnerships create a collaborative environment for addressing the technical and ethical challenges of deep sea mining. For instance, a partnership could support the development of autonomous underwater vehicles (AUVs) equipped with AI-driven navigation and sensing capabilities, enabling more precise and environmentally sensitive resource extraction. These collaborations also provide a platform for aligning public and private interests, ensuring that technological advancements benefit both the economy and the environment.

Research and development (R&D) are essential for driving AI-driven innovation in deep sea mining while addressing the knowledge gaps and technical challenges that remain. Investment in R&D should prioritize areas such as materials science, robotics, environmental monitoring, and energy efficiency. For example, researchers could explore the use of advanced composites and coatings to enhance the durability and performance of mining equipment in corrosive deep ocean environments. Similarly, studies on bio-inspired robotics could lead to the development of extraction systems that mimic the movements and adaptability of marine organisms, reducing ecological disruption.

Environmental monitoring and impact assessment represent key areas for responsible AI-driven innovation. The deep ocean is one of the least understood ecosystems on Earth, and mining activities have the potential to disrupt its delicate balance. AI-powered monitoring systems, integrated with autonomous sensor networks, provide the tools needed to track environmental indicators such as turbidity, water chemistry, and biodiversity. These systems enable operators to detect and mitigate potential impacts in real time, ensuring that operations remain within regulatory thresholds and align with principles of sustainability. For example, an AI system might analyze data from sensors to identify the spread of sediment plumes and recommend adjustments to extraction techniques to minimize harm.

Transparency and accountability are central to fostering responsible AI-driven innovation. Companies engaged in deep sea mining should adopt practices that ensure the public and stakeholders have access to accurate and comprehensive information about their activities. AI technologies can support these efforts by generating real-time data and visualizations that illustrate the environmental and operational performance of mining projects. For instance, a public-facing dashboard powered by AI analytics could display metrics such as sediment plume dispersion, CO_2 emissions, and resource recovery rates, enabling stakeholders to assess the impacts and benefits of mining activities.

Stakeholder engagement is another critical strategy for promoting responsibility and inclusivity in AI-driven deep sea mining. Transparent communication and participatory decision-making processes ensure that diverse perspectives are considered and that governance frameworks reflect the values and priorities of all stakeholders. Mechanisms such as public consultations, advisory committees, and community workshops provide platforms for dialogue and collaboration, fostering trust and legitimacy. For example, coastal communities, environmental organizations, and industry representatives could contribute to discussions about the design and implementation of AI-driven monitoring systems, ensuring that their insights inform project planning and execution.

The ethical dimensions of AI-driven innovation in deep sea mining warrant careful consideration, particularly in terms of equity and justice. Decisions about resource development must balance economic growth with environmental stewardship and social equity, ensuring that the benefits of mining are distributed fairly among stakeholders. For instance, capacity building and knowledge transfer initiatives can empower developing nations and marginalized communities to participate in and benefit from mining activities. A program that provides training in AI technologies and environmental management, for example, could equip local populations with the skills needed to contribute to and share in the economic opportunities created by mining.

International collaboration is essential for fostering responsible AI-driven innovation in deep sea mining, given the global nature of resource demand and the interconnectedness of marine ecosystems. The United States, as a leader in technology and governance, has an opportunity to work with allies and partners to establish global standards for the use of AI in resource extraction. For example, an international consortium could develop guidelines for the ethical deployment of AI in environmental monitoring, ensuring that technologies are used transparently and accountably. These collaborations enhance the collective ability to address the challenges of deep sea mining while promoting shared values and principles.

Despite its promise, the integration of AI into deep sea mining faces challenges that must be addressed to ensure its responsible deployment. The high upfront costs of developing and implementing AI-driven systems, the complexities of operating in extreme environments, and the need for robust data infrastructure all pose significant hurdles. However, these challenges also present opportunities for innovation, collaboration, and leadership. Advances in AI algorithms, sensor technologies, and autonomous systems can reduce costs and improve performance, while interdisciplinary research and stakeholder engagement can provide solutions to ethical and technical concerns.

Public policy and government support are critical for overcoming these challenges and creating an enabling environment for responsible AI-driven innovation. Initiatives such as funding programs, regulatory sandboxes, and international agreements can accelerate the adoption of AI technologies while ensuring that their use aligns with societal values and environmental goals. For example, a regulatory sandbox for AI in deep sea mining could provide a controlled environment for testing new technologies and business models, enabling operators to experiment and innovate without the risk of immediate regulatory penalties.

Epilogue

Navigating the Depths of Innovation and Responsibility

As humanity stands at the threshold of an unprecedented chapter in its relationship with the ocean, the exploration of the deep sea reveals not only immense opportunities but also profound responsibilities. This book has journeyed through the innovations, challenges, and ethical considerations that define the emerging frontier of deep sea mining, with artificial intelligence (AI) as the central enabler. The story of this technology's integration into one of the planet's least understood and most vulnerable environments is a testament to human ingenuity and ambition. Yet, it also underscores the need for balance, between exploration and preservation, between economic gain and ecological integrity, and between innovation and governance.

The deep ocean, with its vast reserves of critical minerals, holds the promise of addressing the urgent demand for resources essential to clean energy, advanced manufacturing, and national security. AI, as a transformative force, equips humanity with the tools to navigate this complexity, enabling precise exploration, efficient extraction, and rigorous environmental stewardship. These capabilities, however, are not an end in themselves. They must serve as instruments for achieving a greater goal: sustainable and equitable resource development that benefits not just the present generation but also those yet to come.

The challenges that remain are as vast as the ocean itself. Unresolved scientific questions about the geological, biological, and chemical processes of the deep sea demand continued research and collaboration. The development of robust engineering solutions capable of operating in extreme conditions requires innovation at the cutting edge of multiple disciplines. The potential ecological impacts of mining activities necessitate vigilance, adaptability, and a commitment to transparency and accountability. And the governance frameworks that will determine the rules of engagement in this arena must reflect shared values of sustainability, inclusivity, and fairness.

Throughout this book, one theme has emerged with clarity: the imperative of collaboration. No single stakeholder, whether a nation, corporation, or institution, can address the complexities of deep sea mining in isolation. The intersection of AI and sustainable resource development demands interdisciplinary approaches, public-private partnerships, and international cooperation. It is through collective effort that humanity can harness the transformative potential of AI while safeguarding the integrity of the ocean, a global commons that connects and sustains us all.

Ethical considerations remain at the heart of this endeavor. The choices made in the pursuit of deep sea mining will shape the ecological and social landscapes of the future, carrying implications far beyond the immediate gains of resource extraction. These choices must prioritize environmental justice, equitable resource distribution, and intergenerational responsibility. They must reflect a recognition of the ocean's intrinsic value, not merely as a repository of minerals but as a vital component of Earth's biosphere and a source of wonder and inspiration.

The deep sea, shrouded in mystery and brimming with potential, challenges humanity to think beyond short-term interests and to act with foresight and integrity. The integration of AI into this domain offers a glimpse of what is possible when technology is harnessed responsibly: a future where innovation and sustainability are not opposing forces but complementary pillars of progress. It is a future that requires courage, humility, and a commitment to shared purpose.

As the pages of this book close, the call to action remains open. Policymakers, industry leaders, researchers, and citizens alike have a role to play in shaping the trajectory of deep sea mining. By embracing the principles of innovation, responsibility, and collaboration, humanity can navigate the depths with wisdom and care, unlocking the potential of the ocean while preserving its legacy for generations to come. The journey ahead is not without its challenges, but it is one that holds the promise of creating a more sustainable, equitable, and harmonious relationship with the vast and mysterious world beneath the waves.